This book deals with a branch of functional analysis that has developed over the last decade. Entropy quantities are connected with the 'degree of compactness' of compact or precompact spaces, and so are appropriate tools for investigating linear and compact operators between Banach spaces. The main intention of this Tract is to study the relations between compactness and other analytical properties, e.g. approximability and eigenvalue sequences, of such operators. The authors present many new and generalized results, some of which have not appeared in the literature before. In the final chapter, the authors demonstrate that, to a certain extent, the geometry of Banach spaces can also be developed on the basis of operator theory.

All mathematicians working in functional analysis and operator theory will welcome this work as a reference or for advanced graduate courses.

T0245348

CAMBRIDGE TRACTS IN MATHEMATICS

General Editors

B. BOLLOBAS, H. HALBERSTAM & C.T.C. WALL

98 *Entropy, compactness and the approximation of operators*

BERND CARL

IRMTRAUD STEPHANI

Entropy, compactness and the approximation of operators

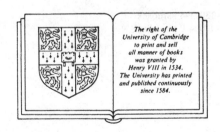

The right of the
University of Cambridge
to print and sell
all manner of books
was granted by
Henry VIII in 1534.
The University has printed
and published continuously
since 1584.

CAMBRIDGE UNIVERSITY PRESS

Cambridge

New York   Port Chester

Melbourne   Sydney

CAMBRIDGE UNIVERSITY PRESS
Cambridge, New York, Melbourne, Madrid, Cape Town, Singapore, São Paulo, Delhi

Cambridge University Press
The Edinburgh Building, Cambridge CB2 8RU, UK

Published in the United States of America by Cambridge University Press, New York

www.cambridge.org
Information on this title: www.cambridge.org/9780521330114

© Cambridge University Press 1990

This publication is in copyright. Subject to statutory exception
and to the provisions of relevant collective licensing agreements,
no reproduction of any part may take place without the written
permission of Cambridge University Press.

First published 1990
This digitally printed version 2008

A catalogue record for this publication is available from the British Library

Library of Congress Cataloguing in Publication data

Carl, Bernd.
    Entropy, compactness, and the approximation of operators / Bernd
Carl, Irmtraud. Stephani.
        p.   cm.
    Includes bibliographical references.
    ISBN 0 521 33011 4
    1. Functional analysis, 2. Entropy (Information theory)
3. Approximation theory, 4. Operator theory. I. Stephani,
Irmtraud. II. Title.
    QC20.7.F84C37   1990
    515.7--dc20   90-31049
                                                        CIP

ISBN 978-0-521-33011-4 hardback
ISBN 978-0-521-09094-0 paperback

Contents

Preface   ix

Introduction   1

1. Entropy quantities   6
   1.1. Entropy numbers of sets   6
   1.2. Entropy moduli of sets   10
   1.3. Entropy numbers of operators   11
   1.4. Entropy moduli of operators   22
   1.5. Entropy classes   26
   1.6. Operator ideals   35

2. Approximation quantities   41
   2.1. Approximation numbers   41
   2.2. Kolmogorov numbers   48
   2.3. Gelfand numbers   54
   2.4. Geometrical parameters   62
   2.5. Duality relations   72
   2.6. Symmetrized approximation numbers   82
   2.7. Local representations of approximation quantities   85

3. Inequalities of Bernstein–Jackson type   95
   3.1. Inequalities of Bernstein type   96
   3.2. Inequalities of Jackson type   101
   3.3. Geometrical parameters   105
   3.4. The Hilbert space setting   115
   3.5. Powers of operators   120

4. A refined Riesz theory   129
   4.1. Main aspects of classical Riesz theory   130
   4.2. Eigenvalues, entropy quantities, and Weyl type inequalities   139

4.3. Generalizations of the classical spectral radius formula 148
4.4. The Hilbert space setting 151

5. Operators with values in $C(X)$ **159**
5.1. Why $C(X)$-valued operators? 159
5.2. Local properties of $C(X)$ 162
5.3. Approximation quantities of $C(X)$-valued operators 168
5.4. The modulus of continuity of functions 170
5.5. The modulus of continuity of $C(X)$-valued operators 174
5.6. Approximation numbers and the modulus of continuity 178
5.7. Entropy numbers and the modulus of continuity 184
5.8. Entropy properties of connected compact metric spaces 188
5.9. Hölder continuous operators 196
5.10. Application of local techniques to Hölder continuous
operators 199
5.11. Integral operators 212
5.12. Hölder continuous integral operators 221
5.13. Operators defined by abstract kernels 225

6. Operator theoretical methods in the local theory of Banach spaces **229**
6.1. Norms of projections 230
6.2. Projection constants 239
6.3. Banach–Mazur distances $d(M, l_2^k)$ of subspaces $M \subseteq l_\infty^n$ 244
6.4. Volumes of convex hulls of finite sets 245
6.5. On a theorem of Pisier 248
6.6. On absolutely 1-summing and 2-summing operators 254
6.7. Tensor product techniques and the little Grothendieck
theorem 259

References **268**
List of symbols **272**
Index **275**

Preface

This book deals with a branch of modern functional analysis which has arisen only in the last 10 years, although it has its origin in a 1932 paper by Pontrjagin and Schnirelman.

In general there is quite a big difference between the level of recent research and the level of lectures as they are given to students. The question arises if this is in the nature of the subject, or if it is mainly a problem of producing an appropriate representation of the subject. Concerning 'Entropy, compactness and the approximation of operators', we came to the opinion that it should be possible to represent the subject at a level which makes reference only to the results of an introductory course on functional analysis. We have tried to write the book in the corresponding style and have listed in the introduction the concepts necessary for an understanding of the book. A few facts beyond the standard elementary knowledge of functional analysis are used without proof. However, a reader who is only interested in the fundamental relations between entropy quantities, approximation quantities, and eigenvalues can leave out the more difficult passages. By reading only sections 1.1, 1.2, 1.3, 1.4 of chapter 1, section 2.1 of chapter 2, section 3.1 of chapter 3, and section 4.2 of chapter 4, he or she will get an impression of the main ideas of the book and will be able to follow the applications of the general results in chapter 5. A course along these lines could be considered complementary to classical functional analysis, in particular to Riesz theory, or to classical approximation theory.

A book is never the work of the author alone, not even if this term stands for a group of two or more people. We had the help of our colleagues Dr Stefan Geiss, Dr Albrecht Hess, Dr Thomas Kühn, and Dr Doris Planer and want to thank them for reading the manuscript and making critical comments. In particular we are obliged to Dr Albrecht Hess who identified himself with the book project and the intentions of the authors. He discovered errors that we had overlooked and improved some earlier drafts. Miss Heike Gierschner typed the text excellently. Many thanks also to her!

Last but not least we wish to express our gratitude to Cambridge University Press. Professor Garling as an adviser to the Press recommended the book for publication in 'The Cambridge Tracts' series and suggested the final title. Publishing director David Tranah was a most reliable contact over a period of three years, meeting our expectations in all respects. We do hope that the reader will be as happy with the final product as we are with all the people who supported us.

Oldenburg   Bernd Carl
Jena, April 1990   Irmtraud Stephani

$T:E \to F$ is denoted by $L(E, F)$. When equipped with the *operator norm*

$$\|T\| = \sup_{\|x\| \leq 1} \|Tx\|$$

the class $L(E, F)$ becomes a Banach space itself.

Compactness properties of operators give rise to subclasses of the class $L(E, F)$. The main intention of the book is to quantify the *'degree' of compactness* of an operator $T:E \to F$ and to study its relation to other analytical properties of T. Among these, *approximability of T by finite rank operators*, a subject dealt with in chapter 2, plays a decisive role.

An operator $T:E \to F$ is called a *finite rank operator*, if its range $R(T)$ is a finite-dimensional subspace of F. If either the Banach space E or the Banach space F is finite-dimensional, the operator T is called *finite-dimensional*.

We recall that any finite-dimensional subspace of a Banach space is closed. In contrast with that an infinite-dimensional linear subspace of a Banach space need not be closed. The subspaces of Banach spaces occurring in the context of this book will in general be both linear and closed. Therefore we shall omit these two adjectives if no confusion is possible.

The *range* $R(T)$ of an operator $T:E \to F$ is a linear subspace of F which is not necessarily closed. In order to have the chance of using a closed linear subspace related to the range $R(T)$ of T we take the closed hull $\overline{R(T)}$ of $R(T)$ if this turns out to be advantageous, for instance for obtaining the *canonical factorization*

$$T = I_{F_0}^F T_0$$

of T through the Banach space $F_0 = \overline{R(T)}$. The operator $T_0:E \to F_0$ then is defined by

$$T_0 x = Tx \quad \text{for } x \in E$$

and is called *the operator induced by T*, while $I_{F_0}^F:F_0 \to F$ is understood to be the *natural* or *canonical embedding* of the 'subspace' F_0 of F into F.

Another *canonical factorization* of an operator $T:E \to F$ refers to the null space $N = N(T)$ of T. By the continuity of T, the *null space*

$$N(T) = \{x \in E : Tx = 0\}$$

or the *kernel* of T, as it is sometimes called, must be a closed subspace of E. This implies that the *quotient space* E/N is complete with respect to the norm

$$\|\bar{x}\| = \inf\{\|x - z\| : z \in N\}$$

of its elements

$$\bar{x} = \{x - z : z \in N\},$$

Introduction

As the title *Entropy, compactness and the approximation of operators* suggests, this book is about entropy, compactness and approximation properties of linear and continuous operators acting between Banach spaces. This indicates that the reader is first of all supposed to be aquainted with the notion of *a Banach space* and the notion of a *linear and continuous operator* $T:E \rightarrow F$ from a Banach space E into a Banach space F. These two notions are closely related to each other.

A *norm* $\|\cdot\|_0$ on a Banach space E is said to be *equivalent* to the original norm $\|\cdot\|$ on E, if there exist constants $c > 0$ and $C > 0$ such that

$$c \cdot \|x\| \leqslant \|x\|_0 \leqslant C \cdot \|x\| \quad \text{for all } x \in E.$$

However, instead of assigning another norm $\|x\|_0$ to the same element $x \in E$ we can also regard x with the new norm $\|x\|_0$ as an element $y = Sx$ of another Banach space E_0. The map $S:E \rightarrow E_0$ defined in this way is a linear and continuous operator from E onto E_0 with a continuous inverse $S^{-1}:E_0 \rightarrow E$. An operator with these properties is called an *isomorphism*. The corresponding Banach spaces E and E_0 are said to be *isomorphic*.

Among the examples of Banach spaces to appear in this book the *Banach spaces $C(X)$* of continuous functions on a compact metric space X and *Hilbert spaces* will take a primary place. In addition the reader is presented with the *spaces l_p* of p-summable sequences, with the corresponding *spaces $L_p(X, \mu)$* of functions f on a compact metric space X whose pth power $|f|^p$ is integrable with respect to a Borel measure μ on X, $1 \leqslant p < \infty$, as well as with the *spaces l_∞ and c_0* of bounded sequences and null sequences, respectively, and with the *space $L_\infty(X, \mu)$* of μ essentially bounded functions on X.

Special linear and continuous operators that will be used are *diagonal operators $D:l_p \rightarrow l_p$* acting in a sequence space l_p, $1 \leqslant p \leqslant \infty$, and *integral operators* from $L_p(X, \mu)$ into $C(X)$ as well as from $C(X)$ into itself.

Given a linear operator T from a Banach space E into a Banach space F the question of continuity of T is normally decided by checking the boundedness of T. Indeed, a linear operator $T:E \rightarrow F$ is continuous if and only if it is bounded. If T is known from the very beginning to be linear and either continuous or bounded we shall in general omit these two adjectives and simply use the notation *operator*. The class of all operators which are called the *cosets* of the elements $x \in E$ with respect to the subspace $N \subseteq E$. The operator $Q_N^E : E \to E/N$ defined by

$$Q_N^E x = \bar{x}$$

is referred to as the *natural* or the *'canonical'* surjection of E onto the quotient space E/N. Quite often we shall also use the expression *quotient map*.

The canonical factorization connected with the null space $N = N(T)$ of $T : E \to F$ that we have in mind is the factorization

$$T = T_0 Q_N^E$$

of T over the quotient space E/N. The operator $T_0 : E/N \to F$ in this situation is defined by

$$T_0 \bar{x} = Tx \quad \text{for } \bar{x} \in E/N$$

and also called *the operator induced by* T. Note that the definition of T_0 in fact makes sense since the value $T_0 \bar{x}$ in F is independent of the choice of a representative x in the coset $\bar{x}$.

Besides the two kinds of canonical factorizations, which apply to arbitrary operators $T : E \to F$, we have specific representations for special operators $T : E \to F$. In particular, if T is a finite rank operator, T can be represented as a finite sum of rank 1 operators. Since a rank 1 operator $T : E \to F$ has a one-dimensional range it allows a representation

$$Tx = A(x) \cdot y_0 \quad \text{for } x \in E$$

with an element $y_0 \in F$ and an operator A from E into the real line $\mathbb{R}$ or the complex plane $\mathbb{C}$. Operators of this kind will be called *functionals* and from now on will be denoted by lower case latin letters $a, b, c, \ldots$ This notation expresses a kind of similarity between elements $x \in E$ and functionals a over E. The notation $\langle x, a \rangle$ for the value of the functional a on the element $x \in E$ emphasizes this idea. It encourages the reader to fix x in E and to let a vary in the Banach space $L(E, \mathbb{R})$ of all linear and continuous functionals over E. Indeed, according to what has been said about $L(E, F)$, the linear space $L(E, \mathbb{R})$ is a Banach space with respect to the norm

$$\|a\| = \sup_{\|x\| \leqslant 1} |\langle x, a \rangle|.$$

It is denoted by E' and called *the dual of* E. As indicated by the notation $\langle x, a \rangle$ any $x \in E$ can be considered as a functional over E'. This functional is obviously linear and, moreover, even continuous, its norm being given by

$$\|x\| = \sup_{\|a\| \leqslant 1} |\langle x, a \rangle|,$$

that is by the norm of x as an element of E. This is a consequence of the famous *Hahn–Banach extension theorem*. The operator that assigns to each $x \in E$ the corresponding functional over E' will be denoted by $K_E : E \to E''$ and called *the canonical embedding of E into its bidual* $E'' = (E')'$.

The agreement on the notation of functionals over a Banach space E secures the possibility of a representation of an operator $A : E \to F$ with rank $(A) \leqslant n$ by a finite sum

$$Ax = \sum_{i=1}^{n} \langle x, a_i \rangle y_i$$

with elements $y_i \in F$ and functionals $a_i \in E'$.

Quite often the investigation of finite rank operators is the basis for the investigation of infinite-dimensional operators, in so far as an infinite-dimensional operator $T : E \to F$ is approximated by finite rank operators $A : E \to F$. In the case of an operator $T : E \to C(X)$ with values in a Banach space $C(X)$ of continuous functions the degree of approximability of T by finite rank operators is essentially determined by the degree of compactness of the underlying compact metric space X. This is proved in detail in chapter 5, section 5.6. On the other hand, one of the central results of the book says that a certain degree of approximability of any operator $T : E \to F$ by finite rank operators implies a certain degree of compactness of T. The precise derivation of this claim is given in section 3.1, 'Inequalities of Bernstein type'.

Another central result concerns the influence of the degree of compactness of an operator $T : E \to E$ on the *rate of decrease of the eigenvalue sequence* $\lambda_1(T), \lambda_2(T), \ldots, \lambda_n(T), \ldots$ The inequalities expressing the corresponding relation between compactness properties and spectral properties of operators $T : E \to E$ are proved in the framework of 'A refined Riesz theory' (chapter 4, section 4.2). They enable us to predict certain summability properties of the eigenvalues $\lambda_n(T)$ under appropriate suppositions about the degree of compactness of T.

But how should we quantify the degree of compactness of operators $T : E \to F$? Among various possibilities for doing this there is a predestinate one which refers to *entropy quantities* of T. What are they?

In chapter 1, sections 1.1 and 1.2, entropy quantities are first introduced for bounded subsets of metric spaces. In the subsequent sections 1.3 and 1.4 these concepts are transferred to operators $T \in L(E, F)$.

The *notion of a metric space*, which is basic for the beginning of the book, is also a central notion for chapter 5 devoted to *operators* $T : E \to C(X)$ *with values in a Banach space* $C(X)$ of continuous functions on a compact metric space X. As already mentioned, the degree of compactness of X or, in other words, the entropy properties of X, imply a certain degree of approximability of T by finite rank operators and hence, because of the Bernstein type inequalities, also certain entropy properties of T. This fact alone would justify the investigation of operators $T: E \rightarrow C(X)$ in a separate chapter. But in addition it turns out that *compactness properties of $C(X)$-valued operators* are in a sense *representatives for compactness properties of operators* between arbitrary Banach spaces. This will be pointed out in section 5.1.

The book ends with chapter 6, 'Operator theoretical methods in the local theory of Banach spaces'. This final chapter seeks to demonstrate that the theory of Banach spaces, which seems to take priority over the theory of operators in Banach spaces, at least up to a certain extent, can also be developed on the basis of the operator theory.

1
Entropy quantities

Entropy quantities, in particular entropy numbers, in their proper sense, are set functions defined for bounded subsets of metric spaces. The theory of linear and continuous operators between Banach spaces involves the consideration of bounded subsets in Banach spaces. Indeed, the image $T(U_E)$ of the closed unit ball U_E of a Banach space E under a linear and continuous operator T from E into a Banach space F is a bounded subset of F and thus possesses well-defined entropy numbers. Having observed this, one can conjecture that entropy numbers might be useful tools for investigating properties of linear and continuous operators between Banach spaces. They *are* in fact and, as we shall see later, in particular prove to be appropriate quantities for reflecting compactness and spectral properties of operators.

1.1. Entropy numbers of sets

Let (X, d) be a metric space and $x_0 \in X$. Then we denote by
$$\mathring{U}(x_0; \varepsilon) = \{x \in X : d(x, x_0) < \varepsilon\}$$
the open ball of radius $\varepsilon > 0$ with centre x_0, and by
$$U(x_0; \varepsilon) = \{x \in X : d(x, x_0) \leqslant \varepsilon\}$$
the corresponding closed ball. A subset $M \subseteq X$ is said to be *bounded* if it is contained in an appropriate ball $U(x_0; \varepsilon)$.

Bounded subsets of metric spaces give rise to covering problems as well as packing problems. The *coverings* we are thinking of are coverings by closed balls $U(x_i; \varepsilon)$ of uniform radius ε so that
$$M \subseteq \bigcup_{i \in I} U(x_i; \varepsilon). \tag{1.1.1}$$

A system of points $\{x_i\}_{i \in I}$ producing a covering (1.1.1) of M in X is called an *ε-net for M in X*. By a *packing* of a bounded subset $M \subseteq X$ we mean a system of points $x_i \in M$, $i \in I$, such that
$$d(x_i, x_k) > 2\rho \quad \text{for } i \neq k \text{ and all } i, k \in I, \tag{1.1.2}$$
where ρ is a certain positive number. A system $\{x_i\}_{i \in I}$ with the property (1.1.2) is also called a *ρ-distant subset of $M \subseteq X$*.

There is a close relation between covering and packing properties of bounded subsets in metric spaces (cf. Kolmogorov and Tichomirov 1959; Mitjagin 1961; Mitjagin and Pełczyński 1966; Lorentz 1966; Triebel 1970; Pietsch 1978). In order to reveal this relation, we introduce entropy numbers $\varepsilon_n(M)$ and capacity numbers or so-called inner entropy numbers $\varphi_n(M)$ for bounded sets $M \subseteq X$. The *nth entropy number* $\varepsilon_n(M)$ is defined by

$$\varepsilon_n(M) = \inf \{ \varepsilon > 0 : \text{there exists an } \varepsilon\text{-net for } M$$
$$\text{in } X \text{ consisting of } q \leqslant n \text{ points} \}.$$

Of course, one can also operate with coverings of M by open balls instead of closed ones, and characterize $\varepsilon_n(M)$ as

$$\varepsilon_n(M) = \inf \left\{ \varepsilon > 0 : \text{there exist } q \leqslant n \text{ points } x_1, x_2, \ldots, x_q \right.$$
$$\left. \text{in } X \text{ such that } M \subseteq \bigcup_{i=1}^{q} \overset{\circ}{U}(x_i; \varepsilon) \right\}.$$

The *nth inner entropy number* $\varphi_n(M)$ is defined by

$$\varphi_n(M) = \sup \{ \rho > 0 : \text{there exists a } \rho\text{-distant subset}$$
$$\text{of } M \text{ consisting of } p > n \text{ points} \}$$

or, equivalently,

$$\varphi_n(M) = \sup \{ \rho > 0 : \text{there exist } p > n \text{ points } x_1, x_2, \ldots, x_p$$
$$\text{in } M \text{ such that } d(x_i, x_k) \geqslant 2\rho \text{ for } i \neq k \}.$$

If M consists of less than $n + 1$ elements we put

$$\varphi_n(M) = 0.$$

Given $\varepsilon > \varepsilon_n(M)$ and ρ with $0 < \rho < \varphi_n(M)$ we can find points $y_1, y_2, \ldots, y_q$ in X and points $x_1, x_2, \ldots, x_p$ in M with $q \leqslant n$ and $p > n$ such that

$$M \subseteq \bigcup_{j=1}^{q} U(y_j; \varepsilon) \text{ and } d(x_i, x_k) > 2\rho \quad \text{for } i \neq k, 1 \leqslant i, k \leqslant p.$$

Since $p > n \geqslant q$ at least one ball $U(y_j; \varepsilon)$ must contain two of the elements $x_i \in M$, say

$$x_i \in U(y_j; \varepsilon) \quad \text{and} \quad x_k \in U(y_j; \varepsilon).$$

Accordingly, we have

$$d(x_i, x_k) \leqslant d(x_i, y_j) + d(y_j, x_k) \leqslant 2\varepsilon.$$

Hence it follows that $2\rho < 2\varepsilon$ which implies

$$\varphi_n(M) \leqslant \varepsilon_n(M). \tag{1.1.3}$$

On the other hand, for any ρ with $\varphi_n(M) < \rho$ there exists a maximal ρ-distant subset $\{x_1, x_2, \ldots, x_q\}$ of M with $q \leqslant n$. The maximality of the subset $\{x_1, x_2, \ldots, x_q\}$ amounts to the fact that one can assign to any $x \in M$ at least one element x_i such that

$$d(x, x_i) \leqslant 2\rho.$$

Thus

$$M \subseteq \bigcup_{i=1}^{q} U(x_i; 2\rho)$$

turns out to be true which shows that $\varepsilon_n(M) \leqslant 2\rho$. The final result is

$$\varepsilon_n(M) \leqslant 2\varphi_n(M). \tag{1.1.4}$$

A subset M of a metric space X is called *precompact* if for every $\varepsilon > 0$ there exists a finite ε-net for M. The minimal number $m = N(M; \varepsilon)$ of elements $x_1, x_2, \ldots, x_m$ in X forming an ε-net for M is called *the entropy function of the precompact set M*. If we are given an arbitrary finite ε-net $\{x_1, x_2, \ldots, x_n\}$ of M so that

$$M \subseteq \bigcup_{i=1}^{n} U(x_i; \varepsilon),$$

we may conclude that $n \geqslant N(M; \varepsilon)$ and

$$\varepsilon_n(M) \leqslant \varepsilon \quad \text{for all } n \geqslant N(M; \varepsilon).$$

Hence we have

$$\lim_{n \to \infty} \varepsilon_n(M) = 0 \tag{1.1.5}$$

for every precompact set $M \subseteq X$. Conversely, the condition (1.1.5) proves to be sufficient for the precompactness of the set M. An equivalent characterization of precompact sets $M \subseteq X$ says that

$$\lim_{n \to \infty} \varphi_n(M) = 0. \tag{1.1.6}$$

Since the sequences $\varepsilon_n(M)$ and $\varphi_n(M)$ are monotonously decreasing, the rate of decrease may be regarded as a measure for the degree of precompactness of the set M.

In a finite-dimensional Banach space every bounded subset is precompact. This can easily be seen by estimating the entropy numbers of the unit ball of a Banach space E with $\dim(E) = m$ from above. Let

$$U_E = \{x \in E : \|x\| \leqslant 1\}$$

stand for the closed unit ball of E and

$$\mathring{U}_E = \{x \in E : \|x\| < 1\}$$

for the open one. We refer to the inner entropy numbers $\varphi_n(U_E)$ of U_E, fix an arbitrary positive number $\rho < \varphi_n(U_E)$, and determine a system of $p > n$ elements $x_1, x_2, \ldots, x_p$ in U_E with

$$\|x_i - x_k\| > 2\rho \quad \text{for } i \neq k. \tag{1.1.7}$$

Then we consider the closed balls

$$U(x_i; \rho) = \{x_i + \rho U_E\}.$$

Since

$$\rho < \varphi_n(U_E) \leqslant \varphi_1(U_E) = 1,$$

we have

$$\|x\| \leqslant \|x - x_i\| + \|x_i\| \leqslant \rho + 1 < 2 \quad \text{for } x \in U(x_i; \rho)$$

and, furthermore,

$$U(x_i; \rho) \cap U(x_k; \rho) = \varnothing \quad \text{for } i \neq k$$

by (1.1.7). Let us regard $U(x_i; \rho)$, $1 \leqslant i \leqslant p$, as a system of p non-intersecting closed subsets of the m-dimensional euclidean space l_2^m. Since these subsets are contained in the closed subset $2U_E$ we may use the Lebesgue measure on l_2^m and carry out a comparison of volumes, namely

$$\sum_{i=1}^{p} \text{vol}_m(U(x_i; \rho)) \leqslant \text{vol}_m(2U_E)$$

which amounts to

$$p \cdot \rho^m \cdot \text{vol}_m(U_E) \leqslant 2^m \text{vol}_m(U_E).$$

Because $\text{vol}_m(U_E) > 0$ and $p > n$ we may conclude that

$$\rho \leqslant 2 \cdot n^{-1/m}.$$

This yields

$$\varphi_n(U_E) \leqslant 2 \cdot n^{-1/m} \tag{1.1.8}$$

and thus confirms the statement $\lim_{n \to \infty} \varphi_n(U_E) = 0$. But, what is more, the sequence $n^{-1/m}$ exactly reflects the asymptotic behaviour of the sequences $(\varphi_n(U_E))$ and $(\varepsilon_n(U_E))$. Indeed, given $\varepsilon > \varepsilon_n(U_E)$ there exist $q \leqslant n$ elements $y_1, y_2, \ldots, y_q$ in E such that

$$U_E \subseteq \bigcup_{i=1}^{q} \{y_i + \varepsilon U_E\}.$$

This time a comparison of volumes in the m-dimensional euclidean space leads us to

$$\text{vol}_m(U_E) \leqslant q \cdot \varepsilon^m \text{vol}_m(U_E)$$

and finally yields

$$\varepsilon_n(U_E) \geqslant n^{-1/m}. \tag{1.1.9}$$

Combining (1.1.9), (1.1.4), and (1.1.8) we recognize that

$$n^{-1/m} \leqslant \varepsilon_n(U_E) \leqslant 4 \cdot n^{-1/m} \quad \text{for dim}(E) = m. \tag{1.1.10}$$

Let us emphasize that the asymptotic behaviour of the entropy numbers $\varepsilon_n(U_E)$ is essentially determined by the dimension m of the underlying Banach space E.

So far E has been tacitly supposed to be a real m-dimensional Banach space. In the case of a complex Banach space of dimension m the comparison of volumes takes place in a real euclidean space of dimension $2m$. Correspondingly, (1.1.10) has to be replaced by

$$n^{-1/2m} \leqslant \varepsilon_n(U_E) \leqslant 4 \cdot n^{-1/2m} \qquad (1.1.11)$$

1.2. Entropy moduli of sets

The comparison of volumes just carried out for coverings of the unit ball U_E of a finite-dimensional Banach space E opens new perspectives for estimating the degree of precompactness even in the general situation of a bounded subset in an arbitrary Banach space.

For the moment let $M \subset E$ be a bounded and Lebesgue-measurable subset of a Banach space E with $\dim(E) = m$. This means that the set M is measurable in the Lebesgue sense when it is considered as a subset of the m-dimensional euclidean space l_2^m. The unit ball U_E of the m-dimensional Banach space E always has this property. A covering

$$M \subseteq \bigcup_{i=1}^{k} \{x_i + \varepsilon U_E\}$$

of the set M, in a similar way as in the case $M = U_E$, then gives rise to an inequality

$$\mathrm{vol}_m(M) \leqslant k\varepsilon^m \, \mathrm{vol}_m(U_E) \qquad (1.2.1)$$

between the volume $\mathrm{vol}_m(M)$ of the set M and the volume $\mathrm{vol}_m(U_E)$ of the unit ball U_E. Replacing ε by the corresponding infimum $\varepsilon_k(M)$ we obtain the inequality

$$\left(\frac{\mathrm{vol}_m(M)}{\mathrm{vol}_m(U_E)} \right)^{1/m} \leqslant k^{1/m} \varepsilon_k(M). \qquad (1.2.2)$$

To obtain an optimal estimation of the so-called *volume ratio* $(\mathrm{vol}_m(M)/\mathrm{vol}_m(U_E))^{1/m}$ from above by the entropy quantities $k^{1/m}\varepsilon_k(M)$ we take the infimum with respect to k on the right-hand side of (1.2.2). Then

$$\left(\frac{\mathrm{vol}_m(M)}{\mathrm{vol}_m(U_E)} \right)^{1/m} \leqslant \inf_{1 \leqslant k < \infty} k^{1/m} \varepsilon_k(M) \qquad (1.2.3)$$

appears. Now we observe that the expression

$$g_n(M) = \inf_{1 \leqslant k < \infty} k^{1/n} \varepsilon_k(M), \quad n = 1, 2, 3, \ldots, \qquad (1.2.4)$$

makes sense for any bounded subset M of an arbitrary Banach space E. It is called *the nth entropy modulus of the set M* (cf. Carl 1982, 1984). The asymptotic behaviour of the sequence of entropy moduli $(g_n(M))$ again is a criterion for the degree of precompactness of the underlying set M.

Indeed, from the definition (1.2.4) of the $g_n(M)$, it immediately follows that

$$\inf_{1 \leqslant k < \infty} \varepsilon_k(M) \leqslant g_n(M) \leqslant n^{1/n} \varepsilon_n(M).$$

This implies

$$\lim_{n \to \infty} g_n(M) = \lim_{n \to \infty} \varepsilon_n(M). \tag{1.2.5}$$

Therefore we have $\lim_{n \to \infty} g_n(M) = 0$ if and only if $M \subset E$ is precompact. In particular, with the help of (1.1.10), we recognize that

$$g_n(U_E) = 0 \tag{1.2.6}$$

if E is a Banach space with $\dim(E) = m < n$.

In the case of a bounded subset $M \subset E$ of a complex Banach space E the analogous definition of the nth entropy modulus $g_n(M)$ is

$$g_n(M) = \inf_{1 \leqslant k < \infty} k^{1/2n} \varepsilon_k(M), \quad n = 1, 2, 3, \ldots \tag{1.2.4}'$$

This is because in a complex Banach space E with $\dim(E) = m$ a comparison of volumes has to take place in a real euclidean space of dimension $2m$.

1.3. Entropy numbers of operators

The term *operator* refers to a continuous linear transformation T from a Banach space E into a Banach space F. The class of all these operators is denoted by $L(E, F)$. The image $T(U_E)$ of the closed unit ball U_E of E, being a bounded subset of F, has entropy numbers $\varepsilon_n(T(U_E))$ as well as inner entropy numbers $\varphi_n(T(U_E))$. Obviously

$$\varepsilon_n(T(U_E)) = \varepsilon_n(T(\mathring{U}_E)) \quad \text{and} \quad \varphi_n(T(U_E)) = \varphi_n(T(\mathring{U}_E)).$$

We introduce the abbreviations

$$\varepsilon_n(T) = \varepsilon_n(T(U_E)) \quad \text{and} \quad \varphi_n(T) = \varphi_n(T(U_E)) \tag{1.3.1}$$

and call $\varepsilon_n(T)$ *the nth entropy number* and $\varphi_n(T)$ *the nth inner entropy number of T*.

First let us list the algebraic properties of the entropy numbers $\varepsilon_n(T)$.

(E1) **Monotonicity:** $\varepsilon_1(T) \geqslant \varepsilon_2(T) \geqslant \cdots \geqslant \varepsilon_n(T) \geqslant \cdots$
and

$$\varepsilon_1(T) = \| T \|.$$

(E2) **Additivity:** $\varepsilon_{kn}(T_1 + T_2) \leqslant \varepsilon_k(T_1) + \varepsilon_n(T_2)$ for $T_1, T_2 \in L(E, F)$, in particular
(E2)' $\varepsilon_n(T_1 + T_2) \leqslant \| T_1 \| + \varepsilon_n(T_2)$.

(E3) **Multiplicativity:** $\varepsilon_{kn}(RS) \leqslant \varepsilon_k(R)\varepsilon_n(S)$
for $S \in L(E, Z)$ and $R \in L(Z, F)$, in particular

(E3) (a) $\varepsilon_n(RS) \leqslant \|R\| \varepsilon_n(S)$
and

(E3) (b) $\varepsilon_k(RS) \leqslant \varepsilon_k(R)\|S\|$.

(ES) **Surjectivity:** $\varepsilon_n(TQ) = \varepsilon_n(T)$
for $T \in L(E, F)$ and any metric surjection $Q : \tilde{E} \to E$.

We recall that *a metric surjection Q from a Banach space $\tilde{E}$ onto a Banach space E* is characterized by the property

$$Q(\mathring{U}_{\tilde{E}}) = \mathring{U}_E. \tag{1.3.2}$$

It implies $(TQ)(\mathring{U}_{\tilde{E}}) = T(\mathring{U}_E)$ so that the surjectivity (ES) of $\varepsilon_n(T)$ becomes obvious.

The monotonicity of the sequence $(\varepsilon_n(T))$ is obvious as well. The inequality $\varepsilon_1(T) \leqslant \|T\|$ is an immediate consequence of $T(U_E) \subseteq \|T\| U_F$. On the other hand, assuming

$$T(U_E) \subseteq \{y_1 + \varepsilon U_F\}$$

we obtain

$$Tx = y_1 + \varepsilon y_+ \quad \text{and} \quad T(-x) = y_1 + \varepsilon y_-$$

for arbitrary $x \in U_E$, where y_+ and y_- are appropriate elements of U_F. It follows that

$$2\|Tx\| = \varepsilon \|y_+ - y_-\| \leqslant 2\varepsilon$$

which tells us that $\|T\| \leqslant \varepsilon_1(T)$.

To verify the additivity we choose $\sigma > 0$ arbitrarily and determine $q \leqslant k$ elements $y_1, y_2, \ldots, y_q$ in F such that

$$T_1(U_E) \subseteq \bigcup_{i=1}^{q} \{y_i + (\varepsilon_k(T_1) + \sigma)U_F\}$$

and simultaneously elements $z_1, z_2, \ldots, z_p$ in F with $p \leqslant n$ and

$$T_2(U_E) \subseteq \bigcup_{j=1}^{p} \{z_j + (\varepsilon_n(T_2) + \sigma)U_F\}.$$

Since

$$(T_1 + T_2)(U_E) \subseteq T_1(U_E) + T_2(U_E)$$

we get a covering

$$(\dot{T}_1 + T_2)(U_E) \subseteq \bigcup_{i=1}^{q} \bigcup_{j=1}^{p} \{y_i + z_j + (\varepsilon_k(T_1) + \varepsilon_n(T_2) + 2\sigma)U_F\}$$

and thus may conclude that

$$\varepsilon_{kn}(T_1 + T_2) \leqslant \varepsilon_k(T_1) + \varepsilon_n(T_2) + 2\sigma$$

which proves (E2).

The multiplicativity can be shown by similar arguments. Given

$S \in L(E, Z)$, $R \in L(Z, F)$ and $\sigma > 0$ we may start from

$$S(U_E) \subseteq \bigcup_{j=1}^{p} \{z_j + (\varepsilon_n(S) + \sigma)U_Z\} \tag{1.3.3}$$

and

$$R(U_Z) \subseteq \bigcup_{i=1}^{q} \{y_i + (\varepsilon_k(R) + \sigma)U_F\} \tag{1.3.4}$$

with $z_1, z_2, \ldots, z_p \in Z$, $p \leqslant n$, and $y_1, y_2, \ldots, y_q \in F$, $q \leqslant k$. Applying the operator R to both sides of (1.3.3) and using (1.3.4) we arrive at

$$(RS)(U_E) \subseteq \bigcup_{i=1}^{q} \bigcup_{j=1}^{p} \{Rz_j + (\varepsilon_n(S) + \sigma)y_i + (\varepsilon_n(S) + \sigma)(\varepsilon_k(R) + \sigma)U_F\}$$

which amounts to $\varepsilon_{kn}(RS) \leqslant \varepsilon_k(R)\varepsilon_n(S)$.

The algebraic properties of the inner entropy numbers $\varphi_n(T)$ differ from those of the $\varepsilon_n(T)$ only in a few details.

(F1) **Monotonicity:** $\varphi_1(T) \geqslant \varphi_2(T) \geqslant \cdots \geqslant \varphi_n(T) \geqslant \cdots$
and

$$\varphi_1(T) = \| T \|.$$

(F2) **Additivity:** $\varphi_{kn}(T_1 + T_2) \leqslant 2(\varphi_k(T_1) + \varphi_n(T_2))$
for $T_1, T_2 \in L(E, F)$, in particular

(F2)' $\varphi_n(T_1 + T_2) \leqslant 2(\| T_1 \| + \varphi_n(T_2))$.

(F3) **Multiplicativity:** $\varphi_{kn}(RS) \leqslant 4\varphi_k(R)\varphi_n(S)$
for $S \in L(E, Z)$ and $R \in L(Z, F)$, furthermore

(F3) (a) $\varphi_n(RS) \leqslant \| R \|\varphi_n(S)$
and

(F3) (b) $\varphi_k(RS) \leqslant \varphi_k(R)\| S \|$.

(FS) **Surjectivity:** $\varphi_n(TQ) = \varphi_n(T)$
for $T \in L(E, F)$ and any metric surjection $Q: \tilde{E} \to E$.

(FI) **Injectivity:** $\varphi_n(JT) = \varphi_n(T)$
for $T \in L(E, F)$ and any metric injection $J: F \to \tilde{F}$.

We recall that a *metric injection J of a Banach space F into a Banach space* $\tilde{F}$ is characterized by the property

$$\| Jy \| = \| y \| \quad \text{for all } y \in F. \tag{1.3.5}$$

The injectivity of the inner entropy numbers of an operator $T \in L(E, F)$ is due to the fact that the value $\varphi_n(T) = \varphi_n(T(U_E))$ – according to its definition – is determined by the metric of the set $M = T(U_E)$ only and does not change if $M \subset F$ is isometrically embedded into a Banach space $\tilde{F}$. The entropy numbers $\varepsilon_n(T) = \varepsilon_n(T(U_E))$, or *outer entropy numbers* as they are called sometimes (see Pietsch 1978), do not share this property because their definition refers to coverings of the set $M = T(U_E)$ by balls of the particular Banach space F which contains the range $T(E)$ of the operator T as a subspace. Nevertheless the outer entropy numbers $\varepsilon_n(T)$ are injective in a weaker sense, namely

$$\varepsilon_n(T) \leqslant 2\varepsilon_n(JT). \qquad (1.3.6)$$

This is due to

$$\varepsilon_n(T) \leqslant 2\varphi_n(T) \leqslant 2\varepsilon_n(T) \qquad (1.3.7)$$

(see (1.1.3) and (1.1.4)) and the injectivity (FI) of the inner entropy numbers $\varphi_n(T)$. It can be shown that the factor 2 in (1.3.6) cannot be reduced (see section 3.5, (3.5.17)).

We give a few comments on the remaining properties of the inner entropy numbers.

The monotonicity of the sequence $(\varphi_n(T))$ again is obvious. So let $\rho > 0$ be such that

$$\| Tx_1 - Tx_2 \| > 2\rho$$

for two elements $x_1, x_2 \in U_E$. Then we automatically have $\rho < \| T \|$ and hence also $\varphi_1(T) \leqslant \| T \|$. Conversely, let $\rho > \varphi_1(T)$. Then

$$\| Tx_1 - Tx_2 \| \leqslant 2\rho$$

turns out to be true for any two elements $x_1, x_2 \in U_E$ and, consequently,

$$\| Tx \| \leqslant 2\rho \quad \text{for all } x \in 2U_E.$$

This implies $\| T \| \leqslant \rho$ and hence $\| T \| \leqslant \varphi_1(T)$.

The additivity (F2) of the inner entropy numbers results from the additivity (E2) of the outer entropy numbers and from (1.3.7).

Similarly the multiplicativity (F3) of the inner entropy numbers rests on the multiplicativity (E3) of the outer entropy numbers and on the relations (1.3.7). The special versions (F3) (a) and (F3) (b) can easily be checked directly.

For the surjectivity (FS) the argument (1.3.2) is sufficient again.

The injectivity (FI) of the inner entropy numbers $\varphi_n(T)$, which has already been discussed, represents an adequate means of investigating the relations between the asymptotic behaviour of the sequences $\varphi_n(T)$ and $\varepsilon_n(T)$ and rank properties of the operator T.

Lemma 1.3.1. *Let E and F be arbitrary real Banach spaces and $T \in L(E, F)$. Then*

$$\varphi_n(T) \leqslant C \cdot n^{-1/m} \quad \text{for } n = 1, 2, 3, \ldots \text{ with } C > 0 \text{ implies } \operatorname{rank}(T) \leqslant m,$$
$$(1.3.8)$$

and

$$\varphi_n(T) \geqslant \tilde{C} \cdot n^{-1/m} \quad \text{for } n = 1, 2, 3, \ldots \text{ with } \tilde{C} > 0 \text{ implies } \operatorname{rank}(T) \geqslant m.$$
$$(1.3.9)$$

Proof. Let us assume rank $(T) > m$. Then there exists a subspace $F_0 \subseteq F$ with dimension dim $(F_0) = d > m$ such that the unit ball $U_{F_0} = F_0 \cap U_F$ of F_0 is absorbed by the set $T(U_E)$. This means

$$U_{F_0} \subseteq \sigma T(U_E)$$

with an appropriate constant $\sigma > 0$. Owing to the injectivity of the inner entropy numbers we may conclude that

$$\varphi_n(U_{F_0}) \leqslant \sigma \varphi_n(T(U_E)) = \sigma \varphi_n(T).$$

Since

$$2\varphi_n(U_{F_0}) \geqslant \varepsilon_n(U_{F_0}) \geqslant n^{-1/d}$$

by (1.1.4) and (1.1.9) the supposition $\varphi_n(T) \leqslant C \cdot n^{-1/m}$ leads us to

$$n^{-1/d} \leqslant 2\sigma C \cdot n^{-1/m} \quad \text{for } n = 1, 2, 3, \dots.$$

But this contradicts the assumption $d > m$ and thus proves (1.3.8).

Now let us assume rank $(T) < m$ and the validity of an estimate (1.3.9) with $\tilde{C} > 0$. The canonical factorization

$$T = J_0 T_0 \tag{1.3.10}$$

of T over its range $F_0 = T(E) = \overline{T(E)}$ with $T_0 : E \to F_0$ as the induced operator and $J_0 : F_0 \to F$ as the embedding map then represents a factorization over a finite-dimensional Banach space F_0 with dim $(F_0) \leqslant m - 1$. Using the multiplicativity (F3)(b) and the injectivity (FI) of φ_n we see that

$$\varphi_n(T) \leqslant \varphi_n(J_0) \| T_0 \| = \varphi_n(I_{F_0}) \| T \| \tag{1.3.11}$$

with I_{F_0} as the identity map of F_0. Since the inner entropy number $\varphi_n(I_{F_0})$ is defined by $\varphi_n(I_{F_0}) = \varphi_n(U_{F_0})$, the estimate (1.1.8) applies, namely

$$\varphi_n(I_{F_0}) \leqslant 2 \cdot n^{-1/(m-1)} \tag{1.3.12}$$

for dim $(F_0) \leqslant m - 1$. The proof of rank $(T) \geqslant m$ is finally completed by the sequence of estimates

$$\tilde{C} \cdot n^{-1/m} \leqslant \varphi_n(T) \leqslant 2 \| T \| \cdot n^{-1/(m-1)}$$

which yields the contradiction

$$n^{1/(m-1)-1/m} \leqslant \frac{2}{\tilde{C}} \| T \| \quad \text{for } n = 1, 2, 3, \dots \qquad \blacksquare$$

Proposition 1.3.1. *An operator $T \in L(E, F)$ acting between arbitrary real Banach spaces E and F is of rank m if and only if there exists a constant $\tilde{C} > 0$ such that*

$$\tilde{C} \cdot n^{-1/m} \leqslant \varphi_n(T) \leqslant 2 \| T \| \cdot n^{-1/m} \quad \text{for } n = 1, 2, 3, \dots \tag{1.3.13}$$

or, equivalently, a constant $C > 0$ such that

$$C \cdot n^{-1/m} \leqslant \varepsilon_n(T) \leqslant 4 \| T \| \cdot n^{-1/m} \quad \text{for } n = 1, 2, 3, \dots \tag{1.3.14}$$

Proof. Let $\text{rank}(T) = m$; then we can employ arguments similar to those in the proof of (1.3.8) and (1.3.9) in Lemma 1.3.1. Indeed, since the range $F_0 = T(E)$ of T is a subspace of dimension m of F, we have

$$n^{-1/m} \leqslant 2\varphi_n(U_{F_0})$$

as well as

$$\varphi_n(U_{F_0}) \leqslant \sigma\varphi_n(T(U_E)) = \sigma\varphi_n(T),$$

for some $\sigma > 0$, because the unit ball U_{F_0} of F_0 is absorbed by the set $T(U_E)$. The final result can be written as

$$\tilde{C} \cdot n^{-1/m} \leqslant \varphi_n(T) \quad \text{with } \tilde{C} = \frac{1}{2\sigma}.$$

On the other hand, the estimate

$$\varphi_n(T) \leqslant 2\|T\| \cdot n^{-1/m}$$

results from (1.3.11) and (1.3.12) if we replace $m - 1$ by m. This proves (1.3.13). Conversely, the supposition (1.3.13) implies $\text{rank}(T) = m$ by Lemma 1.3.1. The statement (1.3.14) is simply an equivalent version of (1.3.13) in terms of outer entropy numbers $\varepsilon_n(T)$. ∎

Both the estimates (1.3.13) and (1.3.14) are characteristic properties of rank m operators T between real Banach spaces. When characterizing rank m operators T acting between complex Banach spaces E and F we have to refer to

$$n^{-1/2m} \leqslant \varepsilon_n(U_{F_0}) \leqslant 2\varphi_n(U_{F_0})$$

and

$$\varphi_n(U_{F_0}) \leqslant 2 \cdot n^{-1/2m}.$$

Thus we obtain

$$\tilde{C} \cdot n^{-1/2m} \leqslant \varphi_n(T) \leqslant 2\|T\| \cdot n^{-1/2m} \qquad (1.3.13)'$$

for $n = 1, 2, 3, \ldots$ with $\tilde{C} > 0$, and

$$C \cdot n^{-1/2m} \leqslant \varepsilon_n(T) \leqslant 4\|T\| \cdot n^{-1/2m} \qquad (1.3.14)'$$

for $n = 1, 2, 3, \ldots$ with $C > 0$ as characteristic properties for rank m operators T in the case of complex Banach spaces E and F.

Proposition 1.3.1 makes clear that, the larger the rank m of an operator T, the smaller the rate of decrease of its entropy numbers. In the case of an operator T with infinite-dimensional range the entropy numbers need not tend to zero at all.

Let us treat as an example a *diagonal operator* $D: l_p \to l_p$ in an l_p space with $1 \leqslant p \leqslant \infty$. The entropy behaviour of D is interesting in itself.

Furthermore, it will be the basis for determining the entropy behaviour of operators T between arbitrary Banach spaces E and F (see chapter 3).

Proposition 1.3.2. *Let $\sigma_1 \geq \sigma_2 \geq \cdots \sigma_k \geq \cdots \geq 0$ be a non-increasing sequence of non-negative numbers and let*

$$Dx = (\sigma_1 \xi_1, \sigma_2 \xi_2, \ldots, \sigma_k \xi_k, \ldots) \qquad (1.3.15)$$

for $x = (\xi_1, \xi_2, \ldots, \xi_k, \ldots) \in l_p$ be the diagonal operator from l_p with $1 \leq p \leq \infty$ into itself, generated by the sequence (σ_i). Then

$$\sup_{1 \leq k < \infty} n^{-1/k}(\sigma_1 \sigma_2 \cdots \sigma_k)^{1/k} \leq \varepsilon_n(D) \leq 6 \cdot \sup_{1 \leq k < \infty} n^{-1/k}(\sigma_1 \sigma_2 \cdots \sigma_k)^{1/k} \qquad (1.3.16)$$

in the case of the real l_p space and

$$\sup_{1 \leq k < \infty} n^{-1/2k}(\sigma_1 \sigma_2 \cdots \sigma_k)^{1/k} \leq \varepsilon_n(D) \leq 6 \cdot \sup_{1 \leq k < \infty} n^{-1/2k}(\sigma_1 \sigma_2 \cdots \sigma_k)^{1/k} \qquad (1.3.16)'$$

in the case of the complex l_p spaces, $n = 1, 2, 3, \ldots$ (see König 1986; Gordon, König and Schütt 1987).

Proof. We confine ourselves to the real case. To prove the estimate from below we fix k and consider the operator $D^{(k)} : l_p^k \to l_p^k$ given by

$$D^{(k)}(\xi_1, \xi_2, \ldots, \xi_k) = (\sigma_1 \xi_1, \sigma_2 \xi_2, \ldots, \sigma_k \xi_k). \qquad (1.3.17)$$

Denoting by $I_k : l_p^k \to l_p$ the natural embedding of l_p^k into l_p and by $P_k : l_p \to l_p^k$ the natural projection of l_p onto l_p^k we can write

$$D^{(k)} = P_k D I_k. \qquad (1.3.18)$$

By (E3) we get

$$\varepsilon_n(D^{(k)}) \leq \varepsilon_n(D) \qquad (1.3.19)$$

from (1.3.18). Now let $\varepsilon > \varepsilon_n(D^{(k)})$. Then there are elements $x_1, x_2, \ldots, x_n$ in l_p^k such that

$$D^{(k)}(U_p^k) \subseteq \bigcup_{i=1}^{n} \{x_i + \varepsilon U_p^k\}, \qquad (1.3.20)$$

where U_p^k stands for the closed unit ball of the space l_p^k. The set $D^{(k)}(U_p^k)$, like the unit ball U_p^k itself, possesses a k-dimensional volume (see sections 1.1 and 1.2), which is related to the volume $\mathrm{vol}_k(U_p^k)$ of U_p^k by

$$\mathrm{vol}_k(D^{(k)}(U_p^k)) = \sigma_1 \sigma_2 \cdots \sigma_k \cdot \mathrm{vol}_k(U_p^k).$$

Accordingly, the inclusion (1.3.20) implies

$$\sigma_1 \sigma_2 \cdots \sigma_k \cdot \mathrm{vol}_k(U_p^k) \leq n \cdot \varepsilon^k \, \mathrm{vol}_k(U_p^k)$$

which amounts to

$$n^{-1/k}(\sigma_1 \sigma_2 \cdots \sigma_k)^{1/k} \leq \varepsilon.$$

Passing from $\varepsilon > \varepsilon_n(D^{(k)})$ to the limit case $\varepsilon = \varepsilon_n(D^{(k)})$ and using (1.3.19) we obtain

$$n^{-1/k}(\sigma_1\sigma_2\cdots\sigma_k)^{1/k} \leqslant \varepsilon_n(D). \qquad (1.3.21)$$

Because the right-hand side of this inequality is independent of k we also have

$$\sup_{1\leqslant k<\infty} n^{-1/k}(\sigma_1\sigma_2\cdots\sigma_k)^{1/k} \leqslant \varepsilon_n(D).$$

Having proved the required estimate for $\varepsilon_n(D)$ from below we now prove the estimate from above. We put

$$\delta(n) = 8\cdot \sup_{1\leqslant k<\infty} n^{-1/k}(\sigma_1\sigma_2\cdots\sigma_k)^{1/k} \qquad (1.3.22)$$

and show that for all n there is an index r with $\sigma_{r+1}\leqslant \delta(n)/4$. For this purpose we choose a natural number r such that $n\leqslant 2^{r+1}$ and thus

$$1 \leqslant 2\cdot n^{-1/(r+1)}. \qquad (1.3.23)$$

We have

$$\sigma_{r+1} \leqslant (\sigma_1\sigma_2\cdots\sigma_{r+1})^{1/(r+1)}$$

because of the monotonicity of the sequence (σ_k) and, hence

$$\sigma_{r+1} \leqslant 2\cdot n^{-1/(r+1)}(\sigma_1\sigma_2\cdots\sigma_{r+1})^{1/(r+1)}$$

in view of (1.3.23). Using the definition (1.3.22) of $\delta(n)$ we now may conclude that

$$\sigma_{r+1} \leqslant \frac{\delta(n)}{4}.$$

If even σ_1 happens to be less than or equal to $\delta(n)/4$ we have

$$\varepsilon_n(D) \leqslant \|D\| = \sigma_1 \leqslant \frac{\delta(n)}{4} \leqslant 6\cdot \sup_{1\leqslant k<\infty} n^{-1/k}(\sigma_1\sigma_2\cdots\sigma_k)^{1/k}$$

for $n = 1, 2, 3, \ldots$ and are done. If this is not so, then there exists an index $r = k$ with

$$\sigma_{k+1} \leqslant \frac{\delta(n)}{4} < \sigma_k.$$

The corresponding sectional operator $D_k : l_p \to l_p$ given by

$$D_k(\xi_1, \xi_2, \ldots, \xi_k, \xi_{k+1}, \ldots) = (\sigma_1\xi_1, \sigma_2\xi_2, \ldots, \sigma_k\xi_k, 0, 0, \ldots) \qquad (1.3.24)$$

is of rank k and the image $D_k(U_p)$ of the closed unit ball U_p of l_p is isometric to the subset $D^{(k)}(U_p^k)$ of l_p^k. In any case $D_k(U_p)$ is a precompact subset of l_p. So let $y_1, y_2, \ldots, y_N$ be a maximal system of elements in $D_k(U_p)$ with

$$\|y_i - y_j\| > \frac{\delta(n)}{2} \quad \text{for } i \neq j$$

(cf. (1.1.6)). The maximality of the system $y_1, y_2, \ldots, y_N$ in $D_k(U_p)$ guarantees that

$$D_k(U_p) \subseteq \bigcup_{j=1}^{N} \left\{ y_j + \frac{\delta(n)}{2} U_p \right\} \tag{1.3.25}$$

and thus

$$\varepsilon_N(D_k) \leqslant \frac{\delta(n)}{2}. \tag{1.3.26}$$

In order to get an estimate for $\varepsilon_N(D)$ we split the operator D into the two items

$$D = (D - D_k) + D_k$$

and apply (E2)', namely

$$\varepsilon_N(D) \leqslant \|D - D_k\| + \varepsilon_N(D_k). \tag{1.3.27}$$

Using $\|D - D_k\| = \sigma_{k+1} \leqslant \delta(n)/4$ and (1.3.26) we arrive at

$$\varepsilon_N(D) \leqslant 3 \cdot \frac{\delta(n)}{4}. \tag{1.3.28}$$

It remains to estimate the number N. This again is achieved by a comparison of volumes. We consider the sets $\{y_j + (\delta(n)/4)U_p^k\}$ as subsets of the space l_p^k which is possible since $y_j \in D_k(U_p)$ and $D_k(U_p) = D^{(k)}(U_p^k)$ (cf. (1.3.17)). They are obviously pairwise disjoint. On the other hand we have

$$\bigcup_{j=1}^{N} \left\{ y_j + \frac{\delta(n)}{4} U_p^k \right\} \subseteq D^{(k)}(U_p^k) + \frac{\delta(n)}{4} U_p^k \subseteq 2D^{(k)}(U_p^k) \tag{1.3.29}$$

because $\delta(n)/4 < \sigma_k \leqslant \sigma_{k-1} \leqslant \cdots \leqslant \sigma_1$. After these remarks a comparison of the euclidean volumes provides

$$N \cdot \left(\frac{\delta(n)}{4} \right)^k \cdot \mathrm{vol}_k(U_p^k) \leqslant 2^k \cdot \sigma_1 \sigma_2 \cdots \sigma_k \cdot \mathrm{vol}_k(U_p^k)$$

which amounts to

$$N \leqslant \left(\frac{8}{\delta(n)} \right)^k \cdot \sigma_1 \sigma_2 \cdots \sigma_k. \tag{1.3.30}$$

Using the definition (1.3.22) of $\delta(n)$ we can estimate the quotient $8/\delta(n)$ by

$$\frac{8}{\delta(n)} \leqslant n^{1/k} \cdot (\sigma_1 \sigma_2 \cdots \sigma_k)^{-1/k}.$$

This implies $N \leqslant n$. From (1.3.28) it now follows that

$$\varepsilon_n(D) \leqslant \tfrac{3}{4} \delta(n) = 6 \cdot \sup_{1 \leqslant k < \infty} n^{-1/k} (\sigma_1 \sigma_2 \cdots \sigma_k)^{1/k}.$$

This completes the proof of (1.3.16). The complex counterpart (1.3.16)' can be treated along the same lines. ∎

An operator T between arbitrary Banach spaces E and F is called *compact*, if $T(U_E)$ is a precompact subset of F. Correspondingly,

$$\lim_{n \to \infty} \varepsilon_n(T) = 0 \qquad (1.3.31)$$

and

$$\lim_{n \to \infty} \varphi_n(T) = 0 \qquad (1.3.32)$$

are characteristic properties of compact operators.

By means of the condition (1.3.31) one can easily check that the diagonal operator $D{:}l_p \to l_p$ studied in Proposition 1.3.2 is compact if and only if $\lim_{n \to \infty} \sigma_n = 0$. Indeed, the left-hand part of the inequality (1.3.16) tells us that

$$n^{-1/n}\sigma_n \leqslant \varepsilon_n(D)$$

and hence

$$\sigma_n \leqslant n^{1/n}\varepsilon_n(D).$$

The assumption $\lim_{n \to \infty} \varepsilon_n(D) = 0$ thus leads us to $\lim_{n \to \infty} \sigma_n = 0$. Conversely, let us suppose $\lim_{n \to \infty} \sigma_n = 0$. Then again we employ the sectional operator $D_k{:}l_p \to l_p$ defined by (1.3.24) and take into consideration that

$$\varepsilon_n(D) \leqslant \|D - D_k\| + \varepsilon_n(D_k) = \sigma_{k+1} + \varepsilon_n(D_k).$$

Given $\varepsilon > 0$ we can achieve $\sigma_{k+1} < \varepsilon/2$ by choosing k sufficiently large. For fixed k, we get $\varepsilon_n(D_k) < \varepsilon/2$ if n is chosen sufficiently large, say $n > n_\varepsilon$, because D_k is of finite rank (cf. Proposition 1.3.1). In this way we obtain the required result

$$\varepsilon_n(D) < \varepsilon \quad \text{for } n > n_\varepsilon.$$

Estimating the degree of compactness of an operator T can be understood as estimating the rate of decrease of its entropy numbers $\varepsilon_n(T)$. By Proposition 1.3.1 the entropy numbers of a compact operator with infinite-dimensional range tend to zero more slowly than any power $n^{-\alpha}$ with $\alpha > 0$. Therefore it is logical to ask if perhaps a certain power $(1 + \log n)^{-\beta}$ with $\beta > 0$ can be used as an upper bound for the $\varepsilon_n(T)$. In fact, the claim

$$\varepsilon_n(T) \leqslant C \cdot (1 + \log n)^{-\beta}, \quad n = 1, 2, 3, \ldots, \qquad (1.3.33)$$

gives rise to an interesting class of compact operators (cf. Triebel 1970). However, it is rather inconvenient to work with these logarithmic terms. They can be removed if we think of $\log n$ as the dyadic logarithm of n

and, instead of $\varepsilon_n(T)$, introduce

$$e_n(T) = \varepsilon_{2^{n-1}}(T) \quad \text{for } n = 1, 2, 3, \ldots \tag{1.3.34}$$

(see Pietsch 1978). Then (1.3.33) yields

$$e_n(T) \leqslant C \cdot n^{-\beta}, \quad n = 1, 2, 3, \ldots \tag{1.3.35}$$

It turns out that (1.3.33) and (1.3.35) are asymptotically equivalent, so that from this point of view there is no loss of information in the transition to the *dyadic entropy numbers*, as the $e_n(T)$ will be called from now on.

Rank properties are also reflected entirely by the dyadic entropy numbers. An operator T acting between real Banach spaces E and F is of rank m if and only if there exists a constant $C > 0$ such that

$$C \cdot 2^{-(n-1)/m} \leqslant e_n(T) \leqslant 4 \| T \| \cdot 2^{-(n-1)/m}, \quad n = 1, 2, 3, \ldots \tag{1.3.36}$$

In the complex case

$$C \cdot 2^{-(n-1)/2m} \leqslant e_n(T) \leqslant 4 \| T \| \cdot 2^{-(n-1)/2m}, \quad n = 1, 2, 3, \ldots \tag{1.3.36'}$$

is a characteristic property of rank m operators T.

For later use we summarize those algebraic properties of the dyadic entropy numbers $e_n(T)$ which formally differ from the algebraic properties of the original entropy numbers $\varepsilon_n(T)$:

(DE2) **Additivity:** $e_{k+n-1}(T_1 + T_2) \leqslant e_k(T_1) + e_n(T_2)$
  for $T_1, T_2 \in L(E, F)$, in particular
(DE2)' $e_n(T_1 + T_2) \leqslant \| T_1 \| + e_n(T_2)$.
(DE3) **Multiplicativity:** $e_{k+n-1}(RS) \leqslant e_k(R)e_n(S)$
  for $S \in L(E, Z)$ and $R \in L(Z, F)$, in particular
(DE3) (a) $e_n(RS) \leqslant \| R \| e_n(S)$
  and
(DE3) (b) $e_k(RS) \leqslant e_k(R) \| S \|$.

For the sake of completeness we introduce the *dyadic inner entropy numbers*

$$f_n(T) = \varphi_{2^{n-1}}(T) \quad \text{for } n = 1, 2, 3, \ldots \tag{1.3.37}$$

(see Pietsch 1978) listing their modified algebraic properties, too:

(DF2) **Additivity:** $f_{k+n-1}(T_1 + T_2) \leqslant 2(f_k(T_1) + f_n(T_2))$
  for $T_1, T_2 \in L(E, F)$, in particular
(DF2)' $f_n(T_1 + T_2) \leqslant 2(\| T_1 \| + f_n(T_2))$.
(DF3) **Multiplicativity:** $f_{k+n-1}(RS) \leqslant 4f_k(R)f_n(S)$
  for $S \in L(E, Z)$ and $R \in L(Z, F)$, furthermore
(DF3) (a) $f_n(RS) \leqslant \| R \| f_n(S)$
  and
(DF3) (b) $f_k(RS) \leqslant f_k(R) \| S \|$.

1.4. Entropy moduli of operators

Given an operator T between arbitrary real Banach spaces E and F we use the abbreviation $g_n(T)$ for the nth entropy modulus $g_n(M)$ of the set $M = T(U_E)$ and call this quantity

$$g_n(T) = \inf_{1 \leq k < \infty} k^{1/n} \varepsilon_k(T), \quad n = 1, 2, 3, \dots, \tag{1.4.1}$$

the nth entropy modulus of T (see Carl 1982, 1984). By (1.2.5) an operator T is compact if and only if $\lim_{n \to \infty} g_n(T) = 0$. Also the algebraic and rank properties of the $g_n(T)$ can easily be derived from the algebraic and rank properties of the $\varepsilon_n(T)$.

(M1) **Monotonicity:** $g_1(T) \geq g_2(T) \geq \cdots \geq g_n(T) \geq \cdots \geq 0$
and

$$g_1(T) = \|T\|.$$

(M2) **Multiplicativity:** $g_n(RS) \leq g_n(R)g_n(S)$
for $S \in L(E, Z)$ and $R \in L(Z, F)$, in particular

(M2) (a) $g_n(RS) \leq \|R\| g_n(S)$
and

(M2) (b) $g_n(RS) \leq g_n(R)\|S\|$.

(M3) **Rank property:** $g_n(T) = 0$ if and only if rank $(T) < n$.

(MS) **Surjectivity:** $g_n(TQ) = g_n(T)$
for $T \in L(E, F)$ and any metric surjection $Q: \tilde{E} \to E$.

Since it is clear that the $g_n(T)$ are monotonous and $g_1(T) \leq \varepsilon_1(T) = \|T\|$, we turn to the proof of $\|T\| \leq g_1(T)$. We remind the reader that the definition of $g_1(T)$ originates from a comparison of volumes in the one-dimensional euclidean space E_1. In order to obtain a one-dimensional operator $S: E_1 \to E_1$ related to the operator $T: E \to F$ in an appropriate way we fix $\eta > 0$ and determine an element $x_0 \in U_E$ as well as a functional $b_0 \in U_{F'}$ such that

$$\|T\| < (1 + \eta)|\langle Tx_0, b_0 \rangle|. \tag{1.4.2}$$

Then we put

$$S\xi = \langle Tx_0, b_0 \rangle \xi \quad \text{for } \xi \in E_1.$$

The image $S(U_{E_1})$ of the one-dimensional 'unit ball' $U_{E_1} = [-1, +1]$ under S possesses the one-dimensional volume $2|\langle Tx_0, b_0 \rangle|$ so that the volume ratio is given by

$$\frac{\text{vol}_1(S(U_{E_1}))}{\text{vol}_1(U_{E_1})} = |\langle Tx_0, b_0 \rangle|.$$

According to (1.2.2) we therefore have

$$|\langle Tx_0, b_0 \rangle| \leq k \cdot \varepsilon_k(S) \quad \text{for } k = 1, 2, 3, \dots.$$

On the other hand, the operator S can be written as a product

$$S = BTA,$$

where $A:E_1 \to E$ and $B:F \to E_1$ are defined by

$$A\xi = \xi x_0 \quad \text{and} \quad By = \langle y, b_0 \rangle,$$

respectively. From this remark it follows by the multiplicativity (E3) of the entropy numbers that

$$\varepsilon_k(S) \leqslant \|B\| \varepsilon_k(T) \|A\|$$

and hence

$$\varepsilon_k(S) \leqslant \varepsilon_k(T)$$

because

$$\|A\| = \|x_0\| \leqslant 1 \quad \text{and} \quad \|B\| = \|b_0\| \leqslant 1.$$

Now the inequality (1.4.2) can be extended to

$$\|T\| < (1 + \eta)k\varepsilon_k(S) \leqslant (1 + \eta)k\varepsilon_k(T)$$

for $k = 1, 2, 3, \ldots$ and thus yields

$$\|T\| \leqslant (1 + \eta) \inf_{1 \leqslant k < \infty} k\varepsilon_k(T) = (1 + \eta)g_1(T).$$

Since $\eta > 0$ can be chosen arbitrarily small, we finally arrive at the desired result

$$\|T\| \leqslant g_1(T).$$

The multiplicativity of the $g_n(T)$ can be proved on the basis of the multiplicativity of the $\varepsilon_k(T)$ by straight-forward computations, namely

$$g_n(RS) = \inf_{1 \leqslant k < \infty} k^{1/n}\varepsilon_k(RS) = \inf_{1 \leqslant k_1, k_2 < \infty} (k_1 k_2)^{1/n}\varepsilon_{k_1 k_2}(RS)$$

$$\leqslant \inf_{1 \leqslant k_1, k_2 < \infty} (k_1^{1/n}\varepsilon_{k_1}(R))(k_2^{1/n}\varepsilon_{k_2}(S)) = g_n(R)g_n(S).$$

For the proof of the rank property (M3) it suffices to refer to Proposition 1.3.1 which says that an operator $T:E \to F$ is of rank $m > 0$ if and only if

$$C \cdot k^{-1/m} \leqslant \varepsilon_k(T) \leqslant 4\|T\| \cdot k^{-1/m} \quad \text{for } k = 1, 2, 3, \ldots \qquad (1.3.14)$$

with $C > 0$.

The surjectivity (MS) of the $g_n(T)$ follows from the surjectivity (ES) of the entropy numbers $\varepsilon_k(T)$.

As an aside, we remark that the injectivity (1.3.6) of the $\varepsilon_k(T)$ up to a factor 2 implies a corresponding weak kind of injectivity for the $g_n(T)$, namely

$$g_n(T) \leqslant 2g_n(JT) \qquad (1.4.3)$$

for any metric injection $J:F \to \tilde{F}$.

We point out that the definition of the nth entropy modulus $g_n(T)$ of an operator T acting between complex Banach spaces E and F is based upon the definition (1.2.4)' of the nth entropy modulus $g_n(M)$ of a bounded subset M of a complex Banach space, namely

$$g_n(T) = \inf_{1 \leq k < \infty} k^{1/2n} \varepsilon_k(T) \qquad (1.4.1)'$$

see (cf. Carl 1982, 1984). The algebraic properties of the $g_n(T)$ are not affected by the change from the real to the complex case. However, some of the proofs have to be modified slightly.

The rank property (M3), the multiplicativity (M2), and the monotonicity (M1) turn out to be norm determining for the entropy moduli $g_n(T)$ in the sense that

$$g_n(I_E) = 1 \quad \text{whenever } \dim(E) \geq n, \qquad (1.4.4)$$

I_E being the identity map of the Banach space E. Indeed, because $\operatorname{rank}(I_E) \geq n$ we may conclude that $g_n(I_E) > 0$ by (M3). Moreover, since $I_E = I_E^2$ we have

$$0 < g_n(I_E) = g_n(I_E^2) \leq g_n(I_E)^2 \leq g_1(I_E)^2 = \|I_E\|^2 = 1$$

by (M2) and (M1) and thus recognize that $g_n(I_E) = 1$.

It should be mentioned that additivity does not even hold in the weak form

$$g_k(T_1 + T_2) \leq \|T_1\| + g_k(T_2).$$

A counterexample can be constructed by means of the finite-dimensional diagonal operator $D^{(k)}: l_p^k \to l_p^k$ considered in section 1.3 (cf. (1.3.17)). For this operator $D^{(k)}$ with the generating sequence $\sigma_1 \geq \sigma_2 \geq \cdots \geq \sigma_k \geq 0$ we obtain

$$(\sigma_1 \sigma_2 \cdots \sigma_k)^{1/k} \leq \inf_{1 \leq n < \infty} n^{1/k} \varepsilon_n(D^{(k)}) = g_k(D^{(k)}). \qquad (1.4.5)$$

as a consequence of (1.3.21). Now let us assume

$$g_k(D^{(k)}) \leq \|D_1^{(k)}\| + g_k(D_2^{(k)}) \qquad (1.4.6)$$

for $D^{(k)} = D_1^{(k)} + D_2^{(k)}$ with the operators $D_1^{(k)}$ and $D_2^{(k)}$ defined by

$$D_1^{(k)}(\xi_1, \xi_2, \ldots, \xi_k) = (0, 0, \ldots, 0, \sigma_k \xi_k)$$

and

$$D_2^{(k)}(\xi_1, \xi_2, \ldots, \xi_k) = (\sigma_1 \xi_1, \ldots, \sigma_{k-1} \xi_{k-1}, 0).$$

Then we have

$$\|D_1^{(k)}\| = \sigma_k \text{ and } g_k(D_2^{(k)}) = 0$$

since $\operatorname{rank}(D_2^{(k)}) < k$. Accordingly, (1.4.6) reduces to

$$g_k(D^{(k)}) \leq \sigma_k.$$

Hence from the monotonicity of the sequence $(\sigma_i)_{1 \leq i \leq k}$ and from (1.4.5) it follows that

$$\sigma_k \leqslant (\sigma_1 \sigma_2 \cdots \sigma_k)^{1/k} \leqslant \sigma_k$$

which implies $\sigma_1 = \sigma_2 = \cdots = \sigma_k = \sigma$. The diagonal operator $D^{(k)}$ has turned out to be a multiple $\sigma I_p^{(k)}$ of the identity map $I_p^{(k)}$ of l_p^k. If $D^{(k)}$ is not of this type the additivity (1.4.6) fails to be valid.

The finite-dimensional diagonal operator $D^{(k)} : l_p^k \to l_p^k$ can also be used for estimating the kth entropy modulus $g_k(D)$ of the infinite-dimensional diagonal operator $D : l_p \to l_p$ generated by the sequence $\sigma_1 \geqslant \sigma_2 \geqslant \cdots \geqslant \sigma_k \geqslant \sigma_{k+1} \geqslant \cdots \geqslant 0$.

Proposition 1.4.1. *Let $D : l_p \to l_p$ for $1 \leqslant p \leqslant \infty$ be the diagonal operator (1.3.15) generated by the non-increasing sequence $\sigma_1 \geqslant \sigma_2 \geqslant \cdots \geqslant \sigma_k \geqslant \sigma_{k+1} \geqslant \cdots \geqslant 0$. Then we have*

$$(\sigma_1 \sigma_2 \cdots \sigma_k)^{1/k} \leqslant g_k(D) \leqslant 6(\sigma_1 \sigma_2 \cdots \sigma_k)^{1/k} \qquad (1.4.7)$$

for $k = 1, 2, 3, \ldots$

Proof. We confine ourselves to diagonal operators $D : l_p \to l_p$ between real l_p spaces so that

$$g_k(D) = \inf_{1 \leqslant n < \infty} n^{1/k} \varepsilon_n(D)$$

is the adequate definition for the kth entropy modulus.

The estimate from below is an immediate consequence of (1.3.16). For estimating $g_k(D)$ from above we apply methods similar to those used for estimating $\varepsilon_n(D)$ from above in the proof of Proposition 1.3.2. First of all we notice that the right-hand side of (1.4.7) yields equality if $\sigma_k = 0$, for this implies $\operatorname{rank}(D) < k$ and hence $g_k(D) = 0$ by (M3). From now on let us assume $\sigma_k > 0$ and determine a maximal system of elements $y_1, y_2, \ldots, y_N$ in $D^{(k)}(U_p^k)$ with

$$\| y_i - y_j \| > 2\rho \quad \text{for } i \neq j, \qquad (1.4.8)$$

where ρ with $0 < \rho \leqslant \sigma_k$ is fixed arbitrarily for the moment. The sets $\{y_j + \rho U_p^k\}$ are obviously pairwise disjoint subsets of $D^{(k)}(U_p^k) + \rho U_p^k$ so that the inclusion

$$\bigcup_{j=1}^{N} \{y_j + \rho U_p^k\} \subseteq D^{(k)}(U_p^k) + \rho U_p^k \qquad (1.4.9)$$

can again be used for a comparison of volumes. As in the case of (1.3.29) we use a series of inclusions

$$\rho U_p^k \subseteq \sigma_k U_p^k \subseteq D^{(k)}(U_p^k)$$

and replace (1.4.9) by

$$\bigcup_{j=1}^{N} \{y_j + \rho U_p^k\} \subseteq 2D^{(k)}(U_p^k).$$

In analogy to (1.3.30) we now may conclude that

$$N \leqslant \left(\frac{2}{\rho}\right)^k \cdot \sigma_1 \sigma_2 \cdots \sigma_k. \tag{1.4.10}$$

This estimate of the cardinality of the system $y_1, y_2, \ldots, y_N$ is the key to the desired estimate of $g_k(D)$ from above. We start with an estimate for $\varepsilon_N(D)$ again using the sectional operator

$$D_k = I_k D^{(k)} P_k \tag{1.4.11}$$

(see (1.3.24)) and

$$\varepsilon_N(D) \leqslant \|D - D_k\| + \varepsilon_N(D_k). \tag{1.3.27}$$

The product formula (1.4.11) gives

$$\varepsilon_N(D_k) \leqslant \varepsilon_N(D^{(k)}).$$

For $\varepsilon_N(D^{(k)})$, we use the maximality of the system $y_1, y_2, \ldots, y_N$ in $D^{(k)}(U_p^k)$ with the property (1.4.8) which implies

$$D^{(k)}(U_p^k) \subseteq \bigcup_{j=1}^{N} \{y_j + 2\rho U_p^k\}.$$

As a result of this we obtain

$$\varepsilon_N(D^{(k)}) \leqslant 2\rho \leqslant 2\sigma_k.$$

Hence it follows that

$$\varepsilon_N(D) \leqslant \|D - D_k\| + 2\sigma_k = \sigma_{k+1} + 2\sigma_k \leqslant 3\sigma_k$$

from (1.3.27) (cf. also (1.3.28)). Taking account of the definition of $g_k(D)$ and of (1.4.10) we recognize that

$$g_k(D) \leqslant N^{1/k} \varepsilon_N(D) \leqslant \frac{2}{\rho} (\sigma_1 \sigma_2 \cdots \sigma_k)^{1/k} \cdot 3\sigma_k.$$

The choice $\rho = \sigma_k$ of ρ completes the proof of (1.4.7). ∎

In contrast to the estimate (1.3.16) for the entropy numbers $\varepsilon_n(D)$ of the diagonal operator D, the estimate (1.4.7) for the entropy moduli $g_k(D)$ remains unchanged if D is considered as an operator from the complex l_p space into itself. This is because the definition of the entropy moduli $g_k(T)$ is itself adapted to the complex situation so long as T acts between complex Banach spaces.

1.5. Entropy classes

In section 1.3 we mentioned the class of operators T characterized by

$$e_n(T) \leqslant C \cdot n^{-\beta} \quad \text{for } n = 1, 2, 3, \ldots \tag{1.3.35}$$

Instead of (1.3.35) we can state

$$\sup_{1\leqslant k<\infty} k^{1/p}e_k(T)<\infty \qquad (1.5.1)$$

where $p=1/\beta$. Another way of selecting classes of compact operators is by imposing certain summability properties on the $e_n(T)$, for instance

$$\sum_{k=1}^{\infty}(k^{\alpha}e_k(T))^q<\infty$$

for $q>0$ and a real number α. Then $\alpha=0$ just means $(e_k(T))\in l_q$. Representing α as $\alpha=1/p-1/q$ with $p>0$ we state the following (cf. Triebel 1970; Carl 1981).

An operator T is *of type* $L_{p,q}^{(e)}$, where $0<p\leqslant\infty$, $0<q<\infty$, if

$$\sum_{k=1}^{\infty}(k^{1/p-1/q}e_k(T))^q<\infty; \qquad (1.5.2)$$

in particular we write $T\in L_{p,q}^{(e)}(E,F)$ if T is an operator of type $L_{p,q}^{(e)}$ from E into F.

Taking into consideration that the sequence space l_q is equipped with the supremum norm when $q=\infty$ we use the symbol $\mathbf{L}_{p,\infty}^{(e)}$ for the class of operators T defined by (1.5.1). For $p=\infty$ the inequality (1.5.1) reduces to

$$\sup_{1\leqslant k<\infty} e_k(T)=\|T\|<\infty$$

thus expressing the coincidence of $\mathbf{L}_{\infty,\infty}^{(e)}$ with the class L of all operators. Furthermore let us emphasize that the definition (1.5.2) of the class $L_{p,q}^{(e)}$ in fact makes sense for $p=\infty$ and $q<\infty$. So we finally make use of the entropy classes $L_{p,q}^{(e)}$ for all p and q with $0<p,q\leqslant\infty$. A unified description of $L_{p,q}^{(e)}$ can obviously be given by

$$T\in L_{p,q}^{(e)}(E,F) \quad \text{if } (k^{1/p-1/q}e_k(T))\in l_q. \qquad (1.5.3)$$

The summability schemes employed here to classify operators with respect to their entropy properties also play an important role in other connections (see chapters 2 and 3). The corresponding classes of sequences are known as *Lorentz sequence spaces*.

A null sequence $x=(\xi_k)$ is said to belong to the Lorentz sequence space $l_{p,q}$ if the non-increasing rearrangement $(s_k(x))$ of its absolute values $|\xi_k|$ satisfies

$$(k^{(1/p-1/q)}s_k(x))\in l_q, \qquad (1.5.4)$$

so that

$$\lambda_{p,q}(x)=\begin{cases}\left(\displaystyle\sum_{k=1}^{\infty}(k^{(1/p-1/q)}s_k(x))^q\right)^{1/q} & \text{for } 0<p\leqslant\infty \text{ and } 0<q<\infty,\\[2em] \displaystyle\sup_{1\leqslant k<\infty} k^{1/p}s_k(x) & \text{for } 0<p<\infty \text{ and } q=\infty\end{cases} \qquad (1.5.5)$$

is finite. The $s_k(x)$ are called *the s-numbers of the null sequence* $x = (\xi_k)$ (see Pietsch 1978). It is quite remarkable that the question whether or not a null sequence $x = (\xi_k)$ belongs to $l_{p,q}$ can be decided by means of the dyadic s-numbers $s_{2^n}(x)$ only, where $n = 0, 1, 2, 3, \ldots$

Lemma 1.5.1. *A null sequence x belongs to $l_{p,q}$ with $q < \infty$ if and only if*

$$\sum_{n=0}^{\infty} (2^{n/p} s_{2^n}(x))^q < \infty, \tag{1.5.6}$$

and $x \in l_{p,\infty}$ if and only if

$$s_{2^n}(x) \leqslant C \cdot 2^{-n/p} \quad \text{for } n = 0, 1, 2, 3, \ldots \tag{1.5.7}$$

Proof. Let $x \in l_{p,q}$ with $0 < p \leqslant \infty$ and $q < \infty$. We then split the infinite series $\sum_{k=1}^{\infty} (k^{(1/p-1/q)} s_k(x))^q$ into finite parts $\sum_{k=2^{n-1}}^{2^n-1} (k^{(1/p-1/q)} s_k(x))^q$ with $n = 1, 2, 3, \ldots$ For $p \leqslant q$ we obtain in this way

$$\sum_{k=1}^{\infty} (k^{(1/p-1/q)} s_k(x))^q \geqslant \sum_{n=1}^{\infty} 2^{n-1} (2^{(n-1)(1/p-1/q)} s_{2^n}(x))^q$$

$$= 2^{-q/p} \cdot \sum_{n=1}^{\infty} (2^{n/p} s_{2^n}(x))^q.$$

Similarly, for $q < p$ it emerges that

$$\sum_{k=1}^{\infty} (k^{(1/p-1/q)} s_k(x))^q \geqslant \sum_{n=1}^{\infty} 2^{n-1} (2^{n(1/p-1/q)} s_{2^n}(x))^q$$

$$= \frac{1}{2} \sum_{n=1}^{\infty} (2^{n/p} s_{2^n}(x))^q.$$

These estimates prove (1.5.6) for $x \in l_{p,q}$ with $0 < p \leqslant \infty$ and $0 < q < \infty$. Conversely, starting from (1.5.6) we obtain

$$\sum_{k=1}^{\infty} (k^{(1/p-1/q)} s_k(x))^q \leqslant \sum_{n=1}^{\infty} 2^{n-1} (2^{n(1/p-1/q)} s_{2^{n-1}}(x))^q$$

$$= 2^{(q/p)-1} \sum_{n=0}^{\infty} (2^{n/p} s_{2^n}(x))^q < \infty \quad \text{for } p \leqslant q$$

and

$$\sum_{k=1}^{\infty} (k^{(1/p-1/q)} s_k(x))^q \leqslant \sum_{n=1}^{\infty} 2^{n-1} (2^{(n-1)(1/p-1/q)} s_{2^{n-1}}(x))^q$$

$$= \sum_{n=0}^{\infty} (2^{n/p} s_{2^n}(x))^q < \infty \quad \text{for } q < p.$$

Now let us suppose $x \in l_{p,\infty}$ with $0 < p < \infty$. Then we have

$$s_k(x) \leqslant C \cdot k^{-1/p} \quad \text{for } k = 1, 2, 3, \ldots,$$

which implies (1.5.7). On the other hand, using (1.5.7) we can estimate the s-numbers $s_k(x)$ for k between 2^{n-1} and 2^n from above by

$$s_k(x) \leqslant s_{2^{n-1}}(x) \leqslant C \cdot 2^{-(n-1)/p} \leqslant C \cdot 2^{1/p} \cdot k^{-1/p}. \qquad \blacksquare$$

As an easy consequence of Lemma 1.5.1 we obtain the *lexicographical order* of the Lorentz sequence spaces $l_{p,q}$.

Lemma 1.5.2. *Let $L_{p_i q_i}$ be Lorentz sequence spaces. Then*

$$l_{p,q_1} \subseteq l_{p,q_2} \quad \text{for } q_1 < q_2 \leqslant \infty \qquad (1.5.8)$$

and

$$l_{p_1,q_1} \subseteq l_{p_2,q_2} \quad \text{for } p_1 < p_2. \qquad (1.5.9)$$

Proof. The inclusion (1.5.8) follows immediately from (1.5.6) and from

$$l_{q_1} \subseteq l_{q_2} \quad \text{for } q_1 < q_2 \leqslant \infty.$$

Hence for the proof of (1.5.9) it suffices to show that

$$l_{p_1,\infty} \subseteq l_{p_2,q_2} \quad \text{for } p_1 < p_2 \text{ and } q_2 < \infty.$$

So let $x \in l_{p_1,\infty}$. Then

$$2^{n/p_2} s_{2^n}(x) \leqslant C \cdot 2^{-n(1/p_1 - 1/p_2)}$$

by (1.5.7), and therefore

$$\sum_{n=0}^{\infty} (2^{n/p_2} s_{2^n}(x))^{q_2} < \infty,$$

which means $x \in l_{p_2,q_2}$. $\qquad \blacksquare$

We supplement Lemma 1.5.2 with the remark that the inclusions between the Lorentz sequence spaces are strict in the sense that

$$l_{p_1,q_1} \neq l_{p_2,q_2} \quad \text{for } (p_1, q_1) \neq (p_2, q_2). \qquad (1.5.10)$$

Indeed, let $p_1 = p_2 = p$ and $q_1 < q_2$ so that (1.5.8) applies. We then consider

$$\xi_k = k^{-1/p}(1 + \log_2 k)^{-1/q_1} \quad \text{for } k = 1, 2, 3, \ldots$$

The corresponding dyadic sequence

$$\xi_{2^{n-1}} = 2^{-(n-1)/p} n^{-1/q_1}$$

shows that $(\xi_k) \in l_{p,q_2}$, but $(\xi_k) \notin l_{p,q_1}$. According to (1.5.9) the sequence (ξ_k) then also fails to belong to $l_{\tilde{p},\tilde{q}}$ for any pair $(\tilde{p}, \tilde{q})$ with $\tilde{p} < p$. This completes the proof of (1.5.10).

The example just given makes it clear that to classify a decreasing null sequence $\xi_1 \geqslant \xi_2 \geqslant \cdots \geqslant \xi_n \geqslant \cdots \geqslant 0$ with respect to the scheme of the Lorentz spaces $l_{p,q}$ one first has to look for a positive power $k^{1/p}$ of k such that $k^{1/p} \xi_k$ is bounded. Of course such a power need not exist as can be demonstrated by the example

$$\xi_k = (1 + \log_2 k)^{-1/q_1}, \quad k = 1, 2, 3, \ldots,$$

derived from the above one by putting $p = \infty$. But according to what has been said above this sequence (ξ_k) is contained in l_{∞,q_2} for any $q_2 > q_1$. In contrast with that there are sequences which do not belong to any of the Lorentz spaces $l_{p,q}$. An example of this kind is given by

$$\xi_k = \frac{1}{1 + \log_2(1 + \log_2 k)}, \quad k = 1, 2, 3, \ldots$$

On the other hand, the sequence

$$\xi_k = 2^{-k}, \quad k = 1, 2, 3, \ldots,$$

possesses the property $\lim_{k\to\infty} k^{1/p}\xi_k = 0$ for any positive power $k^{1/p}$ of k and thus is contained in $l_{p,q}$ for arbitrary (p,q). We denote by

$$s = \bigcap_{p,q>0} l_{p,q}$$

the intersection of all Lorentz spaces and agree to call a null sequence (ξ_k) *rapidly decreasing* if $(\xi_k) \in s$. In other words, a null sequence $x = (\xi_k)$ is rapidly decreasing if and only if the non-increasing rearrangement $(s_k(x))$ of its absolute values $|\xi_k|$ satisfies the condition

$$\lim_{k\to\infty} k^{\alpha} s_k(x) = 0$$

for any positive power k^{α} of k. Note, that this definition differs from the definition

$$\lim_{k\to\infty} k^{\alpha}\xi_k = 0 \quad \text{for all } \alpha > 0$$

of a rapidly decreasing sequence (ξ_k) used in the theory of locally convex spaces (see Pietsch 1965). For instance, the sequence

$$\xi_k = \begin{cases} 2^{-n} & \text{for } k = 2^{2^n}, \\ 0 & \text{otherwise} \end{cases}$$

is rapidly decreasing in the sense of our definition, but is not in the sense of the other one. For monotonously decreasing sequences, however, the two definitions are equivalent, so that the difference between them disappears if we discuss the question, whether or not the sequence of entropy numbers $e_k(T)$ of an operator T is rapidly decreasing.

Having separated the Lorentz sequence spaces from each other we show that even the entropy classes $L_{s,t}^{(e)}$ are subject to a strict lexicographical order. For this purpose we again refer to the diagonal operator $D: l_p \to l_p$ considered in Proposition 1.3.2 and Proposition 1.4.1 and show that D is of type $L_{s,t}^{(e)}$ if and only if the generating sequence (σ_i) belongs to $l_{s,t}$. Since the behaviour of the dyadic entropy numbers $e_n(D)$ is determined by

$$\sup_{1 \leq k < \infty} 2^{-(n-1)/k}(\sigma_1\sigma_2\cdots\sigma_k)^{1/k} \leq e_n(D) \leq 6 \cdot \sup_{1 \leq k < \infty} 2^{-(n-1)/k}(\sigma_1\sigma_2\cdots\sigma_k)^{1/k}$$

$$(1.5.11)$$

(see (1.3.16)) the proof of this fact amounts to proving that the sequence

$$\tau_n = \sup_{1 \leqslant k < \infty} 2^{-(n-1)/k}(\sigma_1\sigma_2\cdots\sigma_k)^{1/k}, \quad n = 1, 2, 3, \ldots,$$

belongs to $l_{s,t}$ if and only if the sequence (σ_i) does. The estimate

$$\sigma_n \leqslant (\sigma_1\sigma_2\cdots\sigma_n)^{1/n} \leqslant 2 \cdot \sup_{1 \leqslant k < \infty} 2^{-(n-1)/k}(\sigma_1\sigma_2\cdots\sigma_k)^{1/k} = 2\tau_n$$

immediately makes it clear that $(\tau_n) \in l_{s,t}$ implies $(\sigma_n) \in l_{s,t}$. The converse conclusion rests upon an inequality connected with the name of Hardy (cf. Hardy, Littlewood and Polya 1964). The essence of this inequality is that given $(\sigma_n) \in l_{s,t}$, where $\sigma_1 \geqslant \sigma_2 \geqslant \cdots \geqslant \sigma_n \geqslant \cdots \geqslant 0$, the sequence $(\sum_{k=1}^n \sigma_k^q/n)^{1/q}$ of the arithmetic means of order q also lies in $l_{s,t}$ so long as $0 < q < \min(s,t)$.

Lemma 1.5.3. (Hardy's inequality). *Let* $\sigma_1 \geqslant \sigma_2 \geqslant \cdots \geqslant \sigma_n \geqslant \cdots \geqslant 0$ *be a non-increasing sequence of non-negative numbers. Then*

$$\sum_{n=1}^N n^{(t/s)-1}\left(\frac{\sum_{k=1}^n \sigma_k^q}{n}\right)^{t/q} \leqslant \left(1 + \frac{s}{s-q}\right)^{t/q} \sum_{k=1}^N k^{(t/s)-1}\sigma_k^t \quad (1.5.12)$$

for $0 < s, t < \infty$, $0 < q < \min(s,t)$,

$$\sum_{n=1}^N n^{-1}\left(\frac{\sum_{k=1}^n \sigma_k^q}{n}\right)^{t/q} \leqslant 2 \sum_{k=1}^N k^{-1}\sigma_k^t \quad (1.5.13)$$

for $0 < q \leqslant t < \infty$, *and*

$$\sup_{1 \leqslant n \leqslant N} n^{1/s}\left(\frac{\sum_{k=1}^n \sigma_k^q}{n}\right)^{1/q} \leqslant \left(\frac{s}{s-q}\right)^{1/q} \sup_{1 \leqslant k \leqslant N} k^{1/s}\sigma_k \quad (1.5.14)$$

for $0 < q < s < \infty$.

Proof. The inequalities in question are based on Hölder's inequality. The crucial point of the proof is to find an appropriate pair of conjugate exponents and to factorize the items σ_k^q of the sum $\sum_{k=1}^n \sigma_k^q$ in an appropriate way. In any of the three cases we have to estimate finite sums $\sum_{k=1}^n k^{-\gamma}$ for $0 \leqslant \gamma < 1$. A comparison of $\sum_{k=1}^n k^{-\gamma}$ with the integral $\int_1^n x^{-\gamma}\,dx$ yields

$$\sum_{k=1}^n k^{-\gamma} \leqslant 1 + \int_1^n x^{-\gamma}\,dx = \frac{n^{1-\gamma}-\gamma}{1-\gamma} \leqslant \frac{n^{1-\gamma}}{1-\gamma}. \quad (1.5.15)$$

Moreover, the proof of (1.5.12) and (1.5.13) requires an estimate of the infinite sections $\sum_{n=k}^\infty n^{-\alpha}$ of the series $\sum_{n=1}^\infty n^{-\alpha}$ convergent for $\alpha > 1$. A comparison with the integral $\int_k^\infty x^{-\alpha}\,dx$ yields

$$\sum_{n=k}^\infty n^{-\alpha} \leqslant k^{-\alpha} + \int_k^\infty x^{-\alpha}\,dx = k^{-\alpha} + \frac{k^{-\alpha+1}}{\alpha-1} \leqslant \left(1 + \frac{1}{\alpha-1}\right)k^{-\alpha+1}. \quad (1.5.16)$$

We start with the case $0 < s, t < \infty$, $0 < q < \min(s,t)$. From Hölder's inequality with the pair of conjugate exponents t/q and $(t/q)' = t/(t-q)$ we obtain

$$\sum_{k=1}^{n} \sigma_k{}^q = \sum_{k=1}^{n} (k^{(q/s)(1-q/t)} \sigma_k{}^q) k^{-(q/s)(1-q/t)}$$

$$\leq \left(\sum_{k=1}^{n} k^{t/s-q/s} \sigma_k{}^t \right)^{q/t} \left(\sum_{k=1}^{n} k^{-q/s} \right)^{1-q/t}.$$

By using (1.5.15) with $\gamma = q/s$ we can continue this estimate to

$$\sum_{k=1}^{n} \sigma_k{}^q \leq \left(\frac{s}{s-q} \right)^{1-q/t} n^{(1-q/s)(1-q/t)} \left(\sum_{k=1}^{n} k^{t/s-q/s} \sigma_k{}^t \right)^{q/t}.$$

The term

$$n^{t/s-1} \left(\frac{\sum_{k=1}^{n} \sigma_k{}^q}{n} \right)^{t/q} = n^{(t/s-t/q)-1} \left(\sum_{k=1}^{n} \sigma_k{}^q \right)^{t/q}$$

appearing in the sum on the left-hand side of (1.5.12) thus satisfies the inequality

$$n^{t/s-1} \left(\frac{\sum_{k=1}^{n} \sigma_k{}^q}{n} \right)^{t/q} \leq \left(1 + \frac{q}{s-q} \right)^{t/q-1} n^{-2+q/s} \sum_{k=1}^{n} k^{(t-q)/s} \sigma_k^t.$$

Hence for the rest of the proof we have to deal with the double sum

$$\sum_{n=1}^{N} n^{-2+q/s} \sum_{k=1}^{n} k^{(t-q)/s} \sigma_k{}^t = \sum_{k=1}^{N} k^{(t-q)/s} \sigma_k{}^t \sum_{n=k}^{N} n^{-2+q/s}.$$

After changing the order of summation we replace the finite section $\sum_{n=k}^{N} n^{-2+q/s}$ by the infinite one, simultaneously applying (1.5.16) with $\alpha = 2 - (q/s) > 1$. In this way we arrive at

$$\sum_{n=1}^{N} n^{t/s-1} \left(\frac{\sum_{k=1}^{n} \sigma_k{}^q}{n} \right)^{t/q} \leq \left(1 + \frac{q}{s-q} \right)^{t/q-1} \left(1 + \frac{s}{s-q} \right) \sum_{k=1}^{N} k^{t/s-1} \sigma_k{}^t$$

$$(1.5.17)$$

which completes the proof of (1.5.12) since

$$\left(1 + \frac{q}{s-q} \right)^{t/q-1} \left(1 + \frac{s}{s-q} \right) \leq \left(1 + \frac{s}{s-q} \right)^{t/q}.$$

For the proof of (1.5.13) we first let $s \to \infty$ on both sides of the inequality (1.5.17) which yields (1.5.13) for $0 < q < t < \infty$. Next we let $q \to t$ and find that (1.5.13) is also true for $q = t$.

For the proof of (1.5.14) we start from

$$\sum_{k=1}^{n} \sigma_k{}^q = \sum_{k=1}^{n} k^{-q/s} k^{q/s} \sigma_k{}^q \leq \sup_{1 \leq k \leq n} k^{q/s} \sigma_k{}^q \sum_{k=1}^{n} k^{-q/s}.$$

Then, using (1.5.15) with $\gamma = q/s$ again enables us to conclude that

$$\sum_{k=1}^{n} \sigma_k{}^q \leqslant \frac{s}{s-q} n^{1-q/s} \sup_{1 \leqslant k \leqslant n} k^{q/s} \sigma_k{}^q.$$

An equivalent statement is

$$n^{1/s} \left(\frac{\sum_{k=1}^{n} \sigma_k{}^q}{n} \right)^{1/q} \leqslant \left(\frac{s}{s-q} \right)^{1/q} \sup_{1 \leqslant k \leqslant n} k^{1/s} \sigma_k.$$

Taking the supremum with respect to $1 \leqslant n \leqslant N$ on both sides of this inequality we finally arrive at (1.5.14). ∎

Corollary. *The monotonous decrease of the sequence $\sigma_1 \geqslant \sigma_2 \geqslant \cdots \geqslant \sigma_n \cdots$ $\geqslant 0$ implies that the sequence $((\sum_{k=1}^{n} \sigma_k{}^q / n)^{1/q})$ of the arithmetic means of order q is also monotonously decreasing for any $q > 0$. Therefore by Lemma 1.5.3 the assumption $(\sigma_i) \in l_{s,t}$ guarantees $((\sum_{k=1}^{n} \sigma_k{}^q / n)^{1/q}) \in l_{s,t}$ for the sequence of the arithmetic means of order q so long as $0 < q < \min(s, t)$.*

Proposition 1.5.1. *The diagonal operator $D: l_p \to l_p$, $1 \leqslant p \leqslant \infty$, defined by a non-increasing sequence $\sigma_1 \geqslant \sigma_2 \geqslant \cdots \geqslant \sigma_n \geqslant \cdots \geqslant 0$ (see (1.3.15)) is of type $L_{s,t}^{(e)}$ if and only if the sequence (σ_i) belongs to $l_{s,t}$.*

Proof. By what was pointed out for (1.5.11) it suffices to estimate the expressions

$$\tau_n = \sup_{1 \leqslant k < \infty} 2^{-(n-1)/k} (\sigma_1 \sigma_2 \cdots \sigma_k)^{1/k}$$

from above by the members of the sequence (σ_i). We consider τ_{n+1}, splitting up the supremum into two terms:

$$\tau_{n+1} \leqslant \sup_{1 \leqslant k \leqslant n} 2^{-n/k} (\sigma_1 \sigma_2 \cdots \sigma_k)^{1/k} + \sup_{k > n} 2^{-n/k} (\sigma_1 \sigma_2 \cdots \sigma_k)^{1/k}.$$

The estimate for each term rests upon the fact that the geometric mean is smaller than the arithmetic mean and hence also

$$(\sigma_1 \sigma_2 \cdots \sigma_k)^{1/k} \leqslant \left(\frac{\sum_{i=1}^{k} \sigma_i{}^q}{k} \right)^{1/q} \quad \text{for all } q > 0.$$

For the first term we obtain

$$\sup_{1 \leqslant k \leqslant n} 2^{-n/k} (\sigma_1 \sigma_2 \cdots \sigma_k)^{1/k} \leqslant \sup_{1 \leqslant k \leqslant n} 2^{-n/k} k^{-1/q} \left(\sum_{i=1}^{k} \sigma_i{}^q \right)^{1/q}$$

$$\leqslant \left(\sum_{i=1}^{n} \sigma_i{}^q \right)^{1/q} \sup_{1 \leqslant k \leqslant n} 2^{-n/k} k^{-1/q}.$$

From the inequality $2^x > x$ for $x > 0$ we get

$$2^{n/k} > q^{1/q} \left(\frac{n}{k} \right)^{1/q}$$

with $x = q(n/k)$ or

$$2^{-n/k}k^{-1/q} < q^{-1/q}n^{-1/q}.$$

Since the right-hand side of this inequality does not depend on k we can write

$$\sup_{1 \leqslant k \leqslant n} 2^{-n/k}(\sigma_1\sigma_2\cdots\sigma_k)^{1/q} \leqslant q^{-1/q}\left(\frac{\sum_{i=1}^n \sigma_i^q}{n}\right)^{1/q}. \qquad (1.5.18)$$

To estimate the supremum of $2^{-n/k}(\sigma_1\sigma_2\cdots\sigma_k)^{1/k}$ with respect to $k > n$ we need only notice that $2^{-n/k} < 1$ and that

$$(\sigma_1\sigma_2\cdots\sigma_k)^{1/k} \leqslant \left(\frac{\sum_{i=1}^k \sigma_i^q}{k}\right)^{1/q} \leqslant \left(\frac{\sum_{i=1}^n \sigma_i^q}{n}\right)^{1/q} \quad \text{for } k > n. \qquad (1.5.19)$$

Adding the two estimates (1.5.18) and (1.5.19) we arrive at the desired estimate

$$\tau_{n+1} \leqslant (1 + q^{-1/q})\left(\frac{\sum_{i=1}^n \sigma_i^q}{n}\right)^{1/q}.$$

To complete the proof we use the result of the above Corollary which enables us to conclude that $(\tau_n) \in l_{s,t}$ on the basis of the assumption $(\sigma_i) \in l_{s,t}$. ∎

In a sense Proposition 1.5.1 says that the rate of decrease of the dyadic entropy numbers $e_n(D)$ of a diagonal operator $D: l_p \to l_p$ is determined by the rate of decrease of the generating sequence (σ_i). However, this statement is true only because we confined ourselves to the scheme of the Lorentz sequence spaces. In the case of a rapidly decreasing sequence (σ_i) the behaviour of the $e_n(D)$ may differ in detail from the behaviour of the σ_i. Let us consider the rapidly decreasing sequence

$$\sigma_i = 2^{-i} \qquad (1.5.20)$$

and estimate the corresponding expressions

$$\tau_{n+1} = \sup_{1 \leqslant k < \infty} 2^{-n/k}\left(\prod_{i=1}^k 2^{-i}\right)^{1/k}$$

$$= \sup_{1 \leqslant k < \infty} 2^{-n/k}2^{-(k+1)/2} = \frac{1}{\sqrt{2}}\sup_{1 \leqslant k < \infty} 2^{-(n/k+k/2)}.$$

An estimate from above results from

$$\left(\frac{n}{k} + \frac{k}{2}\right)^2 = \left(\frac{n}{k} - \frac{k}{2}\right)^2 + 2n \geqslant 2n,$$

namely

$$\tau_{n+1} \leqslant \frac{1}{\sqrt{2}}2^{-\sqrt{2n}}.$$

To estimate τ_{n+1} from below we choose k with

$$\sqrt{2n} \leqslant k < \sqrt{2n} + 1.$$

From

$$\frac{n}{k} + \frac{k}{2} \leqslant \frac{1}{2}\sqrt{2n} + \frac{1}{2}\sqrt{2n} + \frac{1}{2} = \frac{1}{2} + \sqrt{2n}$$

it follows that

$$\tau_{n+1} \geqslant \frac{1}{\sqrt{2}} 2^{-(n/k + k/2)} \geqslant \frac{1}{2} 2^{-\sqrt{2n}}.$$

Thus the general estimate (1.5.11) for the dyadic entropy numbers of the diagonal operator $D:l_p \to l_p$ in this special situation reads as

$$\frac{1}{2} 2^{-\sqrt{2n}} \leqslant e_{n+1}(D) \leqslant 3\sqrt{2} \cdot 2^{-\sqrt{2n}}. \tag{1.5.21}$$

Clearly the sequence $2^{-\sqrt{2n}}$ is also rapidly decreasing. However, it tends to zero more slowly than the generating sequence $\sigma_n = 2^{-n}$.

As far as the general theory of entropy classes is concerned, Proposition 1.5.1 in fact guarantees that their ordering is strict in the sense that

$$L^{(e)}_{p_1,q_1} \neq L^{(e)}_{p_2,q_2} \quad \text{for } (p_1, q_1) \neq (p_2, q_2).$$

1.6. Operator ideals

When listing the algebraic properties of the class $L^{(e)}_{p,q}$ we shall refer to $L^{(e)}_{p,q}$ as A.

(I1) All finite rank operators belong to A.

(I2) If $T_1, T_2 \in A(E, F)$ then $T_1 + T_2 \in A(E, F)$.

(I3) (a) If $S \in A(E, Z)$ and $R \in L(Z, F)$ then $RS \in A(E, F)$.

(I3) (b) If $S \in L(E, Z)$ and $R \in A(Z, F)$ then $RS \in A(E, F)$.

Property (I1) follows from (1.3.36) and the fact that any sequence $(2^{-n\sigma})$ with $\sigma > 0$ belongs to $l_{p,q}$. Properties (I2) and (I3) can easily be derived from the algebraic properties of the $e_n(T)$. We shall sketch the proofs by referring to the corresponding expression

$$\lambda^{(e)}_{p,q}(T) = \lambda_{p,q}((e_n(T)))$$

defined for all T of type $L^{(e)}_{p,q}$.

For $q < \infty$ we have

$$\lambda^{(e)}_{p,q}(T) = \left(\sum_{n=1}^{\infty} (n^{1/p - 1/q} e_n(T))^q \right)^{1/q}. \tag{1.6.1}$$

Given $T_1, T_2 \in L^{(e)}_{p,q}(E, F)$ we make use of

$$e_{2k}(T_1 + T_2) \leqslant e_{2k-1}(T_1 + T_2) \leqslant e_k(T_1) + e_k(T_2)$$

thus concluding that

$$\lambda_{p,q}^{(e)}(T_1 + T_2) \leqslant \left(\sum_{k=1}^{\infty} [(2k-1)^{1/p-1/q}(e_k(T_1) + e_k(T_2))]^q \right.$$
$$\left. + \sum_{k=1}^{\infty} [(2k)^{1/p-1/q}(e_k(T_1) + e_k(T_2))]^q \right)^{1/q}. \qquad (1.6.2)$$

If $p \leqslant q$ this inequality may be extended to

$$\lambda_{p,q}^{(e)}(T_1 + T_2) \leqslant \left(2 \cdot 2^{q/p-1} \cdot \sum_{k=1}^{\infty} [k^{1/p-1/q}(e_k(T_1) + e_k(T_2)]^q \right)^{1/q}$$
$$= 2^{1/p} \cdot \left(\sum_{k=1}^{\infty} [k^{1/p-1/q}(e_k(T_1) + e_k(T_2))]^q \right)^{1/q}. \qquad (1.6.3)$$

If $q < p$ we can replace the terms $(2k-1)^{1/p-1/q}$ and $(2k)^{1/p-1/q}$ on the right-hand side of (1.6.2) by $k^{1/p-1/q}$, so that

$$\lambda_{p,q}^{(e)}(T_1 + T_2) \leqslant 2^{1/q} \cdot \left(\sum_{k=1}^{\infty} [k^{1/p-1/q}(e_k(T_1) + e_k(T_2))]^q \right)^{1/q} \qquad (1.6.4)$$

emerges. Putting

$$s = \min(p, q) \qquad (1.6.5)$$

we obtain

$$\lambda_{p,q}^{(e)}(T_1 + T_2) \leqslant 2^{1/s} \left(\sum_{k=1}^{\infty} [k^{1/p-1/q}(e_k(T_1) + e_k(T_2))]^q \right)^{1/q} \qquad (1.6.6)$$

as a unified version of (1.6.3) and (1.6.4). Next we apply the triangle inequality or quasi-triangle inequality in the space l_q with $q \geqslant 1$ or $q < 1$, respectively, namely

$$\left(\sum_{k=1}^{\infty} |\xi_k + \eta_k|^q \right)^{1/q} \leqslant \kappa_q \left[\left(\sum_{k=1}^{\infty} |\xi_k|^q \right)^{1/q} + \left(\sum_{k=1}^{\infty} |\eta_k|^q \right)^{1/q} \right],$$

where

$$\kappa_q = \max(2^{1/q-1}, 1). \qquad (1.6.7)$$

Then (1.6.6) finally becomes

$$\lambda_{p,q}^{(e)}(T_1 + T_2) \leqslant \kappa_{p,q}^{(e)}[\lambda_{p,q}^{(e)}(T_1) + \lambda_{p,q}^{(e)}(T_2)] \qquad (1.6.8)$$

with

$$\kappa_{p,q}^{(e)} = \max(2^{1/s+1/q-1}, 2^{1/s}) = \begin{cases} 2^{1/p+1/q-1} & \text{for } 0 < p \leqslant q < 1 \\ 2^{2/q-1} & \text{for } 0 < q < 1, 0 < q < p \\ 2^{1/p} & \text{for } 0 < p < q, 1 \leqslant q \\ 2^{1/q} & \text{for } 1 \leqslant q \leqslant p \end{cases} \qquad (1.6.9)$$

This proves (I2) when $A = L_{p,q}^{(e)}$ with $q < \infty$. The case $A = L_{p,\infty}^{(e)}$ can be treated similarly, this time with

$$\kappa_{p,\infty}^{(e)} = 2^{1/p} \qquad (1.6.10)$$

as a possible constant satisfying (1.6.8).

For the proof of (I3)(a) and (I3)(b) we make use of (DE3)(a) and (DE3)(b), respectively, thus concluding that

$$\lambda_{p,q}^{(e)}(RS) \leqslant \| R \| \lambda_{p,q}^{(e)}(S) \qquad (1.6.11)(a)$$

and

$$\lambda_{p,q}^{(e)}(RS) \leqslant \| S \| \lambda_{p,q}^{(e)}(R). \qquad (1.6.11)(b)$$

A class A of operators satisfying properties (I1), (I2), and (I3) is called an *operator ideal* (see Pietsch 1978); the operators belonging to A are sometimes referred to as *operators of type* A. For a fixed pair of Banach spaces E, F the linear space of operators of type A from E into F is referred to as *the component* $A(E, F)$ *of the operator ideal* A.

A function α which assigns a real number $\alpha(T)$ to each operator T of an operator ideal A is called an *ideal quasi-norm* α *on* A if the following properties are satisfied:

(IQ1) $0 \leqslant \alpha(T) < \infty$ for all operators T belonging to A and $\alpha(T) = 0$ if and only if $T = 0$.

(IQ2) There exists a constant $\kappa \geqslant 1$ independent of the particular component $A(E, F)$ of A such that a quasi-triangle inequality

$$\alpha(T_1 + T_2) \leqslant \kappa[\alpha(T_1) + \alpha(T_2)]$$

is valid for any two operators $T_1, T_2 \in A(E, F)$.

(IQ3) (a) $\alpha(RS) \leqslant \| R \| \alpha(S)$
for $S \in A(E, Z)$ and $R \in L(Z, F)$.

(IQ3) (b) $\alpha(RS) \leqslant \| S \| \alpha(R)$
for $S \in L(E, Z)$ and $R \in A(Z, F)$.

Let us emphasize that in contrast to Pietsch (1978) we do not claim $\alpha(I_{\mathbb{K}}) = 1$ for the identity map $I_{\mathbb{K}}$ of the one-dimensional Banach space $\mathbb{K} = E_1$.

An operator ideal A equipped with an ideal quasi-norm α is called *a quasi-normed operator ideal* and denoted by $[A, \alpha]$. A quasi-normed operator ideal $[A, \alpha]$ is said to be *complete* if each component $A(E, F)$ is complete with respect to the quasi-norm α.

To verify properties (I1), (I2), (I3)(a) and (I3)(b) of the entropy classes $L_{p,q}^{(e)}$ we employed the expression

$$\lambda_{p,q}^{(e)}(T) = \lambda_{p,q}((e_n(T))) \qquad (1.6.12)$$

which is properly defined for all operators T of type $L_{p,q}^{(e)}$. We now see that the inequality (1.6.8) corresponds to (IQ2), while the inequalities (1.6.11)(a) and (1.6.11)(b) correspond to (IQ3)(a) and (IQ3)(b), respectively. Thus $\alpha(T) = \lambda_{p,q}^{(e)}(T)$ has turned out to be an ideal quasi-norm on $L_{p,q}^{(e)}$, the property (IQ1) being obvious. Moreover, the completeness of the quasi-normed operator ideal $[L_{p,q}^{(e)}, \lambda_{p,q}^{(e)}]$ is proved by the following proposition.

Proposition 1.6.1. (cf. Triebel 1970). *The entropy ideal $L_{p,q}^{(e)}$ is complete with respect to the ideal quasi-norm $\lambda_{p,q}^{(e)}$.*

Proof. Let $T_m\colon E \to F$ be a Cauchy sequence in $L_{p,q}^{(e)}(E,F)$ with respect to $\lambda_{p,q}^{(e)}$ so that

$$\lambda_{p,q}^{(e)}(T_{m+j} - T_m) < \varepsilon \tag{1.6.13}$$

for sufficiently large $m > N_\varepsilon$. Since

$$\|T\| = e_1(T) \leqslant \lambda_{p,q}^{(e)}(T) \quad \text{for } T \in L_{p,q}^{(e)}(E,F)$$

the sequence T_m turns out to be also a Cauchy sequence with respect to the usual operator norm. Therefore there exists an operator $T_0 \in L(E,F)$ such that

$$\lim_{m \to \infty} \|T_0 - T_m\| = 0.$$

To show that $T_0 \in L_{p,q}^{(e)}(E,F)$ and

$$\lim_{m \to \infty} \lambda_{p,q}^{(e)}(T_0 - T_m) = 0 \tag{1.6.14}$$

we first confine ourselves to a finite partial sum of the infinite series appearing on the left-hand side of (1.6.13) in the case of $q < \infty$. This enables us to conclude that

$$\lim_{j \to \infty} \left(\sum_{n=1}^{k} [n^{1/p - 1/q} e_n(T_{m+j} - T_m)]^q \right)^{1/q} \tag{1.6.15}$$

$$= \left(\sum_{n=1}^{k} \left[n^{1/p - 1/q} \lim_{j \to \infty} e_n(T_{m+j} - T_m) \right]^q \right)^{1/q} \leqslant \varepsilon$$

for $m > N_\varepsilon$ independently of k. But because of (DE2)' we have

$$|e_n(T_{m+j} - T_m) - e_n(T_0 - T_m)| \leqslant \|T_{m+j} - T_0\|$$

which implies

$$\lim_{j \to \infty} e_n(T_{m+j} - T_m) = e_n(T_0 - T_m).$$

Accordingly, (1.6.15) amounts to

$$\left(\sum_{n=1}^{k} [n^{1/p - 1/q} e_n(T_0 - T_m)]^q \right)^{1/q} \leqslant \varepsilon$$

for $m > N_\varepsilon$ independently of k. But then even

$$\left(\sum_{n=1}^{\infty} [n^{1/p - 1/q} e_n(T_0 - T_m)]^q \right)^{1/q} \leqslant \varepsilon$$

turns out to be true for $m > N_\varepsilon$. This proves (1.6.14) and hence in particular $T_0 \in L_{p,q}^{(e)}(E,F)$ for $q < \infty$. The completeness of $L_{p,\infty}^{(e)}(E,F)$ with respect to $\lambda_{p,\infty}^{(e)}$ can be verified by similar arguments. ∎

So far we have not taken account of the surjectivity (ES) of the entropy numbers $e_n(T)$. It implies a corresponding property of the operator ideal $A = L_{p,q}^{(e)}$:

(IS)   If $T \in L(E, F)$ and if $Q: \tilde{E} \to E$ is a metric surjection
            such that $TQ \in A(\tilde{E}, F)$, then $T \in A(E, F)$.

An operator ideal A satisfying the additional condition (IS) is called *surjective* (see Stephani 1972, 1973; Pietsch 1978). In the case of $A = L_{p,q}^{(e)}$ the surjectivity, like the ideal properties (I2) and (I3) or $L_{p,q}^{(e)}$, is based on the behaviour of $\lambda_{p,q}^{(e)}$, namely

$$\lambda_{p,q}^{(e)}(TQ) = \lambda_{p,q}^{(e)}(T) \qquad (1.6.16)$$

for any metric surjection $Q: \tilde{E} \to E$ which changes the operator $T: E \to F$ into an operator $\tilde{T} = TQ$ of type $L_{p,q}^{(e)}$. An ideal quasi-norm α defined on a surjective operator ideal A quite generally is said to be *surjective* if

(IQS) $\alpha(TQ) = \alpha(T)$

for $T \in A(E, F)$ and any metric surjection $Q: \tilde{E} \to E$ (see Stephani 1972, 1973; Pietsch 1978).

In the sense of these definitions $L_{p,q}^{(e)}$ is a surjective operator ideal, $\lambda_{p,q}^{(e)}$ being a surjective ideal quasi-norm on $L_{p,q}^{(e)}$.

What about metric injections $J: F \to \tilde{F}$ which change an operator $T: E \to F$ into an operator $\tilde{T} = JT$ if type A? They quite naturally lead us to the following definition of an *injective operator ideal* (see Stephani 1970; Pietsch 1978):

(II)   If $T \in L(E, F)$ and if $J: F \to \tilde{F}$ is a metric injection
            such that $JT \in A(E, \tilde{F})$, then $T \in A(E, F)$.

Certainly, the entropy numbers $e_n(T)$ are not injective in the proper sense (see (1.3.6)). But the inner entropy numbers $f_n(T)$ are (see section 1.3, (FI) and (1.3.37)). To check the injectivity of the operator ideal $A = L_{p,q}^{(e)}$ it therefore seems appropriate to change from $\lambda_{p,q}^{(e)}(T)$ to

$$\lambda_{p,q}^{(f)}(T) = \lambda_{p,q}((f_n(T))). \qquad (1.6.17)$$

Since

$$f_n(T) \leqslant e_n(T) \leqslant 2f_n(T) \qquad (1.6.18)$$

(see (1.3.7)) we have

$$\lambda_{p,q}^{(f)}(T) \leqslant \lambda_{p,q}^{(e)}(T) \leqslant 2\lambda_{p,q}^{(f)}(T). \qquad (1.6.19)$$

Hence the class $L_{p,q}^{(f)}$ of operators T with $\lambda_{p,q}^{(f)}(T) < \infty$ coincides with the class $A = L_{p,q}^{(e)}$ under consideration. The expression $\lambda_{p,q}^{(f)}$ represents an ideal quasi-norm on $L_{p,q}^{(e)}$ as well, the particular versions (DF3)(a) and (DF3)(b) of the multiplicativity of the inner entropy numbers yielding (IQ3)(a) and (IQ3)(b), respectively. The ideal quasi-norm $\lambda_{p,q}^{(f)}$ is surjective, too, by (FS) and (1.3.37). Moreover, $\lambda_{p,q}^{(f)}$ has the property

$$\lambda_{p,q}^{(f)}(JT) = \lambda_{p,q}^{(f)}(T) \tag{1.6.20}$$

for any metric injection $J:F \to \tilde{F}$ which changes the operator $T:E \to F$ into an operator $\tilde{T} = JT$ of type $L_{p,q}^{(e)}$. This proves that $L_{p,q}^{(e)}$ is an injective operator ideal. An ideal quasi-norm α defined on an injective operator ideal A quite generally is said to be *injective* if

(IQI) $\alpha(JT) = \alpha(T)$ for $T \in A(E, F)$

and any metric injection $J:F \to \tilde{F}$ (see Stephani 1970; Pietsch 1978).

Two ideal quasi-norms α and α_0 on the same operator ideal A are called *equivalent* if there exist positive constants c and C such that

$$c\alpha_0(T) \leqslant \alpha(T) \leqslant C\alpha_0(T) \tag{1.6.21}$$

for all operators T of type A. Quite often there is a favourite ideal quasi-norm on an operator ideal A closely related to the definition of A. In the case of the operator ideal $A = L_{p,q}^{(e)}$ we were led to the surjective ideal quasi-norm $\lambda_{p,q}^{(e)}$ when introducing the class $L_{p,q}^{(e)}$ by condition (1.5.2) or (1.5.1), for $q = \infty$. On the other hand, if one approaches the entropy classes by using the inner entropy numbers $f_n(T)$ one will be motivated to work with the ideal quasi-norm $\lambda_{p,q}^{(f)}$, which is simultaneously surjective and injective.

The class K of all compact operators represents a surjective and injective operator ideal complete with respect to the usual operator norm, where this operator norm is obviously surjective and injective. A proof can be given on the basis of the characteristic property

$$\lim_{n \to \infty} e_n(T) = 0 \quad \text{or} \quad \lim_{n \to \infty} f_n(T) = 0$$

of compact operators T and by using the algebraic properties of the $e_n(T)$ or $f_n(T)$, respectively.

2
Approximation quantities

Entropy and compactness properties of operators are closely related to approximation properties. 'Approximation' means approximation by finite rank operators. Approximation quantities are non-increasing sequences of non-negative numbers $s_n(T)$ defined for arbitrary operators T between Banach spaces and, in some sense, express the degree of approximability of T by finite rank operators. We shall deal with the so-called approximation numbers and with Kolmogorov and Gelfand numbers. In their original meaning both Kolmogorov and Gelfand numbers – like entropy numbers – are set functions. They entered the mathematical literature as certain diameters of sets. The definitive paper by Kolmogorov appeared in 1936 (cf. Kolmogorov 1936). For the diameters in the sense of Gelfand, we do not feel able to date their origin. However, for a detailed and comprehensive survey of the development of the theory of diameters we recommend the book by Pinkus (1985). In the present book, for the sake of economy, we shall confine ourselves to *Kolmogorov and Gelfand numbers of operators* as they were considered within a general theory of so-called s-numbers of operators by Pietsch (see Pietsch 1974, 1978). Nevertheless, we refer to the geometrical meaning of Kolmogorov numbers and Gelfand numbers. The particular definition of Gelfand numbers that we give (see (2.3.5)) has not been used in literature so far (see Stephani 1987). It is thought to emphasize the analogy to the geometrical definition of the Kolmogorov numbers (see (2.2.4)) irrespective of duality arguments. Yet the final results of this chapter are well known.

2.1. Approximation numbers

Given an operator $T \in L(E, F)$ between arbitrary Banach spaces E and F the expression

$$a_n(T) = \inf \{ \| T - A \| : A \in L(E, F) \text{ with rank } (A) < n \} \qquad (2.1.1)$$

is called *the nth approximation number of T* (see Pietsch 1974). The approximation numbers satisfy the following algebraic and rank properties.

(A1) **Monotonicity:** $a_1(T) \geqslant a_2(T) \geqslant \cdots \geqslant a_n(T) \geqslant \cdots$ and
$$a_1(T) = \|T\|.$$

(A2) **Additivity:** $a_{k+n-1}(T_1 + T_2) \leqslant a_k(T_1) + a_n(T_2)$
for $T_1, T_2 \in L(E, F)$, in particular

(A2)' $a_n(T_1 + T_2) \leqslant \|T_1\| + a_n(T_2)$.

(A3) **Multiplicativity:** $a_{k+n-1}(RS) \leqslant a_k(R)a_n(S)$
for $S \in L(E, Z)$ and $R \in L(Z, F)$, in particular

(A3) (a) $a_n(RS) \leqslant \|R\| a_n(S)$
and

(A3) (b) $a_n(RS) \leqslant a_n(R)\|S\|$.

(A4) **Rank property:** $a_n(T) = 0$ if and only if rank $(T) < n$.

(A5) **Norm-determining property:** $a_n(I_E) = 1$
whenever $\dim(E) \geqslant n$, where I_E is the identity map of the Banach space E.

These properties can easily be verified on the basis of the definition (2.1.1). We confine ourselves to the norm-determining property (A5) and to the rank property (A4).

Since $a_n(I_E) \leqslant \|I_E\| = 1$ for all n and E we only have to show that $a_n(I_E) \geqslant 1$ when $\dim(E) \geqslant n$. So let us suppose $\dim(E) \geqslant n$. Then for an arbitrary operator $A \in L(E, E)$ with rank $(A) < n$ there exists an element x_0 in the null space $N(A)$ of A with $\|x_0\| = 1$. Hence it follows that

$$1 = \|x_0\| = \|x_0 - Ax_0\| \leqslant \sup_{\|x\| \leqslant 1} \|(I_E - A)x\| = \|I_E - A\|$$

and thus finally

$$1 \leqslant \inf\{\|I_E - A\| : A \in L(E, E) \text{ with rank } (A) < n\} = a_n(I_E).$$

For the rank property (A4), it is obvious that rank $(T) < n$ implies $a_n(T) = 0$. The converse implication can be proved on the basis of the multiplicativity (A3) and the norm determining property (A5) (cf. Pietsch 1978). However, this requires some preliminaries concerning projections in Banach spaces and representations of identity maps.

An operator $P \in L(E, E)$ mapping a Banach space E into itself is called a *projection* if

$$P^2 = P. \tag{2.1.2}$$

This property causes the range $R(P)$ of a projection P to be a closed subspace of E. Note that an arbitrary closed subspace M of a Banach space E need not necessarily be the range of a projection P. An impressive example of a subspace which is not the range of a projection is given by the space c_0 as a subspace of l_∞ (see Whitley 1966). However, any finite-dimensional subspace of a Banach space is the range of some linear and continuous projection P.

Lemma 2.1.1. *Let $N \subseteq E$ be an n-dimensional subspace of a Banach space E. Then there exists a projection $P \in L(E, E)$ with $R(P) = N$.*

Proof. The space N is spanned by n linearly independent elements x_1, $x_2, \ldots, x_n$, so that every $x \in N$ allows a unique representation

$$x = \sum_{i=1}^{n} \lambda_i x_i.$$

The equations

$$\langle x, a_i^0 \rangle = \lambda_i \quad \text{for } x = \sum_{i=1}^{n} \lambda_i x_i$$

define linear functionals a_i^0 over N which, moreover, are continuous. According to the Hahn–Banach theorem they can be extended to linear and continuous functionals a_i over E. The corresponding operator

$$Px = \sum_{i=1}^{n} \langle x, a_i \rangle x_i \qquad (2.1.3)$$

is a linear and continuous projection of E onto N since

$$P^2 x = \sum_{i=1}^{n} \langle x, a_i \rangle P x_i = \sum_{i=1}^{n} \langle x, a_i \rangle x_i = Px. \qquad \blacksquare$$

A strengthened version of Lemma 2.1.1 results from

Auerbach's lemma. (see Pietsch 1978; Jarchow 1981). *Let N be an n-dimensional Banach space. Then there exist elements $x_1^0, x_2^0, \ldots, x_n^0$ in N and functionals $a_1^0, a_2^0, \ldots, a_n^0$ over N such that*

$$\|x_k^0\| = 1, \quad \|a_k^0\| = 1, \quad \text{and} \quad \langle x_i^0, a_k^0 \rangle = \delta_{ik} \quad \text{for } 1 \leqslant i, k \leqslant n. \quad (2.1.4)$$

Proof. We choose an arbitrary basis $x_1, x_2, \ldots, x_n$ of N and consider the absolute value

$$d(a_1, a_2, \ldots, a_n) = |\det(\langle x_i, a_k \rangle)|$$

of the determinant $\det(\langle x_i, a_k \rangle)$ as a function on the n-fold cartesian product $X = (U_{N'} \times U_{N'} \times \cdots \times U_{N'})$. Obviously this function is continuous. Because of the compactness of X there exists an n-tuple $(a_1^0, a_2^0, \ldots, a_n^0) \in X$ for which $d(a_1, a_2, \ldots, a_n)$ attains its maximum

$$d_0 = d(a_1^0, a_2^0, \ldots, a_n^0) > 0.$$

Since the corresponding matrix $(\langle x_i, a_j^0 \rangle)$ is invertible we can determine elements $x_1^0, x_2^0, \ldots, x_n^0$ in N such that

$$\sum_{j=1}^{n} \langle x_i, a_j^0 \rangle x_j^0 = x_i \quad \text{for } i = 1, 2, \ldots, n.$$

Then we have

$$\sum_{j=1}^{n} \langle x_i, a_j^0 \rangle \langle x_j^0, a_k \rangle = \langle x_i, a_k \rangle, \quad 1 \leqslant i, k \leqslant n, \qquad (2.1.5)$$

for arbitrary $a_k \in U_{N'}$. In particular, with $a_k = a_k^0$ we obtain

$$\sum_{j=1}^{n} \langle x_i, a_j^0 \rangle (\langle x_j^0, a_k^0 \rangle - \delta_{jk}) = 0 \quad \text{for } 1 \leqslant i, k \leqslant n$$

which yields

$$\langle x_j^0, a_k^0 \rangle = \delta_{jk}$$

and hence also

$$1 \leqslant \| x_k^0 \| \cdot \| a_k^0 \| \leqslant \| x_k^0 \|, \quad 1 \leqslant k \leqslant n.$$

In order to get the desired estimate for $\| x_k^0 \|$ from above we once more refer to equations (2.1.5). They tell us that

$$|\det(\langle x_i, a_j^0 \rangle)| \cdot |\det(\langle x_j^0, a_k \rangle)| = |\det(\langle x_i, a_k \rangle)|.$$

But in view of the choice of the n-tuple $(a_1^0, a_2^0, \ldots, a_n^0)$ we have

$$|\det(\langle x_i, a_k \rangle)| \leqslant |\det(\langle x_i, a_k^0 \rangle)|.$$

This implies

$$|\det(\langle x_j^0, a_k \rangle)| \leqslant 1 \quad \text{for arbitrary } a_k \in U_{N'}.$$

Employing an n-tuple $(a_1, a_2, \ldots, a_n) \in X$ with

$$a_k = \begin{cases} a & \text{for } k = i, \\ a_k^0 & \text{for } k \neq i \end{cases}$$

we attain

$$|\det(\langle x_j^0, a_k \rangle)| = |\langle x_i^0, a \rangle| \leqslant 1$$

and thus

$$\| x_i^0 \| = \left| \sup_{\|a\| \leqslant 1} \langle x_i^0, a \rangle \right| \leqslant 1 \quad \text{for } 1 \leqslant i \leqslant n.$$

But now the inequality $1 \leqslant \| x_k^0 \| \cdot \| a_k^0 \| \leqslant \| x_k^0 \|$ above gives $\| a_k^0 \| = 1$. This completes the proof of (2.1.4)   ∎

A system of elements $x_1^0, x_2^0, \ldots, x_n^0$ in an n-dimensional Banach space N satisfying properties (2.1.4) with a corresponding system of functionals $a_1^0, a_2^0, \ldots, a_n^0$ represents a particular basis of N, a so-called *Auerbach basis*. If we choose an Auerbach basis $x_1 = x_1^0, x_2 = x_2^0, \ldots, x_n = x_n^0$ for the n-dimensional subspace $N \subseteq E$ of Lemma 2.1.1 and identify functionals

$a_1, a_2, \ldots, a_n$ in the formula (2.1.3), with extensions of the functionals $a_k^0 \in N'$ to E such that $\| a_k \| = \| a_k^0 \| = 1$, we gain a projection P of E onto N whose norm can be estimated by

$$\| P \| \leqslant n. \qquad (2.1.6)$$

An even stronger result than (2.1.6) will be derived in chapter 6.

Projections are the main tool for extracting identity maps $I_Z : Z \to Z$ from arbitrary operators $T : E \to F$ in the sense that

$$I_Z = BTA,$$

where $A \in L(Z, E)$ and $B \in L(F, Z)$.

Lemma 2.1.2. *Let $T \in L(E, F)$ be an operator between arbitrary Banach space E and F with $\mathrm{rank}(T) \geqslant n$. Then there exist a Banach space Z with $\dim(Z) = n$ and operators $A \in L(Z, E)$, $B \in L(F, Z)$ such that*

$$I_Z = BTA. \qquad (2.1.7)$$

Proof. Because $\mathrm{rank}(T) \geqslant n$ we can find an n-dimensional subspace Z of E whose image under T is an n-dimensional subspace Y of F. By Lemma 2.1.1 the subspace $Y \subseteq F$ is the range of a projection $P \in L(F, F)$. Let $P_0 : F \to Y$ be the operator of F onto Y induced by P and let $I : Z \to E$ be the natural embedding of Z into E. Then the operator

$$S = P_0 T I$$

is an isomorphism of Z onto Y such that

$$S^{-1}S = I_Z.$$

Putting $A = I$ and $B = S^{-1}P_0$ we arrive at the desired representation (2.1.7) of the identity map I_Z. ∎

Now we are in the position to complete the proof of (A4). We assume $\mathrm{rank}(T) \geqslant n$ and choose an n-dimensional Banach space Z as well as operators $A \in L(Z, E)$ and $B \in L(F, Z)$ satisfying (2.1.7). By using (A5) and (A3) we may conclude that

$$1 = a_n(I_Z) \leqslant \| B \| a_n(T) \| A \|$$

and hence $a_n(T) > 0$. If $a_n(T) = 0$ we therefore have $\mathrm{rank}(T) < n$.

The methods just applied also enable us to calculate the approximation numbers $a_n(D)$ of the diagonal operator $D : l_p \to l_p$ studied in sections 1.3, 1.4 and 1.5. Let us recall that

$$Dx = (\sigma_1 \xi_1, \sigma_2 \xi_2, \ldots, \sigma_k \xi_k, \ldots) \qquad (1.3.15)$$

for

$$x = (\xi_1, \xi_2, \ldots, \xi_k, \ldots) \in l_p, \; 1 \leqslant p \leqslant \infty, \text{ with } \sigma_1 \geqslant \sigma_2 \geqslant \cdots \geqslant 0.$$

By D_{n-1} we again denote the $(n-1)$th sectional operator

$$D_{n-1}x = (\sigma_1\xi_1, \sigma_2\xi_2, \dots, \sigma_{n-1}\xi_{n-1}, 0, 0, \dots) \qquad (1.3.24)'$$

of D in l_p and by $D^{(n)}$ the operator

$$D^{(n)}(\xi_1, \xi_2, \dots, \xi_n) = (\sigma_1\xi_1, \sigma_2\xi_2, \dots, \sigma_n\xi_n) \qquad (1.3.17)'$$

acting in l_p^n. Since rank $(D_{n-1}) < n$ and $\|D - D_{n-1}\| = \sigma_n$ we have

$$a_n(D) \leqslant \sigma_n.$$

The case $\sigma_n = 0$ is clear so we assume $\sigma_n > 0$. Then the operator $D^{(n)}$ is invertible, the norm of its inverse $(D^{(n)})^{-1}$ being given by

$$\|(D^{(n)})^{-1}\| = \sigma_n^{-1}.$$

We consider the factorization

$$I_p^{(n)} = (D^{(n)})^{-1} D^{(n)}$$

of the identity map $I_p^{(n)}$ of l_p^n. Applying (A5) and (A3) we obtain

$$1 = a_n(I_p^{(n)}) \leqslant \|(D^{(n)})^{-1}\| a_n(D^{(n)}) = \sigma_n^{-1} \cdot a_n(D^{(n)}).$$

Since

$$D^{(n)} = P_n D I_n \qquad (1.3.18)'$$

we can once more apply the multiplicativity (A3) of the approximation numbers, namely

$$a_n(D^{(n)}) \leqslant \|P_n\| a_n(D) \|I_n\| = a_n(D),$$

so that

$$1 \leqslant \sigma_n^{-1} \cdot a_n(D)$$

appears. The final result is

$$a_n(D) = \sigma_n \qquad (2.1.8)$$

(cf. Pietsch 1978).

An operator T is said to be *approximable* if $\lim_{n \to \infty} a_n(T) = 0$. In the case of operators T acting between Hilbert spaces H and K the approximable operators are in a sense nothing other than diagonal operators with the additional property

$$\lim_{n \to \infty} \sigma_n = 0$$

of the diagonal sequence σ_n. Indeed, *Schmidt's representation theorem* for approximable operators $T \in L(H, K)$ says that there exists finite or at most countable orthonormal systems (x_i) in H and (y_i) in K, as well as a non-increasing sequence $\sigma_1 \geqslant \sigma_2 \geqslant \cdots \geqslant 0$ that is finite or tends to zero, such that

$$Tx = \sum_{i=1}^{\infty} \sigma_i(x, x_i) y_i \qquad (2.1.9)$$

(cf. Pietsch 1965, 1987). Employing the operators $V^*: H \to l_2$ and $U: l_2 \to K$ defined by

$$V^*x = \sum_{i=1}^{\infty} (x, x_i)e_i \quad \text{for } x \in H \quad \text{and} \quad Uz = \sum_{i=1}^{\infty} (z, e_i)y_i \quad \text{for } z \in l_2,$$

respectively, with (e_i) as the standard unit vector basis of l_2, we recognize that

$$T = UDV^*, \qquad (2.1.10)$$

where $D: l_2 \to l_2$ stands for the diagonal operator (1.3.15). Conversely, the operators $U^*: K \to l_2$ and $V: l_2 \to H$ defined by

$$U^*y = \sum_{i=1}^{\infty} (y, y_i)e_i \quad \text{for } y \in K \quad \text{and} \quad Vz = \sum_{i=1}^{\infty} (z, e_i)x_i \quad \text{for } z \in l_2,$$

respectively, enable us to regain the diagonal operator D as

$$D = U^*TV. \qquad (2.1.11)$$

Because $a_n(D) = \sigma_n$ and $\|U\| = \|V^*\| = 1$ we get

$$a_n(T) \leqslant \|U\| a_n(D) \|V^*\| = \sigma_n$$

from (2.1.10) and, on the other hand,

$$\sigma_n = a_n(D) \leqslant \|U^*\| a_n(T) \|V\| = a_n(T)$$

from (2.1.11) and $\|U^*\| = \|V\| = 1$, which yields

$$a_n(T) = \sigma_n. \qquad (2.1.12)$$

Hence, the approximation numbers $a_n(T)$ of an approximable operator T acting between Hilbert spaces H and K coincide with the coefficients σ_n appearing in a non-increasing arrangement in the Schmidt representation (2.1.9) of T (cf. Pietsch 1965, 1978, 1987).

Owing to the general properties of the approximation numbers the class G of approximable operators between arbitrary Banach spaces forms an operator ideal which, moreover, is complete with respect to the usual operator norm (see Pietsch 1978). In accordance with that, the class $L_{p,q}^{(a)}$ of all operators T with $(a_n(T)) \in l_{p,q}$ forms an operator ideal which is complete with respect to the ideal quasi-norm

$$\lambda_{p,q}^{(a)}(T) = \lambda_{p,q}((a_n(T)))$$

(cf. (1.5.5)). We omit the proof, and refer the reader to the proof of the ideal properties of the class $L_{p,q}^{(e)}$ and its completeness with respect to the ideal quasi-norm $\lambda_{p,q}^{(e)}$ in section 1.6. The relation between the approximation ideal $L_{p,q}^{(a)}$ and the entropy ideal $L_{p,q}^{(e)}$ will be disclosed in the next chapter. In the present chapter we shall study among others, the relation between the ideal G of approximable operators and the ideal K

of compact operators which is fundamental to what follows. In this connection we derive two further characterizations of compact operators thus motivating the definition of the so-called Kolmogorov numbers (see section 2.2) and of the so-called Gelfand numbers (see section 2.3).

2.2. Kolmogorov numbers

The following characterization of compact operators is the motive for the introduction of the Kolmogorov numbers.

Proposition 2.2.1. (cf. Pietsch 1965). *An operator $T \in L(E, F)$ between arbitrary Banach spaces E and F is compact if and only if for every $\varepsilon > 0$ there exists a finite-dimensional subspace $N_\varepsilon \subseteq F$ with*

$$T(U_E) \subset N_\varepsilon + \varepsilon U_F. \tag{2.2.1}$$

Proof. Given a compact operator $T: E \to F$, for every $\varepsilon > 0$ we can choose finitely many elements $y_1, y_2, \ldots, y_{n_\varepsilon}$ such that

$$T(U_E) \subseteq \bigcup_{i=1}^{n_\varepsilon} \{y_i + \varepsilon U_F\}.$$

Introducing the finite-dimensional subspace N_ε spanned by the elements $y_1, y_2, \ldots, y_{n_\varepsilon}$ in F we can state (2.2.1). Conversely, let $T \in L(E, F)$ satisfy the condition formulated above. The inclusion (2.2.1) can then be strengthened to

$$T(U_E) \subseteq N_\varepsilon \cap \rho_\varepsilon U_F + \varepsilon U_F \tag{2.2.2}$$

with $\rho_\varepsilon = \| T \| + \varepsilon$ since

$$Tx = z + \varepsilon y \text{ with } y \in U_F, z \in N_\varepsilon \text{ and } \|z\| \leqslant \| T \| + \varepsilon$$

for $x \in U_E$. The set $N_\varepsilon \cap \rho_\varepsilon U_F = \rho_\varepsilon U_{N_\varepsilon}$ is a bounded subset of the finite-dimensional Banach space N_ε and hence even precompact as a subset of F (see section 1.1). Therefore there are finitely many elements $y_i \in F$, $1 \leqslant i \leqslant m_\varepsilon$, such that

$$\rho_\varepsilon U_{N_\varepsilon} \subseteq \sum_{i=1}^{m_\varepsilon} \{y_i + \varepsilon U_F\}. \tag{2.2.3}$$

Combining (2.2.2) and (2.2.3) we obtain

$$T(U_E) \subseteq \bigcup_{i=1}^{m_\varepsilon} \{y_i + 2\varepsilon U_F\}.$$

Because $\varepsilon > 0$ can be chosen arbitrarily we may conclude that T is compact.

∎

The *nth Kolmogorov number* $d_n(T)$ *of an operator* $T \in L(E, F)$ acting between arbitrary Banach spaces E and F is defined to be the infimum of all $\varepsilon > 0$ such that there is a subspace $N_\varepsilon \subseteq F$ with $\dim(N_\varepsilon) < n$ and

$$T(U_E) \subset N_\varepsilon + \varepsilon U_F, \tag{2.2.1}$$

that is to say

$$d_n(T) = \inf\{\varepsilon > 0: T(U_E) \subset N_\varepsilon + \varepsilon U_F, \text{ where } N_\varepsilon \subseteq F \text{ with } \dim(N_\varepsilon) < n\} \tag{2.2.4}$$

(cf. Kolmogorov 1936; Pietsch 1965; Mitjagin and Pełczyński 1966; Triebel 1970).

Let us remark that the value of $d_n(T)$ remains unchanged if the closed balls U_E or U_F are replaced by the corresponding open ones $\mathring{U}_E$ or $\mathring{U}_F$, respectively. That is to say, we can write

$$T(U_E) \subset N_\varepsilon + \varepsilon \mathring{U}_F \tag{2.2.1}'$$

as well as

$$T(\mathring{U}_E) \subset N_\varepsilon + \varepsilon U_F \quad \text{or} \quad T(\mathring{U}_E) \subset N_\varepsilon + \varepsilon \mathring{U}_F \tag{2.2.1}''$$

under the infimum (2.2.4). Moreover, for $\operatorname{rank}(T) \geq n$ the definition (2.2.4) of $d_n(T)$ can be modified by demanding $\dim(N_\varepsilon) = n - 1$ instead of $\dim(N_\varepsilon) < n$. For $\operatorname{rank}(T) < n$ we obviously have $d_n(T) = 0$.

By Proposition 2.2.1 an operator T is compact if and only if $\lim_{n \to \infty} d_n(T) = 0$. Roughly speaking, Proposition 2.2.1 and the definition of the Kolmogorov numbers $d_n(T)$ of an operator $T : E \to F$ deal with parts of the set $T(U_E)$ which lie outside finite-dimensional subspaces $N \subseteq F$ with $\dim(N) < n$. This fact can be brought to light by using the quotient map $Q_N^F : F \to F/N$. Let us recall that the *quotient space* $\bar{F} = F/N$ is the linear space of the cosets $\bar{y} = \{y - z : z \in N\}$ with respect to N equipped with the *quotient norm*

$$\|\bar{y}\| = \inf\{\|y - z\| : z \in N\}, \tag{2.2.5}$$

the quotient map Q_N^F being defined by $Q_N^F y = \bar{y}$. The image $Q_N^F(\mathring{U}_F)$ of the open unit ball $\mathring{U}_F$ under Q_N^F coincides with open unit ball $\mathring{U}_{F/N}$ of F/N,

$$Q_N^F(\mathring{U}_F) = \mathring{U}_{F/N}. \tag{2.2.6}$$

Indeed, an element $\bar{y} \in Q_N^F(\mathring{U}_F)$ is a coset $\bar{y} = \{y - z : z \in N\}$ with $\|y\| < 1$. According to the definition (2.2.5) of the norm in the quotient space F/N we then have $\|\bar{y}\| \leq \|y\| < 1$ which means $\bar{y} \in \mathring{U}_{F/N}$. On the other hand, given $\bar{y} \in F/N$ with $\|\bar{y}\| < 1$ the definition of the norm $\|\bar{y}\|$ in the space F/N tells us that there is an element $y \in F$ with $Q_N^F y = \bar{y}$ and $\|y\| < 1$, that is to say $\bar{y} \in Q_N^F(\mathring{U}_F)$. This completes the proof of (2.2.6) and hence the proof of the fact that any quotient map $Q_N^F : F \to F/N$ is a metric surjection (cf. (1.3.2)).

Proposition 2.2.2. (cf. Pietsch 1978). *The nth Kolmogorov number* $d_n(T)$ *of an operator* $T \in L(E, F)$ *between arbitrary Banach spaces* E *and* F *can be* *expressed as*

$$d_n(T) = \inf\{\|Q_N^F T\| : N \subseteq F, \dim(N) < n\}. \tag{2.2.7}$$

Proof. We use the definition of the nth Kolmogorov number $d_n(T)$ related to (2.2.1)' with a subspace $N = N_\varepsilon$ of F of dimension $\dim(N) < n$. The application of the quotient map $Q_N^F : F \to F/N$ by (2.2.6) then changes the inclusion (2.2.1)' into

$$Q_N^F T(U_E) \subseteq \varepsilon Q_N^F(\mathring{U}_F) = \varepsilon \mathring{U}_{F/N}. \tag{2.2.8}$$

It follows that $\|Q_N^F T\| \leqslant \varepsilon$ and thus

$$\tilde{d}_n(T) \leqslant d_n(T),$$

where

$$\tilde{d}_n(T) = \inf\{\|Q_M^F T\| : M \subseteq F, \dim(M) < n\}$$

stands for the infimum in question. Now let $\delta > 0$ and $N \subseteq F$ with $\dim(N) < n$ be such that

$$\|Q_N^F T\| < \tilde{d}_n(T) + \delta. \tag{2.2.9}$$

This inequality implies

$$Q_N^F T(U_E) \subseteq (\tilde{d}_n(T) + \delta)\mathring{U}_{F/N} = Q_N^F((\tilde{d}_n(T) + \delta)\mathring{U}_F). \tag{2.2.10}$$

Passing to the inverse images in (2.2.10) with respect to Q_N^F we get

$$N + T(U_E) \subseteq N + (\tilde{d}_n(T) + \delta)\mathring{U}_F$$

and thus

$$T(U_E) \subset N + (\tilde{d}_n(T) + \delta)\mathring{U}_F. \tag{2.2.11}$$

This means $d_n(T) \leqslant \tilde{d}_n(T) + \delta$ and hence finally $d_n(T) \leqslant \tilde{d}_n(T)$. ∎

Remark. When $\operatorname{rank}(T) \geqslant n$ the representation (2.2.7) of $d_n(T)$ can obviously be replaced by

$$d_n(T) = \inf\{\|Q_N^F T\| : N \subseteq F, \dim(N) = n - 1\}. \tag{2.2.7'}$$

It turns out that Kolmogorov numbers are closely related to approximation numbers: a basic relation is given by the inequality

$$d_n(T) \leqslant a_n(T). \tag{2.2.12}$$

This can easily be checked: given $T \in L(E, F)$ and $\varepsilon > 0$ there exists an operator $A \in L(E, F)$ with $\operatorname{rank}(A) < n$ such that

$$\|T - A\| < a_n(T) + \varepsilon$$

and, correspondingly,

$$(T - A)(U_E) \subseteq (a_n(T) + \varepsilon)U_F.$$

An inclusion for the set $T(U_E)$ itself is then given by

$$T(U_E) \subseteq A(U_E) + (a_n(T) + \varepsilon)U_F$$

and hence also by

$$T(U_E) \subset N + (a_n(T) + \varepsilon)U_F, \qquad (2.2.13)$$

where $N = A(E)$ denotes the range of A. From (2.2.13) together with $\dim(N) < n$ follows $d_n(T) \leqslant a_n(T) + \varepsilon$ and finally (2.2.12).

The inequality (2.2.12) in particular tells us that every approximable operator is compact. On the other hand, there are compact operators which are not approximable (cf. Köthe 1979). However, a proof of this statement due to Enflo (1973) is beyond the scope of the present book.

For special Banach spaces $\tilde{E}$ and operators T from $\tilde{E}$ into an arbitrary Banach space F the inequality (2.2.12) turns into an equality. These are the *Banach spaces with the so-called metric lifting property*:

$\tilde{E}$ is said to have the metric lifting property if every operator T from $\tilde{E}$ into a quotient space F/F_0 for arbitrary $\varepsilon > 0$ possesses a '*lifting*' $\tilde{T}: \tilde{E} \to F$ such that

$$T = Q_{F_0}^F \tilde{T} \quad \text{and} \quad \| \tilde{T} \| \leqslant (1 + \varepsilon) \| T \|.$$

Proposition 2.2.3. *Let $\tilde{E}$ be a Banach space with the metric lifting property, F an arbitrary Banach space and $T \in L(\tilde{E}, F)$. Then*

$$d_n(T) = a_n(T). \qquad (2.2.14)$$

Proof. Because of (2.2.12) it only remains to show that $a_n(T) \leqslant d_n(T)$. So let $\varepsilon > 0$. The formula (2.2.7) for the nth Kolmogorov number $d_n(T)$ guarantees the existence of a subspace $N \subseteq F$ with $\dim(N) < n$ and $\| Q_N^F T \| < d_n(T) + \varepsilon$. Since $\tilde{E}$ has the metric lifting property the operator $Q_N^F T: \tilde{E} \to F/N$ can be lifted to an operator $\tilde{T}: \tilde{E} \to F$ such that

$$Q_N^F T = Q_N^F \tilde{T} \quad \text{and} \quad \| \tilde{T} \| \leqslant (1 + \varepsilon) \| Q_N^F T \|.$$

Because

$$Q_N^F (T - \tilde{T}) = 0$$

the difference $S = T - \tilde{T}$ is an operator whose range $R(S)$ is contained in N, so $\operatorname{rank}(S) < n$. Therefore we may conclude that

$$a_n(T) \leqslant \| T - S \| = \| \tilde{T} \| \leqslant (1 + \varepsilon) \| Q_N^F T \| < (1 + \varepsilon)(d_n(T) + \varepsilon),$$

and finally arrive at $a_n(T) \leqslant d_n(T)$. ∎

The Banach space $l_1(\Gamma)$ of *summable number families* $\{ \lambda_\gamma \}_{\gamma \in \Gamma}$ over an arbitrary index set Γ has the metric lifting property. The elements $\tilde{x} = \{ \lambda_\gamma \}_{\gamma \in \Gamma}$ of $l_1(\Gamma)$ are characterized by the condition $\sum_{\gamma \in \Gamma} |\lambda_\gamma| < \infty$ which automatically implies that any $\tilde{x}$ has at most countably many components $\lambda_\gamma \neq 0$. The norm on $l_1(\Gamma)$ is defined by

$$\| \tilde{x} \| = \sum_{\gamma \in \Gamma} |\lambda_\gamma|.$$

For details we refer the reader to Pietsch (1965).

For the proof of the metric lifting property of $l_1(\Gamma)$ we first observe that an arbitrary operator $T:l_1(\Gamma) \to Z$ from $l_1(\Gamma)$ into a Banach space Z can be written as

$$T(\{\lambda_\gamma\}) = \sum_{\gamma \in \Gamma} \lambda_\gamma z_\gamma \text{ with } z_\gamma \in Z \text{ and } \|T\| = \sup_{\gamma \in \Gamma} \|z_\gamma\|.$$

In the case of an operator $T:l_1(\Gamma) \to F/F_0$ from $l_1(\Gamma)$ into a quotient space $Z = F/F_0$ the elements $z_\gamma \in F/F_0$ can be gained from elements $y_\gamma \in F$ by $z_\gamma = Q^F_{F_0} y_\gamma$ such that the conditions $\|y_\gamma\| \leq (1 + \varepsilon)\|z_\gamma\|$ are satisfied simultaneously. Accordingly, the operator

$$\tilde{T}(\{\lambda_\gamma\}) = \sum_{\gamma \in \Gamma} \lambda_\gamma y_\gamma$$

represents the desired lifting for

$$T = Q^F_{F_0} \tilde{T} \quad \text{and} \quad \|\tilde{T}\| = \sup_{\gamma \in \Gamma} \|y_\gamma\| \leq (1 + \varepsilon)\|T\|.$$

The Banach spaces $l_1(\Gamma)$ are of universal significance. In fact, every Banach space E appears as a quotient space of an appropriate space $l_1(\Gamma)$. To prove this we choose the closed unit ball U_E of the Banach space E as an index set Γ. In the space $\tilde{E} = l_1(U_E)$ we then consider the linear subspace

$$\tilde{E}_0 = \left\{ \{\lambda_x\}_{x \in U_E} \in \tilde{E} : \sum_{x \in U_E} \lambda_x x = 0 \right\}.$$

The value of the infinite series $\sum_{x \in U_E} \lambda_x x$ for $\{\lambda_x\}_{x \in U_E} \in \tilde{E}$ is always an element of E. Conversely, every element z of E can be expressed in this way. This is obvious for $z = 0$ and for $z \neq 0$ can be achieved by using the number family

$$\lambda_x = \begin{cases} \|z\| & \text{for } x = \dfrac{z}{\|z\|} \in U_E, \\ 0 & \text{everywhere else on } U_E. \end{cases}$$

This means that there is a one-to-one correspondence between the cosets $\overline{\{\lambda_x\}}_{x \in U_E}$ of $\tilde{E}$ with respect to $\tilde{E}_0$ on one hand and the elements of E on the other. Moreover, the norm of an element z in E coincides with the norm

$$\|\overline{\{\lambda_x\}}_{x \in U_E}\| = \inf \left\{ \sum_{x \in U_E} |\lambda_x| : z = \sum_{x \in U_E} \lambda_x x \right\}$$

of the corresponding coset in the quotient space $\tilde{E}/\tilde{E}_0$. This results from the two estimates

$$\inf \left\{ \sum_{x \in U_E} |\lambda_x| : z = \sum_{x \in U_E} \lambda_x x \right\} \leq \|z\|$$

and

$$\|z\| \le \sum_{x \in U_E} |\lambda_x| \text{ or } \|z\| \le \inf\left\{ \sum_{x \in U_E} |\lambda_x| : z = \sum_{x \in U_E} \lambda_x x \right\},$$

respectively. So let us identify the Banach space E with the quotient space $\tilde{E}/\tilde{E}_0$ and let Q_E denote the quotient map from $\tilde{E} = l_1(U_E)$ onto $E = \tilde{E}/\tilde{E}_0$.

We are now in a position to transform an arbitrary operator $T: E \to F$ into an operator $\tilde{T}: \tilde{E} \to F$ defined on a Banach space $\tilde{E}$ with the metric lifting property and thus subject to Proposition 2.2.3.

Theorem 2.2.1. *Let* $T \in L(E, F)$ *be an operator between arbitrary Banach spaces* E *and* F, *and let* Q_E *be the quotient map from* $l_1(U_E)$ *onto* E. *Then*

$$d_n(T) = a_n(TQ_E). \tag{2.2.15}$$

Proof. By Proposition 2.2.3 we have $d_n(TQ_E) = a_n(TQ_E)$. But the definition of the Kolmogorov numbers $d_n(T)$ given on the basis of (2.2.1)'' makes it clear that

$$d_n(TQ) = d_n(T) \tag{2.2.16}$$

for any metric surjection $Q: \tilde{E} \to E$. This proves (2.2.15). ∎

Remark. The property (2.2.16) will be referred to as the property of *surjectivity of the Kolmogorov numbers*.

Theorem 2.2.1 has far-reaching consequences. It shows that an operator T between arbitrary Banach spaces E and F is compact if and only if the operator $TQ_E: l_1(U_E) \to F$ is approximable. Furthermore, the sequence of Kolmogorov numbers $(d_n(T))$ belongs to the Lorentz sequence space $l_{p,q}$ if and only if $(a_n(TQ_E)) \in l_{p,q}$, which means $TQ_E \in L_{p,q}^{(a)}(l_1(U_E), F)$. Denoting the class of operators T with $(d_n(T)) \in l_{p,q}$ by $L_{p,q}^{(d)}$ and putting

$$\lambda_{p,q}^{(d)}(T) = \lambda_{p,q}((d_n(T))) \tag{2.2.17}$$

we can again state that $L_{p,q}^{(d)}$ represents an operator ideal complete with respect to the ideal quasi-norm $\lambda_{p,q}^{(d)}$. The proof rests upon the algebraic and rank properties of the Kolmogorov numbers which, in analogy to the algebraic and rank properties of the approximation numbers, will be referred to as

(K1) **Monotonicity**, (K2) **Additivity**, (K3) **Multiplicativity**,
(K4) **Rank property**, (K5) **Norm determining property**,
(KS) **Surjectivity**.

They can be checked by straight-forward considerations. Let us point out that the norm determining property

(K5) $d_n(I_E) = 1$ for the identity map I_E of a Banach space E
with $\dim(E) \geqslant n$

is an immediate consequence of the representation (2.2.7) since $\|Q_N^E\| = 1$ for any quotient map $Q_N^E : E \to E/N$ if N is properly contained in E.

Even the steps made for the calculation of the approximation numbers $a_n(D)$ of the diagonal operator (1.3.15) may be transferred to the case of the Kolmogorov numbers $d_n(D)$. The result is

$$d_n(D) = \sigma_n.$$

The surjectivity (KS) of the Kolmogorov numbers $d_n(T)$, which has already been given in (2.2.16), implies the surjectivity of the operator ideal $L_{p,q}^{(d)}$, the ideal quasi-norm $\lambda_{p,q}^{(d)}$ itself being surjective. Because $d_n(T) \leqslant a_n(T)$ we have

$$L_{p,q}^{(a)} \subseteq L_{p,q}^{(d)}.$$

On the other hand, an arbitrary surjective operator ideal A containing the approximation ideal $L_{p,q}^{(a)}$ automatically contains all operators $T : E \to F$ with $TQ_E \in L_{p,q}^{(a)}(l_1(U_E), F)$. By Theorem 2.2.1 this means

$$L_{p,q}^{(d)} \subseteq A.$$

Since the class of operators $L_{p,q}^{(d)}$ is itself a surjective operator ideal, one can state that $L_{p,q}^{(d)}$ is the smallest surjective operator ideal containing the approximation ideal $L_{p,q}^{(a)}$. Similarly, the ideal K of compact operators is the smallest surjective operator ideal containing the ideal G of approximable operators.

Within the general theory of operator ideals the concept of *the surjective hull A^S of an arbitrary operator ideal A* has been introduced to describe such situations. The surjective hull A^S of A is understood to be the smallest surjective operator ideal containing A (see Stephani 1972, 1973; Pietsch 1978).

On the basis of these remarks we can write

$$K = G^S \tag{2.2.18}$$

and

$$L_{p,q}^{(d)} = (L_{p,q}^{(a)})^S. \tag{2.2.19}$$

2.3. Gelfand numbers

The characterization of compact operators which we shall derive next will be the starting point for the definition of the so-called Gelfand numbers (see Stephani 1987).

Proposition 2.3.1. *An operator $T \in L(E, F)$ between arbitrary Banach spaces E and F is compact if and only if for every $\varepsilon > 0$ there are finitely many* *functionals $a_i \in E'$, $1 \leqslant i \leqslant n_\varepsilon$, such that*

$$\|Tx\| \leqslant \sup_{1 \leqslant i \leqslant n_\varepsilon} |\langle x, a_i \rangle| + \varepsilon \|x\| \quad \text{for all } x \in E. \tag{2.3.1}$$

Proof. Given a compact operator $T: E \to F$ and $\varepsilon > 0$ we can determine finitely many elements $y_i \in F$, $1 \leqslant i \leqslant n_\varepsilon$, such that

$$T(U_E) \subseteq \bigcup_{i=1}^{n_\varepsilon} \left\{ y_i + \frac{\varepsilon}{2} U_F \right\}. \tag{2.3.2}$$

Furthermore, by the Hahn–Banach theorem for each y_i there exists a functional $b_i \in F'$ with

$$|\langle y_i, b_i \rangle| = \|y_i\| \quad \text{and} \quad \|b_i\| = 1.$$

Passing to the functionals $a_i \in E'$ defined by

$$\langle x, a_i \rangle = \langle Tx, b_i \rangle \quad \text{or} \quad a_i = T' b_i$$

enables us to verify (2.3.1). Though this estimate concerns arbitrary $x \in E$ it suffices to work with $x \in U_E$ because any $x \in E$, $x \neq 0$, can be changed into an element of U_E by taking $x/\|x\|$. The image Tx of an element $x \in U_E$, however, belongs to the union on the right-hand side of (2.3.2), say $Tx \in \{ y_k + (\varepsilon/2) U_F \}$. The norm of the corresponding element y_k admits the estimate

$$\|y_k\| = |\langle y_k, b_k \rangle| \leqslant |\langle y_k - Tx, b_k \rangle| + |\langle Tx, b_k \rangle|$$

$$\leqslant \|y_k - Tx\| \cdot \|b_k\| + \sup_{1 \leqslant i \leqslant n_\varepsilon} |\langle x, a_i \rangle| \leqslant \frac{\varepsilon}{2} + \sup_{1 \leqslant i \leqslant n_\varepsilon} |\langle x, a_i \rangle|.$$

The norm of the image Tx can finally be estimated by

$$\|Tx\| \leqslant \|Tx - y_k\| + \|y_k\| \leqslant \varepsilon + \sup_{1 \leqslant i \leqslant n_\varepsilon} |\langle x, a_i \rangle|,$$

which proves (2.3.1) for arbitrary $x \in E$.

Conversely, let us assume that the operator $T: E \to F$ fulfils the condition (2.3.1) for arbitrary $\varepsilon > 0$ with appropriate functionals $a_i \in E'$. In order to show that $T(U_E)$ is a precompact subset of F we consider a system of elements $y_j \in T(U_E)$ such that

$$\|y_j - y_k\| \geqslant 4\varepsilon \quad \text{for } j \neq k.$$

Making use of a representation $y_j = Tx_j$ with $x_j \in U_E$ we can state that

$$4\varepsilon \leqslant \|Tx_j - Tx_k\| \leqslant \sup_{1 \leqslant i \leqslant n_\varepsilon} |\langle x_j - x_k, a_i \rangle| + 2\varepsilon$$

and thus

$$2\varepsilon \leqslant \sup_{1 \leqslant i \leqslant n_\varepsilon} |\langle x_j - x_k, a_i \rangle| \quad \text{for } j \neq k. \tag{2.3.3}$$

The supremum on the right-hand side of (2.3.3) can be read as the distance of the elements

$$u_j = (\langle x_j, a_i \rangle)_{1 \leqslant i \leqslant n_\varepsilon} \quad \text{and} \quad u_k = (\langle x_k, a_i \rangle)_{1 \leqslant i \leqslant n_\varepsilon}$$

in the Banach space $l^{n_\varepsilon}_\infty$. Since

$$\|u_j\|_\infty = \sup_{1 \leqslant i \leqslant n_\varepsilon} |\langle x_j, a_i \rangle| \leqslant \sup_{1 \leqslant i \leqslant n_\varepsilon} \|a_i\|$$

the system $\{u_j\} \subset l^{n_\varepsilon}_\infty$ is a bounded subset of $l^{n_\varepsilon}_\infty$. But in a finite-dimensional Banach space every bounded subset is precompact (see section 1.1). The elements $u_j \in l^{n_\varepsilon}_\infty$ satisfying the conditions

$$2\varepsilon \leqslant \|u_j - u_k\|_\infty \quad \text{for } j \neq k, \tag{2.3.3}'$$

therefore even form a finite system in $l^{n_\varepsilon}_\infty$. Hence the original system $y_j \in T(U_E)$ is finite as well. Thus $T(U_E)$ has proved to be precompact. ■

The *nth Gelfand number* $c_n(T)$ *of an operator* $T \in L(E, F)$ acting between arbitrary Banach spaces E and F is defined to be the infimum of all $\varepsilon > 0$ such that there are functionals $a_i \in E'$, $1 \leqslant i \leqslant k < n$, which admit an estimate

$$\|Tx\| \leqslant \sup_{1 \leqslant i \leqslant k} |\langle x, a_i \rangle| + \varepsilon \|x\| \quad \text{for all } x \in E, \tag{2.3.4}$$

that is to say

$$c_n(T) = \inf \left\{ \varepsilon > 0 \colon \|Tx\| \leqslant \sup_{1 \leqslant i \leqslant k} |\langle x, a_i \rangle| + \varepsilon \|x\|, \right.$$

$$\left. \text{where } a_i \in E', 1 \leqslant i \leqslant k \text{ with } k < n \right\}. \tag{2.3.5}$$

By Proposition 2.3.1 an operator T is compact if and only if $\lim_{n \to \infty} c_n(T) = 0$. Roughly speaking, Proposition 2.3.1 and the definition of the Gelfand numbers $c_n(T)$ of an operator $T \colon E \to F$ deal with the restrictions of the operator T to subspaces $M \subseteq E$ of finite codimension. Indeed, the subspace

$$M = \{x \in E \colon \langle x, a_i \rangle = 0 \text{ for } 1 \leqslant i \leqslant k < n\}$$

of E gives rise to a quotient space $\bar{E} = E/M$ whose dimension m is at most $n - 1$. But this means that M is of codimension $m \leqslant n - 1$. With I^E_M as the natural embedding of M into E the estimate (2.3.4) says

$$\|TI^E_M\| \leqslant \varepsilon. \tag{2.3.6}$$

Proposition 2.3.2. (cf. Pietsch 1978) *The nth Gelfand number* $c_n(T)$ *of an operator* $T \in L(E, F)$ *between arbitrary Banach spaces* E *and* F *allows the* *representation*

$$c_n(T) = \inf\{\|TI_M^E\|: M \subseteq E, \text{codim}(M) < n\}. \qquad (2.3.7)$$

Proof. Let us abbreviate the infimum on the right-hand side of (2.3.7) by $\tilde{c}_n(T)$. Then $\tilde{c}_n(T) \leqslant c_n(T)$ appears as an immediate consequence of (2.3.6) and the definition (2.3.5) of $c_n(T)$. Now let $\varepsilon > 0$ and $M \subseteq E$ be a subspace with $\text{codim}(M) = m < n$ such that

$$\|TI_M^E\| < \tilde{c}_n(T) + \varepsilon$$

which amounts to

$$\|Tz\| = \|TI_M^E z\| \leqslant (\tilde{c}_n(T) + \varepsilon)\|z\| \quad \text{for } z \in M. \qquad (2.3.8)$$

In order to get the desired estimate for $\|Tx\|$ itself we first choose a basis $\bar{x}_1, \bar{x}_2, \ldots, \bar{x}_m$ in the quotient space E/M. Then every $\bar{x} \in E/M$ can be represented uniquely as

$$\bar{x} = \sum_{i=1}^{m} \lambda_i \bar{x}_i,$$

where the coefficients

$$\lambda_i = \langle \bar{x}, \bar{a}_i \rangle, \quad 1 \leqslant i \leqslant m,$$

are linear and even continuous functionals over E/M because E/M is finite-dimensional. Hence we have

$$|\langle \bar{x}, \bar{a}_i \rangle| \leqslant C \cdot \|\bar{x}\| \quad \text{for } 1 \leqslant i \leqslant m$$

with an appropriate constant $C > 0$. The equations

$$\langle x, a_i \rangle = \langle \bar{x}, \bar{a}_i \rangle, \quad 1 \leqslant i \leqslant m,$$

obviously define linear functionals a_i over E which are also continuous since

$$|\langle x, a_i \rangle| \leqslant C \cdot \|x\| \quad \text{for } 1 \leqslant i \leqslant m,$$

from the definition of the norm $\|\bar{x}\|$ on the quotient space E/M (see (2.2.5)). We want to employ the functionals a_i to obtain a projection $P: E \to E$ with $N(P) = M$ as null space. For this purpose we determine elements $x_i \in E$ such that

$$Q_M^E x_i = \bar{x}_i$$

and then put

$$Px = \sum_{i=1}^{m} \langle x, a_i \rangle x_i. \qquad (2.3.9)$$

The operator P is actually a projection with the desired property

$$Px = 0 \quad \text{if and only if} \quad \langle x, a_i \rangle = 0 \text{ for } 1 \leqslant i \leqslant m.$$

In accordance with that, the operator

$$P_M = I_E - P \qquad (2.3.10)$$

is a projection of E onto M since

$$P_M^2 = I_E - 2P + P^2 = I_E - P = P_M \qquad (2.3.11)$$

and

$$P_M(E) = \{z \in E : z = x - Px\} = N(P) = M. \qquad (2.3.12)$$

The equality

$$I_E = P + P_M$$

enables us to estimate $\|Tx\|$ by

$$\|Tx\| \leqslant \|TPx\| + \|TP_Mx\|. \qquad (2.3.13)$$

To the element $z = P_Mx$ of M we can apply the estimate (2.3.8), giving

$$\|TP_Mx\| \leqslant (\tilde{c}_n(T) + \varepsilon)\|P_Mx\|. \qquad (2.3.14)$$

Now if we use the substitution $P_M = I_E - P$ on the right-hand side of (2.3.14) the desired term $(\tilde{c}_n(T) + \varepsilon)\|x\|$ can be generated, namely

$$(\tilde{c}_n(T) + \varepsilon)\|P_Mx\| \leqslant (\tilde{c}_n(T) + \varepsilon)\|x\| + (\tilde{c}_n(T) + \varepsilon)\|Px\|.$$

In this way (2.3.13) becomes

$$\|Tx\| \leqslant \|TPx\| + (\tilde{c}_n(T) + \varepsilon)\|Px\| + (\tilde{c}_n(T) + \varepsilon)\|x\|.$$

From (2.3.9) the two items $\|TPx\|$ and $(\tilde{c}_n(T) + \varepsilon)\|Px\|$ allow a uniform estimate

$$\|TPx\| + (\tilde{c}_n(T) + \varepsilon)\|Px\| \leqslant \tilde{C} \cdot \sup_{1 \leqslant i \leqslant m} |\langle x, a_i \rangle|.$$

With $\tilde{a}_i = \tilde{C}a_i$ instead of a_i we finally arrive at

$$\|Tx\| \leqslant \sup_{1 \leqslant i \leqslant m} |\langle x, \tilde{a}_i \rangle| + (\tilde{c}_n(T) + \varepsilon)\|x\|.$$

It follows that $c_n(T) \leqslant \tilde{c}_n(T) + \varepsilon$, and furthermore $c_n(T) \leqslant \tilde{c}_n(T)$. ∎

Remark. For $\operatorname{rank}(T) \geqslant n$ the formula (2.3.7) can obviously be replaced by

$$c_n(T) = \inf\{\|TI_M^E\| : M \subseteq E, \operatorname{codim}(M) = n - 1\}. \qquad (2.3.7)'$$

In analogy to the existence of a linear and continuous projection P onto every finite-dimensional subspace N of an arbitrary Banach space E (see Lemma 2.1.1), the existence of a linear and continuous projection P_M onto every subspace $M \subseteq E$ of finite codimension can be stated. This is a byproduct of the proof given for Proposition 2.3.2 (see (2.3.9), (2.3.10), (2.3.11), (2.3.12)).

Also the Gelfand numbers $c_n(T)$ of an operator $T \in L(E, F)$ are related to the approximation numbers $a_n(T)$ by the inequality

$$c_n(T) \leqslant a_n(T). \qquad (2.3.15)$$

For the proof we fix $\varepsilon > 0$ arbitrarily and determine an operator $S \in L(E, F)$ with rank$(S) < n$ such that

$$\| T - S \| < a_n(T) + \varepsilon$$

and, correspondingly,

$$\| (T - S)x \| \leqslant (a_n(T) + \varepsilon) \| x \|.$$

Then an estimate for $\| Tx \|$ itself is given by

$$\| Tx \| \leqslant \| Sx \| + (a_n(T) + \varepsilon) \| x \|. \qquad (2.3.16)$$

A bound for the norm of the finite rank operator S can be obtained in a similar way as for the operator P defined by (2.3.9), namely

$$\| Sx \| \leqslant \sup_{1 \leqslant i \leqslant m} |\langle x, a_i \rangle|$$

with appropriate functionals $a_i \in E'$, $1 \leqslant i \leqslant m < n$. This ensures that we may conclude $c_n(T) \leqslant a_n(T) + \varepsilon$ on the basis of (2.3.16), and hence $c_n(T) \leqslant a_n(T)$.

The inequality (2.3.15) once more tells us that every approximable operator is compact. For special Banach spaces $\tilde{F}$ and arbitrary operators $T: E \to \tilde{F}$ with values in $\tilde{F}$ the inequality (2.3.15) even turns into an equality. These are the *Banach spaces with the* so-called *metric extension property*:

$\tilde{F}$ is said to have the metric extension property if every operator $T: E \to \tilde{F}$ from a Banach space E into $\tilde{F}$ can be extended to any Banach space $\tilde{E}$ containing E as a subspace, where the extension $\tilde{T}: \tilde{E} \to \tilde{F}$ is linear and continuous, and $\| \tilde{T} \| = \| T \|$.

Proposition 2.3.3. *Let $\tilde{F}$ be a Banach space with the metric extension property, E an arbitrary Banach space and $T \in L(E, \tilde{F})$. Then*

$$c_n(T) = a_n(T). \qquad (2.3.17)$$

Proof. Because of (2.3.15) we need only show that $a_n(T) \leqslant c_n(T)$. So let $\varepsilon > 0$. The formula (2.3.7) for the nth Gelfand number $c_n(T)$ then guarantees the existence of a subspace $M \subseteq E$ with codim$(M) < n$ and $\| TI_M^E \| < c_n(T) + \varepsilon$. Since $\tilde{F}$ has the metric extension property the operator $TI_M^E: M \to \tilde{F}$ can be extended to an operator $\tilde{T}: E \to \tilde{F}$ such that

$$TI_M^E = \tilde{T}I_M^E \quad \text{and} \quad \| \tilde{T} \| = \| TI_M^E \|.$$

Since

$$(T - \tilde{T})I_M^E = 0$$

the difference $S = T - \tilde{T}$ is an operator whose kernel $N(S)$ contains M so that rank$(S) < n$. Therefore we may conclude that

$$a_n(T) \leqslant \| T - S \| = \| \tilde{T} \| = \| T I_M^E \| < c_n(T) + \varepsilon$$

and finally obtain $a_n(T) \leqslant c_n(T)$.   ∎

The Banach space $l_\infty(\Gamma)$ of all *bounded number families* $\{\lambda_\gamma\}_{\gamma \in \Gamma}$ over an arbitrary index set Γ has the metric extension property. The norm on $l_\infty(\Gamma)$ is defined by

$$\| \{\lambda_\gamma\} \| = \sup_{\gamma \in \Gamma} |\lambda_\gamma|.$$

An operator T from an arbitrary Banach space E into $l_\infty(\Gamma)$ is given by

$$Tx = \{\langle x, a_\gamma \rangle\}_{\gamma \in \Gamma} \text{ with } a_\gamma \in E' \text{ and } \| T \| = \sup_{\gamma \in \Gamma} \| a_\gamma \|.$$

By the Hahn–Banach theorem each functional a_γ can be extended to any Banach space $\tilde{E}$ containing E as a subspace, the corresponding extension $\tilde{a}_\gamma$ being linear and continuous with $\| \tilde{a}_\gamma \| = \| a_\gamma \|$. Accordingly, the operator

$$\tilde{T}\tilde{x} = \{\langle \tilde{x}, \tilde{a}_\gamma \rangle\}_{\gamma \in \Gamma} \text{ for } \tilde{x} \in \tilde{E}$$

represents the desired extension of T to $\tilde{E}$, for

$$T = \tilde{T} I_E^{\tilde{E}} \text{ and } \| \tilde{T} \| = \sup_{\gamma \in \Gamma} \| \tilde{a}_\gamma \| = \| T \|.$$

Hence the metric extension property of $l_\infty(\Gamma)$ is secured.

Like the Banach spaces $l_1(\Gamma)$, the Banach spaces $l_\infty(\Gamma)$ are also of universal significance. Indeed, every Banach space F can be regarded as a subspace of an appropriate space $l_\infty(\Gamma)$. To prove this we choose the closed unit ball $U_{F'}$ of the dual space F' of F as an index set Γ. In the space $\tilde{F} = l_\infty(U_{F'})$ we then consider the special number families $\{\langle y, b \rangle\}_{b \in U_{F'}}$ generated by elements $y \in F$. They obviously form a subspace $\tilde{F}_0$ of $l_\infty(U_{F'})$. Moreover, $\{\langle y, b \rangle\}_{b \in U_{F'}} = 0$ if and only if $y = 0$, and

$$\sup_{b \in U_{F'}} |\langle y, b \rangle| = \| y \|.$$

Therefore we can identify the subspace $\tilde{F}_0$ of $\tilde{F} = l_\infty(U_{F'})$ with the Banach space F. The corresponding embedding of $\tilde{F}_0 = F$ into $\tilde{F} = l_\infty(U_{F'})$ will be denoted by $J_F : F \to l_\infty(U_{F'})$.

We are now in a position to transform an arbitrary operator $T : E \to F$ into an operator $\tilde{T} : E \to \tilde{F}$ with values in a Banach space $\tilde{F}$ with the metric extension property and thus subject to Proposition 2.3.3.

Theorem 2.3.1. *Let $T \in L(E, F)$ be an operator between arbitrary Banach spaces E and F and let J_F be the embedding of F into $l_\infty(U_{F'})$. Then*

$$c_n(T) = a_n(J_F T). \qquad (2.3.18)$$

Proof. By Proposition 2.3.3 we have $c_n(J_F T) = a_n(J_F T)$. But the definition (2.3.5) of the Gelfand numbers $c_n(T)$ makes it clear that

$$c_n(JT) = c_n(T) \qquad (2.3.19)$$

for any metric injection $J: F \to \tilde{F}$. This proves (2.3.18). ∎

Remark. The property (2.3.19) will be referred to as the property of *injectivity of the Gelfand numbers*.

Theorem 2.3.1, like Theorem 2.2.1, has far-reaching consequences. It shows that an operator T between arbitrary Banach spaces E and F is compact if and only if the operator $J_F T: E \to l_\infty(U_{F'})$ is approximable. Furthermore, the relation (2.3.18) implies that the sequence of Gelfand numbers $(c_n(T))$ belongs to the Lorentz sequence space $l_{p,q}$ if and only if $(a_n(J_F T)) \in l_{p,q}$, which means $J_F T \in L_{p,q}^{(a)}(E, l_\infty(U_{F'}))$. We shall denote the class of operators T with $(c_n(T)) \in l_{p,q}$ by $L_{p,q}^{(c)}$ and put

$$\lambda_{p,q}^{(c)}(T) = \lambda_{p,q}((c_n(T))). \qquad (2.3.20)$$

$L_{p,q}^{(c)}$ turns out to be an operator ideal complete with respect to the ideal quasi-norm $\lambda_{p,q}^{(c)}$. This fact is again due to the algebraic and rank properties of the Gelfand numbers $c_n(T)$. We shall refer to them in the usual manner as

(G1) **Monotonicity**, (G2) **Additivity**, (G3) **Multiplicativity**,
(G4) **Rank property**, (G5) **Norm determining property**, (GI) **Injectivity**.
We leave out the proofs and confine ourselves to mentioning that the norm determining property,

(G5) $c_n(I_E) = 1$ for the identity map I_E of a Banach space E with $\dim(E) \geqslant n$,

is this time a consequence of the representation (2.3.7) since $\| I_M^E \| = 1$ for any natural embedding $I_M^E: M \to E$ of a subspace $M \subseteq E$ with $\dim(M) \geqslant 1$.

As a consequence of the algebraic and rank properties of the $c_n(T)$ we again obtain

$$c_n(D) = \sigma_n$$

for the Gelfand numbers of the diagonal operator (1.3.15).

The injectivity (GI) of the Gelfand numbers $c_n(T)$, which has already been given in (2.3.19), implies the injectivity of the operator ideal $L_{p,q}^{(c)}$, the ideal quasi-norm $\lambda_{p,q}^{(c)}$ itself being injective. Because $c_n(T) \leqslant a_n(T)$ we have

$$L_{p,q}^{(a)} \subseteq L_{p,q}^{(c)}.$$

On the other hand, an arbitrary injective operator ideal A containing the approximation ideal $L_{p,q}^{(a)}$ automatically contains all operators $T: E \to F$ with $J_F T \in L_{p,q}^{(a)}(E, l_\infty(U_{F'}))$. By Theorem 2.3.1 this means

$$L_{p,q}^{(c)} \subseteq A.$$

Since the class of operators $L_{p,q}^{(c)}$ is itself an injective operator ideal, one can state that $L_{p,q}^{(c)}$ is the smallest injective operator ideal containing the approximation ideal $L_{p,q}^{(a)}$. Similarly, the ideal K of compact operators is the smallest injective operator ideal containing the ideal G of approximable operators.

Within the general theory of operator ideals the concept of *the injective hull A^1 of an arbitrary operator ideal A* has been introduced to describe such situations. The injective hull A^1 of A is understood to be the smallest injective operator ideal containing A (see Stephani 1970; Pietsch 1978).

We conclude these observations with the two formulas

$$K = G^1 \tag{2.3.21}$$

and

$$L_{p,q}^{(c)} = (L_{p,q}^{(a)})^1. \tag{2.3.22}$$

2.4. Geometrical parameters

In this section we shall derive converse inequalities to the inequalities

$$c_n(T) \leqslant a_n(T) \quad \text{and} \quad d_n(T) \leqslant a_n(T)$$

between approximation numbers $a_n(T)$ on the one side and Gelfand $c_n(T)$ or Kolmogorov numbers $d_n(T)$ on the other side (see (2.3.15) and (2.2.12)). We remind the reader of the situation where F is a Banach space with the metric extension property, and

$$a_n(T) = c_n(T)$$

turns out to be true for all operators T mapping an arbitrary Banach space E into F (see Proposition 2.3.3). Similarly, in the case of a Banach space E with the metric lifting property we could prove

$$a_n(T) = d_n(T)$$

for all operators T mapping E into an arbitrary Banach space F (see Proposition 2.2.3). To obtain general estimates for the approximation numbers $a_n(T)$ from above by the Gelfand numbers $c_n(T)$ and the Kolmogorov numbers $d_n(T)$ we introduce certain quantities $p_n(E, F)$ and $q_n(E, F)$, respectively, dependent on the pair of Banach spaces (E, F). Since these quantities are determined by extension and lifting properties of the Banach spaces involved we shall refer to them as *geometrical parameters*.

An operator T defined on a subspace M of a Banach space E and having its values in a Banach space F is said to admit an *extension* to E if there exists a linear and continuous operator $\tilde{T}: E \to F$ such that

$$T = \tilde{T} I_M^E, \qquad (2.4.1)$$

where I_M^E denotes the natural embedding of M into E. For a fixed triple of Banach spaces $(M; E, F)$ with $M \subseteq E$ *the extension constant* $p(M; E, F)$ is defined by

$$p(M; E, F) = \inf\{\rho > 0: \text{ any } T \in L(M, F) \text{ admits an}$$
$$\text{extension } \tilde{T} \text{ to } E \text{ with } \|\tilde{T}\| \leqslant \rho \|T\|\}. \qquad (2.4.2)$$

Of course, the infimum (2.4.2) need not be finite. The very situation $p(M; E, F) = \infty$ can be described as follows: for arbitrary $\rho > 0$ there exists an operator $T \in L(M, F)$ such that any extension $\tilde{T} \in L(E, F)$ of T to E – provided an extension does exist at all – possesses a norm $\|\tilde{T}\| > \rho \|T\|$.

Since the norm of an extension $\tilde{T}$ of T cannot be less than the norm of T itself we have

$$p(M; E, F) \geqslant 1 \qquad (2.4.3)$$

in any case. If $\tilde{F}$ is a Banach space with the metric extension property then

$$p(M; E, \tilde{F}) = 1$$

is true for arbitrary Banach spaces E and subspaces $M \subseteq E$ (see section 2.3).

Next we consider the case $F = M$ and suppose $p(M; E, M) < \infty$. Then in particular the identity map I_M of the Banach space M possesses an extension $\tilde{I}_M: E \to M$ to E. The corresponding operator

$$P_M = I_M^E \tilde{I}_M$$

mapping the Banach space E into itself is a projection of E onto M. Indeed, on the basis of

$$I_M = \tilde{I}_M I_M^E$$

the characteristic property

$$P^2 = P \qquad (2.1.2)$$

of a projection P can easily be verified for $P = P_M$. On the other hand, if $M \subseteq E$ is known to be the range of a projection $P_M: E \to E$ we can use the canonical factorization

$$P_M = I_M^E (P_M)_0$$

of P_M through its range M, with $(P_M)_0: E \to M$ as the operator of E onto M induced by P_M, and then define an extension $\tilde{T}: E \to F$ for an arbitrary operator $T: M \to F$ by

$$\tilde{T} = T(P_M)_0. \qquad (2.4.4)$$

Because

$$I_M = (P_M)_0 I_M^E$$

the property (2.4.1) required for an extension $\tilde{T}$ of T is actually satisfied. Furthermore we obtain the norm estimate

$$\|\tilde{T}\| \leqslant \|T\| \cdot \|(P_M)_0\| = \|T\| \cdot \|P_M\|$$

as a result of (2.4.4). Thus the extension constant $p(M; E, F)$ can be estimated by

$$p(M; E, F) \leqslant \inf\{\|P_M\| : P_M \in L(E, E) \text{ a projection with } R(P_M) = M\}. \qquad (2.4.5)$$

The infimum on the right-hand side of (2.4.5) is called *the relative projection constant of $M \subseteq E$ with respect to E* and is denoted by $\lambda(M, E)$ (see König 1986). Since $\lambda(M, E)$ can also be interpreted as

$$\lambda(M, E) = \inf\{\|\tilde{I}_M\| : \tilde{I}_M \text{ an extension of } I_M \text{ to } E\}$$

we recognize that

$$\lambda(M, E) \leqslant p(M; E, M). \qquad (2.4.6)$$

Combining (2.4.6) with (2.4.5) for $F = M$ we get

$$\lambda(M, E) = p(M; E, M).$$

We formulate the final result as follows:

  $p(M; E, F) < \infty$ *for all Banach spaces F if and only if M is*

  the range of a projection $P_M \in L(E, E)$.

If $E = H$ is a Hilbert space then there exists a projection P_M with $\|P_M\| = 1$ onto every subspace M of H so that

$$\lambda(M, H) = 1.$$

By (2.4.3) we therefore have

$$p(M; H, F) = 1 \text{ for arbitrary Banach spaces } F. \qquad (2.4.7)$$

What we are mainly interested in is the situation of a subspace M of finite codimension in an arbitrary Banach space E. A subspace M of this kind always admits a projection. In section 2.3 for the proof of Proposition 2.3.2 we constructed a projection $P_M \in L(E, E)$ onto a subspace $M \subseteq E$ of finite codimension (see (2.3.10)). The complementary situation of a subspace $N \subseteq E$ of finite dimension was dealt with in section 2.1. In Lemma 2.1.1 we proved that there is a projection onto any finite-dimensional subspace N of a Banach space E and afterwards, by using Auerbach's Lemma, showed the existence of a projection $P \in L(E, E)$ onto an arbitrary subspace $N \subseteq E$ of dimension n with

$$\|P\| \leqslant n. \qquad (2.1.6)$$

It turns out that the norm estimate (2.1.6) can be improved further. We formulate stronger results along these lines which will help us to estimate the geometrical parameters to be studied in this section, as well as those to be introduced in section 3.3. The proofs of the following two lemmas, however, will not be given until chapter 6 (see Theorems 6.1.1 and 6.1.2).

Lemma 2.4.1. *Let $N \subseteq E$ be an n-dimensional subspace of a Banach space E. Then there exists a projection P of E onto N which can be factorized as*

$$P = RS \text{ with } S \in L(E, l_2^n) \text{ and } R \in L(l_2^n, E)$$

over l_2^n such that

$$\|P\| \leqslant \|R\| \|S\| \leqslant \sqrt{n}.$$

Lemma 2.4.2. *Let $M \subseteq E$ be a subspace of a Banach space E with* $\text{codim}(M) = n$. *Then for every $\varepsilon > 0$ there exists a projection $P \in L(E, E)$ with* $N(P) = M$ *which can be factorized as*

$$P = RS \text{ with } S \in L(E, l_2^n) \text{ and } R \in L(l_2^n, E)$$

over l_2^n such that

$$\|P\| \leqslant \|R\| \|S\| \leqslant (1 + \varepsilon)\sqrt{n}.$$

Let us consider the extension constant $p(M; E, F)$ for a subspace $M \subseteq E$ of codimension m. By (2.4.5) we have

$$p(M; E, F) \leqslant \|P_M\|$$

for any projection $P_M \in L(E, E)$ with $R(P_M) = M$. If we first choose a projection P in E according to Lemma 2.4.2 and then put

$$P_M = I_E - P$$

we in fact obtain a projection of E onto M (cf. (2.3.10), (2.3.11), (2.3.12)). The norm of P_M is subject to the estimate

$$\|P_M\| \leqslant 1 + \|P\| \leqslant 1 + (1 + \varepsilon)\sqrt{m}.$$

This finally leads us to

$$p(M; E, F) \leqslant 1 + \sqrt{m}. \tag{2.4.8}$$

Letting M vary in the class of all subspaces $M \subseteq E$ of codimension $n - 1$ we define *the nth extension constant $p_n(E, F)$ for a fixed pair of Banach spaces (E, F)* by

$$p_n(E, F) = \sup\{p(M; E, F): M \subseteq E, \text{codim}(M) = n - 1\}. \tag{2.4.9}$$

In view of (2.4.8) the supremum (2.4.9) is bounded by

$$p_n(E, F) \leqslant 1 + (n - 1)^{1/2}. \tag{2.4.10}$$

The exact value $p_n(E, F)$ is a geometrical parameter dependent on the pair of Banach spaces (E, F). It is this very parameter which is sufficient for the estimation of the nth approximation number $a_n(T)$ from above by the nth Gelfand number $c_n(T)$, as follows.

Proposition 2.4.1. *Let $T \in L(E, F)$ be an operator between arbitrary Banach spaces E and F. Then*

$$a_n(T) \leqslant p_n(E, F)c_n(T). \qquad (2.4.11)$$

Proof. If $\text{rank}(T) < n$ we have $a_n(T) = c_n(T) = 0$. So let us assume that $\text{rank}(T) \geqslant n$. Then we proceed in the same way as for the proof of $a_n(T) \leqslant c_n(T)$ in Proposition 2.3.3. Namely, given $\varepsilon > 0$ we consider a subspace $M \subseteq E$ with $\text{codim}(M) = n - 1$ and $\| TI_M^E \| < c_n(T) + \varepsilon$ (see (2.3.7)′). According to the definition of the extension constant $p(M; E, F)$ the restriction TI_M^E of the operator $T: E \to F$ to M can again be extended to an operator $\tilde{T}: E \to F$ such that

$$TI_M^E = \tilde{T} I_M^E \quad \text{and} \quad \| \tilde{T} \| \leqslant (p(M; E, F) + \varepsilon) \| TI_M^E \|.$$

Since the difference $S = T - \tilde{T}$ is an operator with $\text{rank}(S) < n$ we may conclude that

$$a_n(T) \leqslant \| T - S \| = \| \tilde{T} \| \leqslant (p(M; E, F) + \varepsilon) \| TI_M^E \|$$
$$\leqslant (p(M; E, F) + \varepsilon)(c_n(T) + \varepsilon).$$

Taking the supremum (2.4.9) with respect to all subspaces $M \subseteq E$ of $\text{codim}(M) = n - 1$ we obtain

$$a_n(T) \leqslant (p_n(E, F) + \varepsilon)(c_n(T) + \varepsilon)$$

and hence finally the inequality (2.4.11). ∎

The result (2.4.11) implies the result

$$a_n(T) = c_n(T) \qquad (2.3.17)$$

for operators T from an arbitrary Banach space E into a Banach space $\tilde{F}$ with the metric extension property since

$$p_n(E, \tilde{F}) = 1$$

in this case. For a Hilbert space H in the role of E and arbitrary Banach spaces F we have

$$p_n(H, F) = 1$$

because of (2.4.7). Therefore (2.3.17) turns out to be true even for operators T mapping a Hilbert space H into an arbitrary Banach space F. This statement gives rise to a geometrical representation for the approximation numbers $a_n(T)$ of operators $T: H \to F$ (cf. Pietsch 1978).

Proposition 2.4.2. *Let H be a Hilbert space, F an arbitrary Banach space, and $T \in L(H, F)$. Then*

$$a_n(T) = \inf\{ \| T - TP \| : P \in L(H, H) \text{ is an orthogonal}$$
$$\text{projection with } \text{rank}(P) < n\}. \qquad (2.4.12)$$

Proof. Since $S = TP$ is an operator with $\text{rank}(S) < n$ we have

$$a_n(T) \leqslant \| T - TP \|$$

and therefore also a corresponding inequality between $a_n(T)$ and the infimum of $\| T - TP \|$ with respect to all orthogonal projections $P: H \to H$ of $\text{rank}(P) < n$. For the converse estimate we choose $\varepsilon > 0$ and again determine a subspace $M \subseteq H$ with $\text{codim}(M) < n$ such that

$$\| T I_M^H \| < c_n(T) + \varepsilon.$$

But since M is the range of an orthogonal projection P_M we may write

$$\| T I_M^H \| = \| T P_M \|.$$

Next we introduce the orthogonal projection

$$P = I_H - P_M$$

complementary to P_M. Then $\| T I_M^H \|$ takes the shape

$$\| T I_M^H \| = \| T - TP \|.$$

Realizing that $\text{rank}(P) < n$ we now can take the infimum of $\| T - TP \|$ with respect to all orthogonal projections $P: H \to H$ of $\text{rank}(P) < n$. The result is

$$\inf\{ \| T - TP \| : P \in L(H, H) \text{ is an orthogonal projection}$$
$$\text{with } \text{rank}(P) < n\} \leqslant \| T I_M^H \| < c_n(T) + \varepsilon. \qquad (2.4.13)$$

Because $c_n(T) = a_n(T)$ the inequality (2.4.13) actually reduces to the desired estimate for $a_n(T)$ from below as ε tends to zero. ∎

Having considered the Hilbert space situation $T \in L(H, F)$ we turn again to the case of operators T between arbitrary Banach spaces E and F, and formulate a general estimate which follows immediately from (2.4.10) and (2.4.11).

Proposition 2.4.3. *Let $T \in L(E, F)$ be an operator between arbitrary Banach spaces E and F. Then*

$$a_n(T) \leqslant (2n)^{1/2} c_n(T). \qquad (2.4.14)$$

The idea for stating (2.4.14) was to express the growth of the approximation numbers $a_n(T)$ relative to the Gelfand numbers $c_n(T)$ by a multiple ρn^α of a certain power of n. This could be achieved on the basis of (2.4.10) by using

$$1 + (n-1)^{1/2} \leqslant (2n)^{1/2}.$$

For the estimation for the nth approximation number $a_n(T)$ from above by the nth Kolmogorov number $d_n(T)$ we now introduce geometrical parameters which are determined by lifting properties of the underlying pair of Banach spaces (E, F).

An operator T mapping a Banach space E into a quotient space F/N of a Banach space F is said to possess a *lifting* to F if there exists a linear and continuous operator $\tilde{T}: E \to F$ such that

$$T = Q_N^F \tilde{T}, \qquad (2.4.15)$$

where Q_N^F denotes the quotient map of F onto F/N (see section 2.2). For a fixed triple of Banach spaces $(E, F; N)$ with $N \subseteq F$ the *lifting constant* $q(E, F; N)$ is defined by

$$q(E, F; N) = \inf\{\rho > 0 \colon \text{any } T \in L(E, F/N) \text{ possesses a}$$
$$\text{lifting } \tilde{T} \text{ to } F \text{ with } \|\tilde{T}\| \leqslant \rho \|T\|\}. \qquad (2.4.16)$$

We have

$$q(E, F; N) \geqslant 1 \qquad (2.4.17)$$

in any case. If $\tilde{E}$ is a Banach space with the metric lifting property then

$$q(\tilde{E}, F; N) = 1$$

for arbitrary Banach spaces F and $N \subseteq F$. In general, however, the infimum (2.4.16) need not even be finite.

For the following we identify E with the quotient space F/N and suppose $q(F/N, F; N) < \infty$. Then in particular the identity map $I_{F/N}$ of the quotient space F/N possesses a lifting $\tilde{I}_{F/N} : F/N \to F$ so that

$$I_{F/N} = Q_N^F \tilde{I}_{F/N}. \qquad (2.4.18)$$

Transposing the order of the two factors $\tilde{I}_{F/N}$ and Q_N^F in the product (2.4.18) we obtain a projection

$$P = \tilde{I}_{F/N} Q_N^F$$

in the Banach space F. The characteristic property $P^2 = P$ can easily be checked:

$$P^2 = \tilde{I}_{F/N}(Q_N^F \tilde{I}_{F/N})Q_N^F = \tilde{I}_{F/N} Q_N^F = P$$

by (2.4.18). Moreover, the subspace $N \subseteq F$ turns out to be the null space $N(P)$ of P. Namely $y \in N$ is equivalent to $Q_N^F y = 0$ and from $Q_N^F y = 0$ it follows that

$$Py = \tilde{I}_{F/N} Q_N^F y = 0.$$

Conversely, $Py = 0$ implies

$$Q_N^F y = (Q_N^F \tilde{I}_{F/N})Q_N^F y = 0$$

by (2.4.18). Having recognized the subspace $N \subseteq F$ as the null space $N(P) = N$ of the projection P we can immediately construct a projection $P_N \in L(F, F)$ with $R(P_N) = N$, namely

$$P_N = I_F - P$$

(see (2.3.10), (2.3.11), (2.3.12)). Hence under the assumption $q(F/N, F; N) < \infty$ the subspace $N \subseteq F$ can be expressed as the range of a projection. On the other hand, if $N \subseteq F$ is known to be the range of a projection $P_N \in L(F, F)$ we take the complementary projection

$$P = I_F - P_N$$

whose range $R(P)$ is given by $R(P) = N(P_N)$ and whose null space $N(P)$ coincides with $N = R(P_N)$. Then we define an operator $J: F/N \to F$ by

$$J\bar{y} = Py \quad \text{with} \quad \bar{y} = Q_N^F y.$$

This definition makes sense since the image $J\bar{y}$ of the coset

$$\bar{y} = \{u \in F : u = y - z \text{ with } z \in N\}$$

of $y \in F$ with respect to N is independent of the choice of a representative $u \in \bar{y}$ because $N(P) = N$. The operator J is obviously linear and moreover continuous. The continuity of J is a consequence of the inequality

$$\|J\bar{y}\| = \|Py\| = \|P(y - z)\| \leqslant \|P\| \, \|y - z\| \tag{2.4.19}$$

valid for all $z \in N$ and of the estimate

$$\|J\bar{y}\| \leqslant \|P\| \inf_{z \in N} \|y - z\| = \|P\| \, \|\bar{y}\| \tag{2.4.20}$$

which results from (2.4.19). Combining the operator J with the quotient map $Q_N^F : F \to F/N$ we obtain

$$Q_N^F J \bar{y} = Q_N^F P y = \overline{Py} = \bar{y},$$

that is to say

$$Q_N^F J = I_{F/N}. \tag{2.4.21}$$

On the basis of this representation of the identity map $I_{F/N}$ of the quotient space F/N we are in a position to construct a lifting $\tilde{T}: E \to F$ for an arbitrary operator $T: E \to F/N$, namely

$$\tilde{T} = JT. \tag{2.4.22}$$

The characteristic equation (2.4.15) for the relation between an operator $T: E \to F/N$ and a lifting $\tilde{T}: E \to F$ is obtained from (2.4.21) by multiplying throughout by T on the right. Furthermore, from (2.4.22) and (2.4.20) we obtain the norm estimate

$$\|\tilde{T}\| \leqslant \|J\| \, \|T\| \leqslant \|P\| \, \|T\|.$$

Thus the lifting constant $q(E, F; N)$ can be estimated by

$$q(E, F; N) \leqslant \inf\{\|P\| : P \in L(F, F) \text{ is a projection with } N(P) = N\}. \tag{2.4.23}$$

We formulate the final result as follows:

> $q(E, F; N) < \infty$ *for all Banach spaces E if and only if N is the range or the null space of a projection* $P_N \in L(F, F)$ *or* $P \in L(F, F)$, *respectively.*

In the case of a Hilbert space $F = H$ we have

$$q(E, H; N) = 1 \text{ for arbitrary Banach spaces } E \qquad (2.4.24)$$

because each subspace $N \subseteq H$ is the null space of a projection $P \in L(H, H)$ with $\|P\| = 1$.

What we are mainly interested in is the situation of a finite-dimensional subspace N in an arbitrary Banach space F. Lemma 2.4.1 guarantees the existence of a projection $P_N \in L(F, F)$ with

$$\|P_N\| \leqslant \sqrt{m}$$

onto any m-dimensional subspace $N \subseteq F$. The operator

$$P = I_F - P_N$$

then represents a projection $P \in L(F, F)$ with null space $N(P) = N$ (see (2.3.10), (2.3.11), (2.3.12)), subject to the norm estimate

$$\|P\| \leqslant 1 + \|P_N\| \leqslant 1 + \sqrt{m}.$$

In the case of a subspace $N \subseteq F$ with $\dim(N) = m$ the estimate (2.4.23) for $q(E, F; N)$ can thus be extended to

$$q(E, F; N) \leqslant 1 + \sqrt{m}. \qquad (2.4.25)$$

Letting N vary in the class of all subspaces $N \subseteq F$ of dimension $n - 1$ we now define *the nth lifting constant* $q_n(E, F)$ *for a fixed pair of Banach spaces* (E, F) by

$$q_n(E, F) = \sup\{q(E, F; N): N \subseteq F, \dim(N) = n - 1\}. \qquad (2.4.26)$$

From (2.4.25) the supremum (2.4.26) is bounded by

$$q_n(E, F) \leqslant 1 + \sqrt{n - 1}. \qquad (2.4.27)$$

The nth lifting constant $q_n(E, F)$ is the very geometrical parameter sufficient for the estimation of the nth approximation number $a_n(T)$ of an operator $T: E \to F$ from above by the nth Kolmogorov number $d_n(T)$.

Proposition 2.4.4. *Let* $T \in L(E, F)$ *be an operator between arbitrary Banach spaces E and F. Then*

$$a_n(T) \leqslant q_n(E, F)d_n(T). \qquad (2.4.28)$$

Proof. If $\operatorname{rank}(T) < n$ we have $a_n(T) = d_n(T) = 0$. So let us assume $\operatorname{rank}(T) \geqslant n$. We then proceed in the same way as for the proof of

$a_n(T) \le d_n(T)$ in Proposition 2.2.3. Namely, given $\varepsilon > 0$ we consider a subspace $N \subseteq F$ with $\dim(N) = n - 1$ and $\| Q_N^F T \| < d_n(T) + \varepsilon$ (see (2.2.7)'). According to the definition of the lifting constant $q(E, F; N)$ the operator $Q_N^F T : E \to F/N$ can be lifted to an operator $\tilde{T} : E \to F$ such that

$$Q_N^F T = Q_N^F \tilde{T} \text{ and } \| \tilde{T} \| \le (q(E, F; N) + \varepsilon) \| Q_N^F T \|.$$

Since the difference $S = T - \tilde{T}$ is an operator of $\mathrm{rank}(S) < n$ we may conclude that

$$a_n(T) \le \| T - S \| = \| \tilde{T} \| \le (q(E, F; N) + \varepsilon) \| Q_N^F T \|$$
$$\le (q(E, F; N) + \varepsilon)(d_n(T) + \varepsilon).$$

Taking the supremum (2.4.26) with respect to all subspaces $N \subseteq F$ of $\dim(N) = n - 1$ we obtain

$$a_n(T) \le (q_n(E, F) + \varepsilon)(d_n(T) + \varepsilon)$$

and hence finally the inequality (2.4.28). ∎

If $\tilde{E}$ is a Banach space with the metric lifting property we have

$$q_n(\tilde{E}, F) = 1$$

for arbitrary Banach spaces F so that the statement (2.4.28) of Proposition 2.4.4 implies the statement

$$a_n(T) = d_n(T) \qquad (2.2.14)$$

of Proposition 2.2.3. For a Hilbert space H in place of F and arbitrary Banach spaces E we have

$$q_n(E, H) = 1$$

from (2.4.24). Therefore (2.2.14) turns out to be true also for operators T from an arbitrary Banach space E into a Hilbert space H. From this remark we obtain the following geometrical representation for the approximation numbers $a_n(T)$ of operators $T : E \to H$ (cf. Pietsch 1978).

Proposition 2.4.5. *Let H be a Hilbert space, E an arbitrary Banach space, and $T \in L(E, H)$. Then*

$$a_n(T) = \inf\{ \| T - PT \| : P \in L(H, H) \text{ is an orthogonal}$$
$$\text{projection with } \mathrm{rank}(P) < n \}. \qquad (2.4.29)$$

Proof. Since $\mathrm{rank}(PT) < n$ it is sufficient to estimate the infimum on the right-hand side of (2.4.29) from above by $a_n(T)$. So let $\varepsilon > 0$ and $N \subseteq H$ be a subspace with $\dim(N) < n$ such that

$$\| Q_N^H T \| < d_n(T) + \varepsilon. \qquad (2.4.30)$$

But the quotient space H/N of a Hilbert space H with respect to a subspace N is isometrically isomorphic to the orthogonal complement $N^\perp = M$ of

N, and the quotient map $Q_N^H: H \to H/N$ respresents the orthogonal projection P_M of H onto $M = N^\perp$. Therefore we may write

$$\|P_M T\| < d_n(T) + \varepsilon$$

instead of (2.4.30). If we now replace P_M by

$$P_M = I_H - P_N$$

with P_N as the orthogonal projection of H onto N we obtain

$$\|T - P_N T\| < d_n(T) + \varepsilon.$$

Realizing that $\operatorname{rank}(P_N) < n$ we can now take the infimum of $\|T - PT\|$ with respect to all orthogonal projections $P: H \to H$ of $\operatorname{rank}(P) < n$. In this way we arrive at

$$\inf\{\|T - PT\| : P \in L(H, H) \text{ is an orthogonal}$$
$$\text{projection with } \operatorname{rank}(P) < n\} < d_n(T) + \varepsilon. \tag{2.4.31}$$

Since $d_n(T) = a_n(T)$ the inequality (2.4.31) reduces to the desired estimate for $a_n(T)$ from below as ε tends to zero. ∎

In the case of operators $T: E \to F$ between arbitrary Banach spaces E and F we can give a general estimate corresponding to (2.4.14).

Proposition 2.4.6. *Let $T \in L(E, F)$ be an operator between arbitrary Banach spaces E and F. Then*

$$a_n(T) \leqslant (2n)^{1/2} d_n(T). \tag{2.4.32}$$

The proof of (2.4.32) rests upon (2.4.27) and (2.4.28).

2.5. Duality relations

The *dual* $T': F' \to E'$ of an operator $T: E \to F$ is defined by

$$\langle x, T'b \rangle = \langle Tx, b \rangle \text{ for } x \in E \text{ and } b \in F'.$$

Like T, it is a linear and continuous operator with norm $\|T'\| = \|T\|$. Which other properties does T' inherit from T? Asking this question we think of approximation and compactness properties as they are reflected by approximation numbers, Kolmogorov numbers, Gelfand numbers, and entropy numbers.

First we consider the approximation numbers $a_n(T)$ of operators $T: E \to F$. Given $\varepsilon > 0$ let $A \in L(E, F)$ with $\operatorname{rank}(A) < n$ be such that

$$\|T - A\| < a_n(T) + \varepsilon.$$

Then we also have

$$\|T' - A'\| < a_n(T) + \varepsilon.$$

This implies

$$a_n(T') < a_n(T) + \varepsilon$$

because the dual A' of A is of rank$(A') < n$ as well. Hence it follows that

$$a_n(T') \leqslant a_n(T). \tag{2.5.1}$$

Equality, however, in general fails to hold. A counterexample is given by the identity map $I_0: l_1 \to c_0$ of l_1 into c_0 (see Hutton 1974). Since $\| I_0 \| = 1$ we have

$$a_n(I_0) \leqslant 1 \quad \text{for } n = 1, 2, 3, \ldots$$

On the other hand, we can verify that

$$a_n(I_0) \geqslant 1 \quad \text{for } n = 1, 2, 3, \ldots \tag{2.5.2}$$

by proving

$$\| I_0 - A \| \geqslant 1 \tag{2.5.3}$$

for an arbitrary finite rank operator $A: l_1 \to c_0$. The proof (see Hutton 1974; Pietsch 1978) rests upon the fact that any finite rank operator is compact. Thus for $A \in L(l_1, c_0)$ with rank$(A) < \infty$ and for $\varepsilon > 0$ we can determine finitely many elements $y_i = (\eta_k^{(i)})$ in $c_0, 1 \leqslant i \leqslant m$, such that

$$A(U_1) \subseteq \bigcup_{i=1}^{m} \{ y_i + \varepsilon U_0 \}, \tag{2.5.4}$$

where U_1 and U_0 are the closed unit balls in l_1 and c_0, respectively. Because $\lim_{k \to \infty} \eta_k^{(i)} = 0$ for $1 \leqslant i \leqslant m$ a natural number n can be chosen such that

$$|\eta_n^{(i)}| \leqslant \varepsilon \quad \text{for } 1 \leqslant i \leqslant m. \tag{2.5.5}$$

Let us estimate the norm $\| (I_0 - A)e_n \|$ of the image of the unit vector $e_n \in U_1$ under $I_0 - A$ in c_0. In view of (2.5.4) there exists an element y_j with

$$\| Ae_n - y_j \| \leqslant \varepsilon. \tag{2.5.6}$$

Using (2.5.6), the definition of the norm in the space c_0, and (2.5.5) we obtain

$$\| (I_0 - A)e_n \| \geqslant \| I_0 e_n - y_j \| - \| Ae_n - y_j \|$$

$$\geqslant \sup_{1 \leqslant k < \infty} |\delta_{kn} - \eta_k^{(j)}| - \varepsilon \geqslant 1 - |\eta_n^{(j)}| - \varepsilon \geqslant 1 - 2\varepsilon.$$

But since

$$\| I_0 - A \| \geqslant \| (I_0 - A)e_n \|$$

we also have

$$\| I_0 - A \| \geqslant 1 - 2\varepsilon$$

and get (2.5.3) by letting $\varepsilon \to 0$. This way we arrive at (2.5.2). The final result is

$$a_n(I_0) = 1 \quad \text{for } n = 1, 2, 3, \ldots \tag{2.5.7}$$

Next we consider the dual operator $I'_0 : c'_0 \to l'_1$. The defining equality

$$\langle x, I'_0 b \rangle = \langle I_0 x, b \rangle = \sum_{i=1}^{\infty} \xi_i \beta_i$$

for $I'_0 : l_1 \to l_\infty$ with $x = (\xi_i) \in l_1$ and $b = (\beta_i) \in l_1$ makes it clear that I'_0 is nothing other than the identity map I_∞ of l_1 into l_∞. Although

$$a_1(I'_0) = \| I'_0 \| = \| I_0 \| = 1$$

the approximation numbers $a_n(I'_0) = a_n(I_\infty)$ differ from the $a_n(I_0)$ for $n \geqslant 2$. We first show that

$$a_n(I_\infty) \leqslant \tfrac{1}{2} \quad \text{for } n = 2, 3, 4, \ldots \tag{2.5.8}$$

For this purpose we employ the one-dimensional operator $A_1 : l_1 \to l_\infty$ defined by

$$A_1 x = \left(\frac{1}{2} \sum_{i=1}^{\infty} \xi_i, \frac{1}{2} \sum_{i=1}^{\infty} \xi_i, \ldots \right) \in l_\infty \quad \text{for } x = (\xi_i) \in l_1.$$

For the operator

$$(I_\infty - A_1) x = \left(\frac{1}{2}\xi_1 - \frac{1}{2} \sum_{i \neq 1} \xi_i, \frac{1}{2}\xi_2 - \frac{1}{2} \sum_{i \neq 2} \xi_i, \ldots \right)$$

we carry out the norm estimate

$$\| (I_\infty - A_1) x \| = \sup_{1 \leqslant k < \infty} \frac{1}{2} \left| \xi_k - \sum_{i \neq k} \xi_i \right| \leqslant \frac{1}{2} \sum_{i=1}^{\infty} |\xi_i|.$$

and get

$$\| I_\infty - A_1 \| \leqslant \tfrac{1}{2}$$

which proves (2.5.8). To show that

$$a_n(I_\infty) = \tfrac{1}{2} \quad \text{for } n = 2, 3, 4, \ldots \tag{2.5.9}$$

we suppose that $a_n(I_\infty) < \tfrac{1}{2}$ for some $n \geqslant 2$. This implies the existence of a finite rank operator $A : l_1 \to l_\infty$ such that

$$\| I_\infty - A \| < \tfrac{1}{2}. \tag{2.5.10}$$

The conclusions that follow will culminate in a contradiction to the compactness of A. This contradiction will appear in the shape of an inequality

$$\| A e_i - A e_k \| \geqslant \varepsilon_0 \tag{2.5.11}$$

valid for any two unit vectors $e_i \neq e_k$ of l_1 and some positive number $\varepsilon_0 > 0$. Let

$$A e_m = (\alpha_j^{(m)})$$

be the image of e_m under A in l_∞. Then obviously we have

$$\| A e_i - A e_k \| \geqslant |\alpha_k^{(i)} - \alpha_k^{(k)}|. \tag{2.5.12}$$

For the desired result (2.5.11) we still need the two estimates

$$\|I_\infty - A\| \geqslant |1 - \alpha_k^{(k)}| \quad \text{and} \quad \|I_\infty - A\| \geqslant |\alpha_k^{(i)}| \quad \text{for } i \neq k \qquad (2.5.13)$$

of the operator norm $\|I_\infty - A\|$ which are immediate consequences of

$$\|I_\infty - A\| \geqslant \|(I_\infty - A)e_k\| \quad \text{and} \quad \|I_\infty - A\| \geqslant \|(I_\infty - A)e_i\|,$$

respectively. Combining (2.5.10), (2.5.13) and (2.5.12) we obtain

$$0 < 1 - 2\|I_\infty - A\| \leqslant 1 - |1 - \alpha_k^{(k)}| - |\alpha_k^{(i)}| \leqslant |\alpha_k^{(i)} - \alpha_k^{(k)}| \leqslant \|Ae_i - Ae_k\|,$$

which is the contradiction (2.5.11) we have been looking for, with $\varepsilon_0 = 1 - 2\|I_\infty - A\|$.

The result

$$a_n(I_0) = 1 \quad \text{and} \quad a_n(I_\infty) = \tfrac{1}{2} \text{ for } n \geqslant 2$$

is interesting from another point of view. Since

$$I_\infty = J_0 I_0, \qquad (2.5.14)$$

with $J_0 : c_0 \to l_\infty$ as the natural embedding of c_0 into l_∞, we learn that the approximation numbers are not injective. Furthermore we recognize that

$$c_n(I_\infty) = a_n(I_\infty) \qquad (2.5.15)$$

because l_∞ is a Banach space with the metric extension property (cf. Proposition 2.3.3).

The difference between $a_n(I_0')$ and $a_n(I_0)$ is due to the non-compactness of the operator I_0. Indeed, we have

$$a_n(T') = a_n(T) \qquad (2.5.16)$$

for a compact operator T (see Hutton 1974; Pietsch 1978). We shall derive (2.5.16) for compact operators by using (2.5.1) and an estimate for $a_n(T)$ from above by $a_n(T'')$ valid for arbitrary operators $T : E \to F$. This estimate is based upon the so-called *principle of local reflexivity*.

The principle of local reflexivity deals with the finite-dimensional subspaces of the *bidual* $(F')' = F''$ of an arbitrary Banach space F. In this connection the Banach space F itself is considered as a subspace of its bidual F'' by ascribing to a fixed element $y \in F$ the values $\langle y, b \rangle$ where b varies in F'. Hence $y \in F$ can be expressed as a linear functional y'' over F' whose norm $\|y''\|$ coincides with $\|y\|$ since

$$\|y''\| = \sup_{\|b\| \leqslant 1} |\langle y, b \rangle| = \|y\|.$$

The isometric embedding of F into F'' defined in this way is called *the canonical embedding* and denoted by $K_F : F \to F''$. If K_F is an isometric isomorphism of F onto F'' the Banach space F is called *reflexive*. An arbitrary Banach space F is *locally reflexive* in the following sense.

Principle of local reflexivity. *Let F be an arbitrary Banach space. Then for every finite-dimensional subspace $M \subseteq F''$ and every $\varepsilon > 0$ there exists an isomorphism $S_0: M \to F_0$ mapping M onto a finite-dimensional subspace $F_0 \subseteq F$ such that*

$$S_0 y = y \text{ for all } y \in M \cap F \quad \text{and} \quad \|S_0\| \, \|S_0^{-1}\| \leqslant 1 + \varepsilon. \qquad (2.5.17)$$

The principle of local reflexivity was discovered by Lindenstrauss and Rosenthal (1969) and proved in a slightly stronger version by Johnson, Rosenthal and Zippin in 1971. Since the proof is beyond the scope of this book we omit it.

If we start from F' and carry out the canonical embedding $K_{F'}: F' \to F'''$ we can afterwards regain the identity map $I_{F'}$ of F' as the composition

$$I_{F'} = (K_F)' K_{F'}. \qquad (2.5.18)$$

To verify (2.5.18) we apply the operator $(K_F)' K_{F'}$ to an arbitrary functional $b \in F'$ which yields

$$\langle y, (K_F)' K_{F'} b \rangle = \langle K_F y, K_{F'} b \rangle$$

for any $y \in F$ according to the definition of the dual operator $(K_F)'$, and thus

$$\langle y, (K_F)' K_{F'} b \rangle = \langle y, b \rangle \quad \text{for all } y \in F \text{ and } b \in F'.$$

This proves the assertion (2.5.18).

The decisive step in the proof of $a_n(T') = a_n(T)$ for a compact operator T will be the transition to *the bidual operator* $T'': E'' \to F''$. The bidual T'' of an arbitrary operator $T: E \to F$ is related to T by

$$K_F T = T'' K_E. \qquad (2.5.19)$$

Indeed, given $x \in E$ and $b \in F'$ we have

$$\langle b, T'' K_E x \rangle = \langle T'b, K_E x \rangle.$$

Passing from $K_E x$ as a functional over E' to x in its original role as an element of E we may write

$$\langle T'b, K_E x \rangle = \langle x, T'b \rangle = \langle Tx, b \rangle.$$

By analogous arguments we obtain

$$\langle b, K_F T x \rangle = \langle Tx, b \rangle$$

and hence finally

$$\langle b, T'' K_E x \rangle = \langle b, K_F T x \rangle \quad \text{for all } x \in E \text{ and } b \in F'.$$

This proves (2.5.19).

The relation (2.5.19), the principle of local reflexivity, and the *entropy measure of non-compactness*

$$e(T) = \lim_{n \to \infty} \varepsilon_n(T) \qquad (2.5.20)$$

are the main tools for estimating $a_n(T)$ from above by $a_n(T'')$.

Proposition 2.5.1. (see Hutton 1974; Pietsch 1978; Edmunds and Tylli 1986). *Let $T \in L(E, F)$ be an operator between arbitrary Banach spaces E and F. Then*

$$a_n(T) \leqslant 2e(T) + a_n(T'').$$ (2.5.21)

Proof. Given $\varepsilon > 0$ we determine an operator $A \in L(E'', F'')$ with rank$(A) < n$ such that

$$\| T'' - A \| < a_n(T'') + \varepsilon.$$ (2.5.22)

Furthermore we fix $\delta > e(T)$ and choose a natural number m with $\varepsilon_m(T) < \delta$. Then we can find elements $y_1, y_2, \ldots, y_m$ in F producing a covering

$$T(U_E) \subseteq \bigcup_{i=1}^{m} \{ y_i + \delta U_F \}$$ (2.5.23)

of the set $T(U_E)$. The elements $K_F y_i \in F''$ and the range $R(A)$ of the finite rank operator $A : E'' \to F''$ together generate a finite-dimensional subspace M in the bidual F'' of F. For $M \subseteq F''$ the principle of local reflexivity guarantees the existence of an operator S_0 mapping M onto a subspace $F_0 \subseteq F$ such that

$$S_0 y_i = y_i \quad \text{for } 1 \leqslant i \leqslant m \text{ and } \| S_0 \| \leqslant 1 + \varepsilon.$$ (2.5.24)

This is a consequence of (2.5.17) since S_0 and S_0^{-1} act as identities on $M \cap F$ so that $\| S_0 \| \geqslant 1$ and $\| S_0^{-1} \| \geqslant 1$. Since the range $R(A)$ of A is contained in M the operator A admits a factorization over M, say

$$A = J A_0,$$

where $J : M \to F''$ stands for the identity map of M into F''. On the other hand, the operator $S_0 : M \to F_0$ can then be regarded as an operator from M into F by applying the identity map J_0 of F_0 into F, say

$$S = J_0 S_0.$$

Now we are able to transform $A : E'' \to F''$ into an operator $B : E \to F$ by putting

$$B = S A_0 K_E.$$

Because rank$(A) = $ rank$(A_0) < n$, we have rank$(B) < n$ as well, and therefore

$$a_n(T) \leqslant \| T - B \|.$$

To get an upper bound for $\| T - B \|$ we estimate $\| Tx - Bx \|$ for $x \in U_E$ using an appropriate element y_i of the covering (2.5.23), namely

$$\| Tx - Bx \| \leqslant \| Tx - y_i \| + \| y_i - Bx \| \leqslant \delta + \| y_i - S A_0 K_E x \|.$$

Because of (2.5.24) we can estimate the norm $\| y_i - S A_0 K_E x \|$ by

$$\| y_i - S A_0 K_E x \| = \| S y_i - S A_0 K_E x \| \leqslant (1 + \varepsilon) \| y_i - A_0 K_E x \|,$$

where $\| y_i - A_0 K_E x \|$ is meant to be the norm in $M \subseteq F''$. Operating in F'' instead and considering (2.5.23) as a covering of the set $K_F T(U_E)$ in F'' we may conclude that

$$\| y_i - A_0 K_E x \| = \| K_F y_i - A K_E x \|$$
$$\leqslant \| K_F y_i - K_F T x \| + \| K_F T x - A K_E x \|$$
$$\leqslant \delta + \| K_F T x - A K_E x \|.$$

By using (2.5.19) we are now able to switch over from T to T'' and thus to employ (2.5.22), namely

$$\| K_F T x - A K_E x \| = \| T'' K_E x - A K_E x \| \leqslant \| T'' - A \| < a_n(T'') + \varepsilon.$$

Let us summarize the result of our computations,

$$\| T x - B x \| < \delta + (1 + \varepsilon)(\delta + a_n(T'') + \varepsilon) \quad \text{for } x \in U_E.$$

Hence it follows that

$$a_n(T) \leqslant \| T - B \| \leqslant \delta + (1 + \varepsilon)(\delta + a_n(T'') + \varepsilon).$$

In view of the fact that $\varepsilon > 0$ can be chosen arbitrarily small and $\delta > e(T)$ arbitrarily close to $e(T)$ we finally get (2.5.21). ∎

Proposition 2.5.2. (see Hutton 1974). *Let $T \in K(E, F)$ be a compact operator between arbitrary Banach spaces E and F. Then*

$$a_n(T') = a_n(T). \tag{2.5.16}$$

Proof. Since the operator T is compact we have $e(T) = 0$ and hence obtain

$$a_n(T) \leqslant a_n(T'') \leqslant a_n(T')$$

from (2.5.21) and (2.5.1). This complements the inequality (2.5.1) to give the desired equality (2.5.16). ∎

As shown in (2.5.7) and (2.5.9) for $T = I_0$ and $T' = I_\infty$, the approximation numbers $a_n(T')$ of the dual T' of T may differ from the approximation numbers $a_n(T)$ of T itself if T is non-compact, but there is a universal relation between $a_n(T)$ and $a_n(T')$. Before deriving a corresponding estimate we present two auxiliary lemmas both dealing with the 'entropy measure of non-compactness' $e(T)$ (see (2.5.20)) of an arbitrary operator T.

Lemma 2.5.1. *Let $T \in L(E, F)$ be an operator between arbitrary Banach spaces E and F. Then*

$$e(T) \leqslant 2e(T''). \tag{2.5.25}$$

Proof. Since the outer entropy numbers $\varepsilon_n(T)$ are injective up to a factor 2 (see (1.3.6)), we have

$$\varepsilon_n(T) \leqslant 2\varepsilon_n(K_F T)$$

which yields

$$\varepsilon_n(T) \leqslant 2\varepsilon_n(T''K_E) \leqslant 2\varepsilon_n(T'') \qquad (2.5.26)$$

by (2.5.19). The desired inequality (2.5.25) results from (2.5.26) by letting $n \to \infty$. ∎

Lemma 2.5.2. *Let $T \in L(E, F)$ be an operator between arbitrary Banach spaces E and F. Then*

$$e(T) \leqslant a_n(T), \quad n = 1, 2, 3, \dots \qquad (2.5.27)$$

Proof. We fix n and choose $\delta > a_n(T)$ arbitrarily. Then we determine an operator $A: E \to F$ with rank$(A) < n$ such that

$$\| T - A \| < \delta.$$

Because

$$T(U_E) \subseteq \| T - A \| U_F + A(U_E) \subseteq \delta U_F + A(U_E)$$

a covering of the set $T(U_E)$ can be gained via a covering of the set $A(U_E)$. Since the operator A is of finite rank it is compact. Therefore, given $\varepsilon > 0$, there exist finitely many elements $y_i \in F$, $1 \leqslant i \leqslant m$, producing a covering

$$A(U_E) \subseteq \bigcup_{i=1}^{m} \{ y_i + \varepsilon U_F \}$$

of the set $A(U_E)$. This implies

$$T(U_E) \subseteq \bigcup_{i=1}^{m} \{ y_i + \varepsilon U_F + \delta U_F \} \subseteq \bigcup_{i=1}^{m} \{ y_i + (\delta + \varepsilon) U_F \}.$$

It follows that

$$e(T) \leqslant \varepsilon_m(T) \leqslant \varepsilon + \delta.$$

Letting $\varepsilon \to 0$ we obtain $e(T) \leqslant \delta$ and finally $e(T) \leqslant a_n(T)$. ∎

Though we are mainly interested in the relations between $a_n(T)$ and $a_n(T')$ we first develop another estimate for $a_n(T)$ from above by $a_n(T'')$ which, in a sense, is primary and, moreover, will be used in the next section.

Proposition 2.5.3. *Let $T \in L(E, F)$ be an operator between arbitrary Banach spaces E and F. Then*

$$a_n(T) \leqslant 5a_n(T''). \qquad (2.5.28)$$

Proof. By (2.5.25) we may replace the term $2e(T)$ on the right-hand side of (2.5.21) by $4e(T'')$ so that we have

$$a_n(T) \leqslant 4e(T'') + a_n(T'').$$

Applying (2.5.27) to T'' we recognize the validity of (2.5.28). ∎

On the basis of (2.5.28) and (2.5.1) we can now state the universal relation between $a_n(T)$ and $a_n(T')$ announced above.

Proposition 2.5.4. (see Edmunds and Tylli 1986). *Let $T \in L(E, F)$ be an operator between arbitrary Banach spaces E and F. Then*

$$a_n(T') \leqslant a_n(T) \leqslant 5a_n(T').$$ (2.5.29)

∎

Having studied the approximation numbers we turn to the Kolmogorov and Gelfand numbers of the dual operator T' of $T: E \to F$. Let us recall that

$$d_n(T) = a_n(TQ_E) \quad \text{and} \quad c_n(T) = a_n(J_F T)$$

(see (2.2.15), (2.3.18)) and, correspondingly,

$$d_n(T') = a_n(T'Q_{F'})$$ (2.5.30)

and

$$c_n(T') = a_n(J_{E'} T').$$ (2.5.31)

The canonical surjections $Q_{F'}: l_1(U_{F'}) \to F'$ and injections $J_{E'}: E' \to l_\infty(U_{E''})$ appearing in the representation formulas (2.5.30) and (2.5.31), respectively, are closely related to the duals J'_F and Q'_E of the canonical injections $J_F: F \to l_\infty(U_{F'})$ and surjections $Q_E: l_1(U_E) \to E$, respectively. Indeed, since $J_F: F \to l_\infty(U_{F'})$ is given by

$$J_F y = \{\langle y, b \rangle\}_{b \in U_{F'}} \quad \text{for } y \in F$$

the dual operator $J'_F: l_\infty(U_{F'})' \to F'$ in particular assigns to every element $c = \{\lambda_b\}_{b \in U_{F'}} \in l_1(U_{F'})$ in its role as a functional over $l_\infty(U_{F'})$ the functional $J'_F c$ over F defined by

$$\langle y, J'_F c \rangle = \langle J_F y, c \rangle = \sum_{b \in U_{F'}} \lambda_b \langle y, b \rangle = \left\langle y, \sum_{b \in U_{F'}} \lambda_b b \right\rangle.$$

This makes it clear that the restriction of J'_F to the subspace $l_1(U_{F'})$ of $l_\infty(U_{F'})' = l_1(U_{F'})''$ coincides with the canonical surjection

$$Q_{F'}\{\lambda_b\} = \sum_{b \in U_{F'}} \lambda_b b$$

of $l_1(U_{F'})$ onto F' (see section 2.2). Abbreviating the canonical embedding $K_{l_1(U_{F'})}$ of $l_1(U_{F'})$ into $l_1(U_{F'})'' = l_\infty(U_{F'})'$ by K we may thus write

$$Q_{F'} = J'_F K.$$ (2.5.32)

On the other hand, the operator $Q'_E: E' \to l_1(U_E)' = l_\infty(U_E)$, dual to the canonical surjection

$$Q_E\{\lambda_x\} = \sum_{x \in U_E} \lambda_x x$$

of $l_1(U_E)$ onto E, transforms any functional $a \in E'$ into a functional $Q'_E a$

over $l_1(U_E)$ according to the rule

$$\langle \{\lambda_x\}, Q'_E a \rangle = \langle Q_E \{\lambda_x\}, a \rangle = \left\langle \sum_{x \in U_E} \lambda_x x, a \right\rangle = \sum_{x \in U_E} \lambda_x \langle x, a \rangle$$

which amounts to

$$Q'_E a = \{\langle x, a \rangle\}_{x \in U_E}.$$

Hence, in contrast to the canonical injection $J_{E'}$ which embeds E' isometrically into the Banach space $l_\infty(U_{E''})$ of bounded functions on $U_{E''}$, the operator Q'_E maps E' isometrically into the Banach space $l_\infty(U_E)$ of bounded functions on U_E. Denoting by $P_0 : l_\infty(U_{E''}) \to l_\infty(U_E)$ the operator which restricts any bounded function on $U_{E''}$ to the subset $K_E(U_E) \subseteq U_{E''}$ we can describe Q'_E by

$$Q'_E = P_0 J_{E'}. \qquad (2.5.33)$$

Q'_E is an isometric injection as well. This statement, the relation (2.5.32) between $Q_{F'}$ and J'_F, and (2.5.1) are the crucial facts for the calculation or estimation of $d_n(T')$ and $c_n(T')$.

Proposition 2.5.5. (see Pietsch 1974, 1978). *Let $T \in L(E, F)$ be an operator between arbitrary Banach spaces E and F. Then*

$$c_n(T') \le d_n(T) \qquad (2.5.34)$$

and

$$d_n(T') = c_n(T). \qquad (2.5.35)$$

Proof. To estimate $c_n(T')$ from above we insert Q'_E making use of the injectivity of the Gelfand numbers, namely

$$c_n(T') = c_n(Q'_E T') = c_n((TQ_E)').$$

The proof of (2.5.34) is completed by the series of inequalities

$$c_n((TQ_E)') \le a_n((TQ_E)') \le a_n(TQ_E) = d_n(T).$$

On the other hand, if we insert $Q_{F'} = J'_F K$ into the representation formula (2.5.30) for $d_n(T')$ we get

$$d_n(T') = a_n(T'J'_F K) \le a_n(T'J'_F) = a_n((J_F T)')$$

and further

$$a_n((J_F T)') \le a_n(J_F T) = c_n(T).$$

To check the equality of $d_n(T')$ and $c_n(T)$ we once more use the injectivity of the Gelfand numbers, namely

$$c_n(T) = c_n(K_F T).$$

By using (2.5.19) and (2.5.34) we obtain in this way

$$c_n(T) = c_n(T'' K_E) \le c_n(T'') \le d_n(T')$$

and hence (2.5.35). ∎

As an immediate consequence of (2.5.35) we can formulate

Schauder's theorem. *An operator T between arbitrary Banach spaces E and F is compact if and only if its dual $T': F' \to E'$ is compact.*

In the case of a compact operator $T: E \to F$ the inequality (2.5.34) can be strengthened to an equality.

Proposition 2.5.6. (see Pietsch 1974, 1978). *Let $T \in K(E, F)$ be a compact operator between arbitrary Banach spaces E and F. Then*

$$d_n(T) = c_n(T').$$ (2.5.36)

Proof. The compactness of T implies the compactness of TQ_E and therefore

$$d_n(T) = a_n(TQ_E) = a_n((TQ_E)') = a_n(Q'_E T')$$

by Proposition 2.5.2. The range $l_1(U_E)' = l_\infty(U_E)$ of $Q'_E : E' \to l_\infty(U_E)$ is a Banach space with the metric extension property so that we have

$$a_n(Q'_E T') = c_n(Q'_E T')$$

by Proposition 2.3.3. Because of the injectivity of the Gelfand numbers we finally arrive at

$$d_n(T) = c_n(T').$$ ∎

The identity map $I_0 : l_1 \to c_0$, which served as an example of an operator T with $a_n(T') < a_n(T)$ for $n \geq 2$, can also be used to demonstrate that (2.5.36) need not hold for a non-compact operator. Indeed, since l_1 is a Banach space with the metric lifting property we have

$$d_n(I_0) = a_n(I_0) = 1 \quad \text{for } n = 1, 2, 3, \dots$$

by Proposition 2.2.3 and (2.5.7). On the other hand, the Gelfand numbers $c_n(I'_0)$ of the dual operator $I'_0 = I_\infty$ are given by

$$c_n(I'_0) = a_n(I'_0) = \tfrac{1}{2} \quad \text{for } n \geq 2$$

(see (2.5.9) and (2.5.15)).

2.6. Symmetrized approximation numbers

Apart from a possible difference between $d_n(T)$ and $c_n(T')$ when T is a non-compact operator it is in a sense unsatisfactory that, when comparing the degree of compactness of T and T', one has to use the Gelfand numbers for T and the Kolmogorov numbers for T' or vice versa. For this reason we introduce *symmetrized approximation numbers* $t_n(T)$ for operators T between arbitrary Banach spaces E and F by

$$t_n(T) = a_n(J_F T Q_E)$$ (2.6.1)

(see Pietsch 1978). The definition (2.6.1) is obviously equivalent to

$$t_n(T) = d_n(J_F T) \qquad (2.6.2)$$

as well as to

$$t_n(T) = c_n(T Q_E). \qquad (2.6.3)$$

The relation (2.6.2) makes it clear that the symmetrized approximation numbers $t_n(T)$ are surjective, while (2.6.3) shows that they are injective. The main advantage of the symmetrized approximation numbers $t_n(T)$ in comparison with Kolmogorov and Gelfand numbers, however, is their complete symmetry.

Proposition 2.6. (see Pietsch 1978). *Let $T \in L(E, F)$ be an operator between arbitrary Banach spaces E and F. Then*

$$t_n(T') = t_n(T). \qquad (2.6.4)$$

Proof. On the basis of the representation formula (2.6.3) we obtain

$$t_n(T') = c_n(T' Q_{F'}) = c_n(T' J'_F K) \leqslant c_n((J_F T)')$$

by using (2.5.32). For $c_n((J_F T)')$ the estimate (2.5.34) applies:

$$c_n((J_F T)') \leqslant d_n(J_F T) = t_n(T).$$

To estimate $t_n(T)$ from above we again start with (2.6.3) and then apply (2.5.35), namely

$$t_n(T) = c_n(T Q_E) = d_n((T Q_E)') = d_n(Q'_E T').$$

We then make use of (2.5.33) which yields

$$d_n(Q'_E T') = d_n(P_0 J_{E'} T') \leqslant d_n(J_{E'} T') = t_n(T').$$

This completes the proof of (2.6.4).   ∎

Monotonicity (T1), additivity (T2), and the rank property (T4) can be derived immediately for the symmetrized approximation numbers from the corresponding properties (A1), (A2), and (A4) of the usual approximation numbers. The multiplicativity, from (2.6.2) and (2.6.3), can at least be maintained in two weaker versions, namely

(T3)(a) $t_{k+n-1}(RS) \leqslant t_k(R) d_n(S)$

and

(T3)(b) $t_{k+n-1}(RS) \leqslant c_k(R) t_n(S)$,

for $S \in L(E, Z)$ and $R \in L(Z, F)$. The class $L^{(t)}_{p,q}$ of operators T with $(t_n(T)) \in l_{p,q}$ forms a complete quasi-normed operator ideal with respect to the ideal quasi-norm

$$\lambda^{(t)}_{p,q}(T) = \lambda_{p,q}((t_n(T))$$

which, moreover, is surjective and injective, the ideal quasi-norm $\lambda^{(t)}_{p,q}$ itself being surjective and injective.

The norm-determining property

$$t_n(I_E) = 1 \text{ for the identity map } I_E \text{ of a Banach}$$
(T5)　space E with $\dim(E) \geqslant n$

also holds. The proof, however, rests on a rather deep result concerning pairs of finite-dimensional subspaces in an arbitrary Banach space due to Krein, Krasnoselskij and Milman (1948) (cf. also Przeworska-Rolewicz and Rolewicz 1968; Pinkus 1985). When adapted to the present situation of a Banach space E with $\dim(E) \geqslant n$ considered as a subspace of $\tilde{E} = l_\infty(U_{E'})$ this result says that, given $\varepsilon > 0$ and an arbitrary subspace $\tilde{N} \subset \tilde{E}$ with $\dim(\tilde{N}) < n$, there always exists an element $x_\varepsilon \in E$ with $\|x_\varepsilon\| = 1$ such that $\|x_\varepsilon - \tilde{z}\| > 1/(1 + \varepsilon)$ for all $\tilde{z} \in \tilde{N}$. Hence an inclusion

$$J_E(U_E) \subset \tilde{N} + \delta \cdot U_{\tilde{E}}$$

turns out to be impossible whenever $\delta < 1$, which means

$$d_n(J_E) \geqslant 1 \quad \text{and} \quad t_n(I_E) \geqslant 1$$

according to the definition of the nth Kolmogorov number $d_n(J_E)$ and the nth symmetrized approximation number $t_n(I_E)$, respectively. The converse inequality $t_n(I_E) \leqslant 1$ again follows from the monotonicity $t_n(T) \leqslant t_1(T) = \|T\|$.

In a similar way to approximation numbers, Kolmogorov numbers, and Gelfand numbers, the norm determining property (T5) of the symmetrized approximation numbers is basic for calculating the values $t_n(D)$ of the diagonal operator $D: l_p \to l_p$ given by (1.3.15). That is, considering the identity map $I_p^{(n)}: l_p^n \to l_p^n$ and using the same notation as in (1.3.17)' and (1.3.18)' (see section 2.1) we get

$$1 = t_n(I_p^{(n)}) \leqslant \|(D^{(n)})^{-1}\| t_n(D^{(n)}) = \sigma_n^{-1} t_n(D^{(n)})$$

and, furthermore,

$$t_n(D^{(n)}) \leqslant \|P_n\| t_n(D) \|I_n\| = t_n(D) \leqslant a_n(D) = \sigma_n.$$

This yields

$$t_n(D) = a_n(D) = \sigma_n. \tag{2.6.5}$$

To end this section let us emphasize that the symmetrized approximation numbers give rise to the following compactness theorem.

Refined version of Schauder's theorem. *An operator T between arbitrary Banach spaces E and F is compact if and only if $\lim_{n \to \infty} t_n(T) = 0$ and, moreover,*

$$t_n(T') = t_n(T), \tag{2.6.4}$$

that is to say the degree of compactness of T and T' is the same in so far as it is measured by the symmetrized approximation numbers t_n.

It is an open problem whether the degree of compactness reflected by the rate of decrease of the entropy numbers is also, in some sense, the same for T and T'.

2.7. Local representations of approximation quantities

The term *local* refers to the behaviour of an operator $T:E \to F$ on the finite-dimensional subspaces M of the Banach space E. By *local representations of approximation quantities* we mean representations of the approximation quantities $s_n(T)$ under consideration, as approximation numbers $a_n(T)$, Gelfand numbers $c_n(T)$, Kolmogorov numbers $d_n(T)$, and symmetrized approximation numbers $t_n(T)$, which are based on the corresponding approximation quantities $s_n(TI_M^E)$ of the finite-dimensional restrictions TI_M^E of T. We have

$$s_1(T) = \|T\|$$

for all approximation quantities $s_n(T)$. The operator norm itself admits a local representation

$$\|T\| = \sup \{ \|TI_M^E\| : M \subseteq E, \dim(M) < \infty \}$$

since the definition of $\|T\|$ says

$$\|T\| = \sup \{ \|Tx\| : x \in E, \|x\| \leqslant 1 \}.$$

For $n \geqslant 2$, however, the approximation quantities $s_n(T)$ in general differ from the corresponding *local quantities*

$$\hat{s}_n(T) = \sup \{ s_n(TI_M^E) : M \subseteq E, \dim(M) < \infty \} \qquad (2.7.1)$$

in the sense that only

$$\hat{s}_n(T) \leqslant s_n(T) \qquad (2.7.2)$$

is valid. Yet in special situations equality

$$\hat{s}_n(T) = s_n(T), \quad n = 1, 2, 3, \ldots,$$

holds. In the case of the approximation numbers $s_n(T) = a_n(T)$ this is so if T maps a Banach space E into a dual space F'.

Proposition 2.7.1. *Let $T \in L(E, F')$ be an operator mapping a Banach space E into a dual space F'. Then*

$$\hat{a}_n(T) = a_n(T), \quad n = 1, 2, 3, \ldots \qquad (2.7.3)$$

Proof. Obviously it suffices to consider the nth approximation numbers $a_n(TI_M^E)$ of the restrictions TI_M^E of T to finite-dimensional subspaces $M \subseteq E$ with $\dim(M) \geqslant n$ since $a_n(TI_M^E) = 0$ whenever $\dim(M) < n$. Given an arbitrary finite-dimensional subspace $M \subseteq E$ with $\dim(M) \geqslant n$ and $\varepsilon > 0$

we determine an operator $S_M : M \to F'$ with rank $(S_M) < n$ such that

$$\| T I_M^E - S_M \| \leqslant (1 + \varepsilon) a_n (T I_M^E). \qquad (2.7.4)$$

Let $N(M) \subseteq F'$ be an $(n-1)$-dimensional subspace of F' containing the range of S_M. According to Auerbach's Lemma (see section 2.1) there exists a projection $P_{N(M)} \in L(F', F')$ of F' onto $N(M)$ which can be represented as

$$P_{N(M)} b = \sum_{i=1}^{n-1} \langle b, v_i(M) \rangle b_i(M) \qquad (2.7.5)$$

in terms of a basis $b_i(M)$ of $N(M) \subseteq F'$ and corresponding biorthogonal functionals $v_i(M) \in F''$ such that

$$\| b_i(M) \| = 1 \quad \text{and} \quad \| v_i(M) \| = 1. \qquad (2.7.6)$$

Inserting $b = S_M x$ into (2.7.5) and taking into consideration that $S_M(M) \subseteq N(M) = R(P_{N(M)})$ we obtain

$$S_M x = \sum_{i=1}^{n-1} \langle S_M x, v_i(M) \rangle b_i(M)$$

and hence

$$S_M x = \sum_{i=1}^{n-1} \langle x, u_i(M) \rangle b_i(M) \quad \text{for } x \in M,$$

where the functionals $u_i(M) = S_M' v_i(M)$ over M are subject to the norm estimate

$$\| u_i(M) \| \leqslant \| S_M' \| \, \| v_i(M) \| = \| S_M \|.$$

Moreover, from (2.7.4) we have

$$\| S_M \| \leqslant \| S_M - T I_M^E \| + \| T I_M^E \| \leqslant (1 + \varepsilon) a_n (T I_M^E) + \| T \|$$
$$\leqslant (2 + \varepsilon) \| T \|$$

and thus also

$$\| u_i(M) \| \leqslant (2 + \varepsilon) \| T \|$$

independently of i and M. According to the Hahn–Banach theorem the functionals $u_i(M) \in M'$ can be extended to linear functionals $\tilde{u}_i(M)$ over E with the same norm so that

$$\| \tilde{u}_i(M) \| \leqslant (2 + \varepsilon) \| T \| \qquad (2.7.7)$$

remains true even for the extended functionals $\tilde{u}_i(M) \in E'$. Let

$$\tilde{S}_M x = \sum_{i=1}^{n-1} \langle x, \tilde{u}_i(M) \rangle b_i(M) \quad \text{for } x \in E \qquad (2.7.8)$$

denote the corresponding operators $\tilde{S}_M : E \to F'$ of rank $(\tilde{S}_M) < n$ indexed by the finite-dimensional subspaces M with dim $(M) \geqslant n$.

Since the system of finite-dimensional subspaces $M \subseteq E$ is upwards directed with respect to the set-theoretical inclusion, the system of operators $\tilde{S}_M \in L(E, F')$ forms a so-called *generalized sequence* (see Dunford and Schwartz 1958). In what follows we shall show that this generalized sequence $\tilde{S}_M$ possesses a generalized subsequence $\tilde{S}_{M_\kappa}$ convergent to an operator $\tilde{S} \in L(E, F')$ with rank $(\tilde{S}) < n$ in some kind of weak operator topology. The proof rests on the theorem of Alaoglu–Bourbaki (cf. Dunford and Schwartz 1958; Köthe 1960; Taylor and Lay 1980) which says that in a dual space E' every closed ball is compact in the $\sigma(E', E)$-topology generated on E' by the elements of E. Compactness of a set in this context means that every generalized sequence belonging to the set has a convergent generalized subsequence, the limit again belonging to the set. If we apply the Alaoglu–Bourbaki theorem to the closed ball $(2 + \varepsilon)\|T\|U_{E'}$ of the dual E' we are led to a convergent generalized subsequence of the generalized sequence $\{\tilde{u}_i(M)\}$ with a limit point $\tilde{u}_i \in (2 + \varepsilon)\|T\|U_{E'}$. By the same argument we obtain a functional $b_i \in U_{F'}$ as the limit of a generalized subsequence of $\{b_i(M)\} \subseteq U_{F'}$. Moreover, the selection of the generalized subsequences can be carried out subsequently for $\{\tilde{u}_1(M)\}, \ldots, \{\tilde{u}_{n-1}(M)\}$, and for $\{b_1(M)\}, \ldots, \{b_{n-1}(M)\}$. The final result is a common index set $\{M_\kappa\}$, upwards directed, such that simultaneously

$$\lim_\kappa \tilde{u}_i(M_\kappa) = \tilde{u}_i \quad \text{for } 1 \leqslant i \leqslant n-1$$

and

$$\lim_\kappa b_i(M_\kappa) = b_i \quad \text{for } 1 \leqslant i \leqslant n-1$$

hold in the $\sigma(E', E)$-topology and the $\sigma(F', F)$-topology, respectively. We shall make use of these limit relations by stating that for arbitrary $x_0 \in E$, $y_0 \in F$, and $\varepsilon > 0$ there exists a subspace $M(\varepsilon, x_0, y_0)$ such that

$$|\langle x_0, \tilde{u}_i(M_\kappa) - \tilde{u}_i \rangle| \leqslant \frac{\varepsilon}{n-1} \tag{2.7.9}$$

and

$$|\langle y_0, b_i(M_\kappa) - b_i \rangle| \leqslant \frac{\varepsilon}{n-1} \tag{2.7.10}$$

for $1 \leqslant i \leqslant n-1$ if $M_\kappa \supseteq M(\varepsilon, x_0, y_0)$. On the basis of (2.7.9) and (2.7.10) one can easily prove that the generalized subsequence $\{\tilde{S}_{M_\kappa}\}$ of the original sequence (2.7.8) of operators $\tilde{S}_M$ converges to the operator

$$\tilde{S}x = \sum_{i=1}^{n-1} \langle x, \tilde{u}_i \rangle b_i$$

in the sense that $|\langle y_0, (\tilde{S}_{M_\kappa} - \tilde{S})x_0 \rangle|$ becomes arbitrarily small for sufficiently large M_κ. Indeed, we have

$$|\langle y_0, (\tilde{S}_{M_\kappa} - \tilde{S})x_0 \rangle| = \left| \sum_{i=1}^{n-1} \langle x_0, \tilde{u}_i(M_\kappa) - \tilde{u}_i \rangle \langle y_0, b_i(M_\kappa) \rangle \right.$$

$$+ \left. \sum_{i=1}^{n-1} \langle x_0, \tilde{u}_i \rangle \langle y_0, b_i(M_\kappa) - b_i \rangle \right|$$

$$\leqslant \| y_0 \| \sum_{i=1}^{n-1} |\langle x_0, \tilde{u}_i(M_\kappa) - \tilde{u}_i \rangle|$$

$$+ (2 + \varepsilon) \| T \| \| x_0 \| \sum_{i=1}^{n-1} |\langle y_0, b_i(M_\kappa) - b_i \rangle|$$

from (2.7.6) and (2.7.7). Accordingly we have

$$|\langle y_0, (\tilde{S}_{M_\kappa} - \tilde{S})x_0 \rangle| \leqslant \varepsilon \| y_0 \| + \varepsilon(2 + \varepsilon) \| T \| \| x_0 \| \qquad (2.7.11)$$

for $M_\kappa \supseteq M(\varepsilon, x_0, y_0)$.

Next we estimate the operator norm $\| T - \tilde{S} \|$ from above by determining $x_0 \in U_E$ for fixed $\varepsilon > 0$ such that

$$\| T - \tilde{S} \| \leqslant (1 + \varepsilon) \| (T - \tilde{S})x_0 \|. \qquad (2.7.12)$$

Similarly, the norm of the functional $b_0 = (T - \tilde{S})x_0$ over F can be bounded from above by

$$\| (T - \tilde{S})x_0 \| \leqslant (1 + \varepsilon) |\langle y_0, (T - \tilde{S})x_0 \rangle| \qquad (2.7.13)$$

with an appropriate element $y_0 \in U_F$ depending on ε. Combining (2.7.12) and (2.7.13) and then inserting $\tilde{S}_{M_\kappa} x_0$ we arrive at

$$\| T - \tilde{S} \| \leqslant (1 + \varepsilon)^2 |\langle y_0, (T - \tilde{S}_{M_\kappa})x_0 \rangle| + (1 + \varepsilon)^2 |\langle y_0, (\tilde{S}_{M_\kappa} - \tilde{S})x_0 \rangle|.$$

If we still suppose $M_\kappa \supseteq M(\varepsilon, x_0, y_0)$ we can apply the estimate (2.7.11) to the term $|\langle y_0, (\tilde{S}_{M_\kappa} - \tilde{S})x_0 \rangle|$ with $\| x_0 \| \leqslant 1$ and $\| y_0 \| \leqslant 1$. On the other hand, we may conclude that

$$|\langle y_0, (T - \tilde{S}_{M_\kappa})x_0 \rangle| \leqslant \| (T - \tilde{S}_{M_\kappa}) I_{M_\kappa}^E \|$$

so far as $x_0 \in M_\kappa$. The two claims $M_\kappa \supseteq M(\varepsilon, x_0, y_0)$ and $x_0 \in M_\kappa$, however, can be satisfied simultaneously owing to the general properties of a directed subsystem of a given directed system. In this way we obtain

$$\| T - \tilde{S} \| \leqslant (1 + \varepsilon)^2 \| T I_{M_\kappa}^E - S_{M_\kappa} \| + \varepsilon(1 + \varepsilon)^2 + \varepsilon(1 + \varepsilon)^2 (2 + \varepsilon) \| T \|.$$

Since rank $(\tilde{S}) < n$ we have

$$a_n(T) \leqslant \| T - \tilde{S} \|.$$

Furthermore, from (2.7.4) and the definition of the local approximation number $\hat{a}_n(T)$ we can replace the term $\| T I_{M_\kappa}^E - S_{M_\kappa} \|$ by $(1 + \varepsilon)\hat{a}_n(T)$ which yields

$$a_n(T) \leqslant (1 + \varepsilon)^3 \hat{a}_n(T) + \varepsilon(1 + \varepsilon)^2 + \varepsilon(1 + \varepsilon)^2 (2 + \varepsilon) \| T \|.$$

Taking the limit as $\varepsilon \to 0$ we finally obtain the inequality

$$a_n(T) \leqslant \hat{a}_n(T)$$

required for the proof of (2.7.3). ∎

We shall see later that in the case of an operator T mapping a Banach space E into a Banach space F, which is not a dual space, the *local approximation numbers* $\hat{a}_n(T)$ need not coincide with the actual approximation numbers $a_n(T)$. But the local approximation numbers $\hat{a}_n(T)$ always coincide with the actual approximation numbers $a_n(T'')$ of the bidual T'' of T.

Proposition 2.7.2. *Let* $T \in L(E, F)$ *be an operator between arbitrary Banach spaces E and F. Then*

$$\hat{a}_n(T) = a_n(T'') = a_n(K_F T). \tag{2.7.14}$$

Proof. Starting from $K_F T = T'' K_E$ (see (2.5.19)) we recognize that

$$a_n(K_F T) \leqslant a_n(T'').$$

For estimating $a_n(T'')$ from above we use the representation

$$I_{F''} = K'_{F'}(K_F)''$$

of the identity map of F'' which results from (2.5.18) by taking the dual operators $I'_{F'} = I_{F''}$ and making use of $((K_F)'K_{F'})' = K'_{F'}(K_F)''$, namely

$$a_n(T'') = a_n(K'_{F'}(K_F)''T'') \leqslant a_n((K_F)''T'') = a_n((K_F T)'').$$

From (2.5.1) we obtain

$$a_n((K_F T)'') \leqslant a_n(K_F T).$$

Having thus proved the equality $a_n(K_F T) = a_n(T'')$ we refer to Proposition 2.7.1 and remark that

$$a_n(K_F T) = \hat{a}_n(K_F T) \leqslant \hat{a}_n(T).$$

Next we shall estimate $\hat{a}_n(T)$ from above. For this purpose we analyse the terms $s_n(T I_M^E) = a_n(T I_M^E)$ over which the supremum on the right-hand side of (2.7.1) runs. Because $\dim(M) < \infty$ the operators $T I_M^E$ are of finite rank, and therefore Proposition 2.5.2 tells us that

$$a_n(T I_M^E) = a_n((T I_M^E)'') = a_n(T'' I_M^{E''}).$$

It follows that

$$a_n(T I_M^E) \leqslant a_n(T'')$$

and finally

$$\hat{a}_n(T) \leqslant a_n(T'').$$

This completes the proof of (2.7.14). ∎

Let us once more consider the identity map $I_0 : l_1 \to c_0$. Since c_0 is not a dual space Proposition 2.7.1 does not apply. However, the local approximation numbers $\hat{a}_n(I_0)$ can be calculated by using Proposition 2.7.2. We observe that in the present situation $K_F = K_{c_0}$ is the natural embedding of c_0 into l_∞ so that

$$K_{c_0} I_0 = I_\infty$$

(see (2.5.14)). According to (2.5.7) we have $a_n(I_0) = 1$ for $n = 1, 2, 3, \ldots$ and, on the other hand,

$$a_n(I_\infty) = \tfrac{1}{2} \quad \text{for } n = 2, 3, 4, \ldots, \tag{2.5.9}$$

which can be read as

$$\hat{a}_n(I_0) = \tfrac{1}{2} \quad \text{for } n \geqslant 2.$$

In section 2.5 we explained that the difference between $a_n(I_0)$ and $a_n(I_\infty) = a_n(I_0')$ was due to the non-compactness of the operator $I_0 : l_1 \to c_0$. This, in fact, is also a possible explanation for the difference between $a_n(I_0)$ and $\hat{a}_n(I_0)$ since $\hat{a}_n(T) = a_n(T)$ for compact operators T.

Proposition 2.7.3. *Let $T \in K(E, F)$ be a compact operator between arbitrary Banach spaces E and F. Then*

$$\hat{a}_n(T) = a_n(T). \tag{2.7.15}$$

Proof. Proposition 2.7.2 tells us that $\hat{a}_n(T) = a_n(T'')$. By Schauder's theorem (see section 2.5) the compactness of T transfers to T' and hence also to T''. By Proposition 2.5.2 we obtain in this way $a_n(T'') = a_n(T)$ and thus (2.7.15). ∎

Though the local approximation numbers $\hat{a}_n(T)$ may differ from the actual approximation numbers $a_n(T)$ there is a universal relation between the two quantities.

Proposition 2.7.4. *Let $T \in L(E, F)$ be an operator between arbitrary Banach spaces E and F. Then*

$$\hat{a}_n(T) \leqslant a_n(T) \leqslant 5\hat{a}_n(T). \tag{2.7.16}$$

Proof. We recall that $a_n(T) \leqslant 5a_n(T'')$ (see Proposition 2.5.3) and refer to (2.7.14). ∎

With the *local Gelfand numbers*

$$\hat{c}_n(T) = \sup\{c_n(TI_M^E) : M \subseteq E, \dim(M) < \infty\} \tag{2.7.17}$$

the situation is more satisfactory since $\hat{c}_n(T) = c_n(T)$ is always valid (see Pietsch 1986).

Proposition 2.7.5. *Let* $T \in L(E, F)$ *be an operator between arbitrary Banach spaces* E *and* F. *Then*

$$\hat{c}_n(T) = c_n(T). \tag{2.7.18}$$

Proof. From the representation formula $c_n(T) = a_n(J_F T)$ of the Gelfand numbers $c_n(T)$ (see Theorem 2.3.1) the local Gelfand numbers $\hat{c}_n(T)$ of T coincide with the local approximation numbers $\hat{a}_n(J_F T)$ of the operator $J_F T$ mapping the Banach space E into the dual space $l_\infty(U_{F'}) = l_1(U_{F'})'$. Therefore Proposition 2.7.1 yields the desired result

$$\hat{c}_n(T) = a_n(J_F T) = c_n(T). \qquad \blacksquare$$

The treatment of the *local Kolmogorov numbers*

$$\hat{d}_n(T) = \sup\{d_n(TI_M^E): M \subseteq E, \dim(M) < \infty\} \tag{2.7.19}$$

requires some knowledge of the local structure of the spaces $l_1(\Gamma)$:

For every finite-dimensional subspace $M \subseteq l_1(\Gamma)$ and every $\varepsilon > 0$ there exists a finite-dimensional subspace $\tilde{M} \subseteq l_1(\Gamma)$ containing M and an isomorphism A of $\tilde{M}$ onto l_1^m, where $m = \dim(\tilde{M})$, such that

$$\|A\| \, \|A^{-1}\| < 1 + \varepsilon. \tag{2.7.20}$$

A proof of this will be presented in a more general context in section 5.2. For the moment we take it as valid and prove a fundamental inequality between the local Kolmogorov numbers $\hat{d}_n(T)$ and the local approximation numbers $\hat{a}_n(TQ_E)$.

Proposition 2.7.6. *Let* $T \in L(E, F)$ *be an operator between arbitrary Banach spaces* E *and* F. *Then*

$$\hat{a}_n(TQ_E) \leqslant \hat{d}_n(T). \tag{2.7.21}$$

Proof. We consider the restriction $TQ_E I_M^{l_1(U_E)}$ of the operator $TQ_E: l_1(U_E) \to F$ to a finite-dimensional subspace $M \subset l_1(U_E)$ and embed M into a finite-dimensional subspace $\tilde{M} \subset l_1(U_E)$ which admits an isomorphism A onto l_1^m with the property (2.7.20) stated above. The natural embedding $I_M^{l_1(U_E)}: M \to l_1(U_E)$ can then be expressed as the composition

$$I_M^{l_1(U_E)} = I_{\tilde{M}}^{l_1(U_E)} A^{-1} A I_M^{\tilde{M}}.$$

This gives rise to the estimate

$$a_n(TQ_E I_M^{l_1(U_E)}) \leqslant a_n(TQ_E I_{\tilde{M}}^{l_1(U_E)} A^{-1}) \|A\|$$

of the nth approximation number of the restriction $TQ_E I_M^{l_1(U_E)}$. Since A^{-1} acts on l_1^m we may apply Proposition 2.2.3 and conclude that

$$a_n(TQ_E I_{\tilde{M}}^{l_1(U_E)} A^{-1}) = d_n(TQ_E I_{\tilde{M}}^{l_1(U_E)} A^{-1}).$$

The operator $S = Q_E I_M^{l_1(U_E)} A^{-1}$ maps l_1^m into E. The canonical factorization

$$S = I_N^E S_0$$

of S through its range $N = S(l_1^m)$ enables us to estimate $d_n(T Q_E I_M^{l_1(U_E)} A^{-1})$ by

$$d_n(T Q_E I_M^{l_1(U_E)} A^{-1}) \leqslant d_n(T I_N^E) \|S_0\|.$$

Because

$$\|S_0\| = \|S\| \leqslant \|Q_E\| \ \|I_M^{l_1(U_E)}\| \ \|A^{-1}\| = \|A^{-1}\|$$

this estimate can be extended to

$$d_n(T Q_E I_M^{l_1(U_E)} A^{-1}) \leqslant d_n(T I_N^E) \|A^{-1}\|.$$

Altogether we obtain

$$a_n(T Q_E I_M^{l_1(U_E)}) \leqslant d_n(T I_N^E) \|A\| \ \|A^{-1}\| < (1 + \varepsilon) d_n(T I_N^E). \qquad (2.7.22)$$

Taking the supremum with respect to all finite-dimensional subspaces $N \subseteq E$ on the right-hand side of (2.7.22) and the supremum with respect to all finite-dimensional subspaces $M \subseteq l_1(U_E)$ on the left-hand side leads us to

$$\hat{a}_n(T Q_E) \leqslant (1 + \varepsilon) \hat{d}_n(T),$$

and finally yields (2.7.21) by letting $\varepsilon \to 0$. ∎

We are now in a position to derive the relation between the local Kolmogorov numbers $\hat{d}_n(T)$ and the actual Kolmogorov numbers $d_n(T)$ by straight-forward considerations.

Proposition 2.7.7. *Let* $T \in L(E, F')$ *be an operator mapping a Banach space* E *into a dual space* F'. *Then*

$$\hat{d}_n(T) = d_n(T). \qquad (2.7.23)$$

Proof. The proof of $d_n(T) \leqslant \hat{d}_n(T)$ is based on the representation $d_n(T) = a_n(T Q_E)$ for the Kolmogorov numbers $d_n(T)$ (see Theorem 2.2.1) and on Proposition 2.7.1 which tells us that

$$a_n(T Q_E) = \hat{a}_n(T Q_E)$$

under the present assumptions. It is completed by referring to (2.7.21). ∎

Proposition 2.7.8. *Let* $T \in L(E, F)$ *be an operator between arbitrary Banach spaces* E *and* F. *Then*

$$\hat{d}_n(T) = d_n(T'') = d_n(K_F T). \qquad (2.7.24)$$

Proof. On the one hand we have

$$d_n(K_F T) = d_n(T'' K_E) \leqslant d_n(T'')$$

and, on the other hand,

$$d_n(T'') = d_n(K'_{F'}(K_F)''T'') \leqslant d_n((K_FT)'')$$

since

$$I_{F''} \stackrel{.}{=} (I_{F'})' = K'_{F'}(K_F)''$$

by (2.5.18). From Proposition 2.5.5 it follows that

$$d_n((K_FT)'') = c_n((K_FT)') \leqslant d_n(K_FT).$$

For the local Kolmogorov numbers $\hat{d}_n(T)$, we see that $d_n(K_FT) = \hat{d}_n(K_FT)$ by Proposition 2.7.7 and thus

$$d_n(K_FT) \leqslant \hat{d}_n(T).$$

As an upper bound for $\hat{d}_n(T)$ we again get $d_n(T'')$ if we apply Proposition 2.5.6 and Proposition 2.5.5 to the individual terms $d_n(TI_M^E)$ involved in the supremum (2.7.19), namely

$$d_n(TI_M^E) = c_n((TI_M^E)') = d_n((TI_M^E)'') = d_n(T''I_M^{E''}) \leqslant d_n(T'').$$

This completes the proof of (2.7.24).   ∎

Obviously the identity map $I_0 : l_1 \to c_0$ can again serve as an example to demonstrate the possibility of a difference between $d_n(T)$ and $\hat{d}_n(T)$ since

$$d_n(I_0) = a_n(I_0) = 1 \quad \text{for } n = 1, 2, 3, \dots$$

(see Proposition 2.2.3 and (2.5.7)), but

$$\hat{d}_n(I_0) = d_n(K_{c_0}I_0) = d_n(I_\infty) = a_n(I_\infty) = \tfrac{1}{2} \quad \text{for } n \geqslant 2.$$

Proposition 2.7.9. *Let* $T \in K(E, F)$ *be a compact operator between arbitrary Banach spaces E and F. Then*

$$\hat{d}_n(T) = d_n(T). \tag{2.7.25}$$

Proof. The consecutive applications of Propositions 2.7.8, 2.5.5, and 2.5.6 yields

$$\hat{d}_n(T) = d_n(T'') = c_n(T') = d_n(T). \qquad \blacksquare$$

Proposition 2.7.10. *For an operator* $T \in L(E, F)$ *between arbitrary Banach spaces E and F the universal relation*

$$\hat{d}_n(T) \leqslant d_n(T) \leqslant 5\hat{d}_n(T) \tag{2.7.26}$$

holds between the local and the actual Kolmogorov numbers.

Proof. We once more recall that $d_n(T) = a_n(TQ_E)$ and make use of $a_n(TQ_E) \leqslant 5\hat{a}_n(TQ_E)$ (see Proposition 2.7.4) and $\hat{a}_n(TQ_E) \leqslant \hat{d}_n(T)$ (see Proposition 2.7.6).   ∎

It has turned out that the local Kolmogorov numbers $\hat{d}_n(T)$ are related to the actual Kolmogorov numbers $d_n(T)$ in the same way as the local approximation numbers $\hat{a}_n(T)$ to the actual approximation numbers $a_n(T)$.

The behaviour of the *local symmetrized approximation numbers*

$$\hat{t}_n(T) = \sup\{t_n(TI_M^E): M \subseteq E,\ \dim(M) < \infty\}, \qquad (2.7.27)$$

like the behaviour of the local Gelfand numbers, is determined by the fact that the operator $J_F T Q_E$, which enters the definition $t_n(T) = a_n(J_F T Q_E)$ of the $t_n(T)$, has its range in a dual space.

Proposition 2.7.11. *Let $T \in L(E, F)$ be an operator between arbitrary Banach spaces E and F. Then*

$$\hat{t}_n(T) = t_n(T). \qquad (2.7.28)$$

Proof. Since $t_n(T) = d_n(J_F T)$ (see (2.6.2)) we also have $\hat{t}_n(T) = \hat{d}_n(J_F T)$. To $\hat{d}_n(J_F T)$ we can apply Proposition 2.7.7 which proves (2.7.28). ∎

3
Inequalities of Bernstein–Jackson type

The classical inequalities of Bernstein–Jackson express the relations between structural properties of continuous real-valued functions on an interval $[a, b]$ or of continuous real-valued periodic functions and approximation properties of these functions. Approximation means approximation by algebraic or trigonometric polynomials (see Natanson 1955). The term *structural properties* means properties such as *Hölder continuity* with a certain positive exponent $\alpha, 0 < \alpha \leqslant 1$. In this connection the so-called *modulus of continuity*

$$\omega(f; \delta) = \sup_{|x' - x''| \leqslant \delta} |f(x') - f(x'')| \tag{3.0.1}$$

plays a decisive role, in the sense that Hölder continuity of $f \in C[a, b]$ with an exponent $\alpha \leqslant 1$ is equivalent to

$$\omega(f; \delta) \leqslant C\delta^{\alpha}, \tag{3.0.2}$$

where C is an appropriate constant.

The minimal error

$$E_n(f) = \inf\{\|f - p\| : \deg(p) \leqslant n\}$$

with respect to the approximation of a function $f \in C[a, b]$ by polynomials p of degree $\deg(p) \leqslant n$ shows an obvious similarity to the $(n + 1)$th approximation number

$$a_{n+1}(T) = \inf\{\|T - A\| : A \in L(E, F) \text{ with rank}(A) \leqslant n\} \tag{2.1.1'}$$

of an operator $T \in L(E, F)$. In view of this fact we take the well-known *Bernstein inequality* for functions

$$\omega\left(f; \frac{b-a}{2n}\right) \leqslant \frac{C}{n} \sum_{k=1}^{n} E_k(f), \quad n = 1, 2, \ldots, \tag{3.0.3}$$

as a model for establishing corresponding inequalities between approximation quantities and entropy quantities of operators. We actually obtain an inequality of the form

$$e_n(T) \leqslant \frac{C}{n} \sum_{k=1}^{n} a_k(T), \quad n = 1, 2, \ldots, \tag{3.0.4}$$

for operators $T \in L(E, F)$ between arbitrary Banach spaces E and F as a special version of a more general inequality (see (3.1.9)). This in a sense justifies a parallel consideration of entropy quantities of operators on one side, and of the modulus of continuity of functions on the other side, as measures for structural properties of operators and functions, respectively.

Furthermore, *Jackson's inequality*

$$E_n(f) \leqslant C \cdot \omega\left(f; \frac{b-a}{2n}\right), \qquad (3.0.5)$$

for functions $f \in C[a, b]$ has an adequate counterpart for operators. Indeed, let T be an operator from an arbitrary Banach space E into a Banach space $C(X)$ of continuous functions on a compact metric space (X, d). Then the definition

$$\omega(T; \delta) = \sup_{\|x\| \leqslant 1} \omega(Tx; \delta), \qquad (3.0.6)$$

of a *modulus of continuity of T* makes sense. It refers to the classical modulus of continuity

$$\omega(f; \delta) = \sup_{d(t', t'') \leqslant \delta} |f(t') - f(t'')| \qquad (3.0.7)$$

of the images $f = Tx$ of $x \in U_E$ in $C(X)$. In analogy to (3.0.5) an estimate

$$a_{n+1}(T) \leqslant \omega(T; \varepsilon_n(X)) \qquad (3.0.8)$$

for the $(n + 1)$th approximation number $a_{n+1}(T)$ of a compact operator $T: E \to C(X)$ turns out to be true. A proof of (3.0.8), however, is beyond the scope of this chapter and will be given in chapter 5 only (see Theorem 5.6.1). In the present chapter a Jackson-type inequality will be derived for operators T acting between Hilbert spaces (see section 3.4), a weakened version of this inequality being

$$a_n(T) \leqslant 2e_n(T). \qquad (3.0.9)$$

3.1. Inequalities of Bernstein type

The inequalities of Bernstein type to be derived for arbitrary operators $T: E \to F$ rest on the entropy behaviour of finite rank operators (see section 1.3).

Theorem 3.1.1. (see Carl 1981). *Let $0 < p < \infty$ and let $T \in L(E, F)$ be an operator between arbitrary Banach spaces E and F. Then*

$$\sup_{1 \leqslant k \leqslant m} k^{1/p} e_k(T) \leqslant c_p \sup_{1 \leqslant k \leqslant m} k^{1/p} a_k(T) \quad \text{for } m = 1, 2, 3, \ldots \qquad (3.1.1)$$

where
$$c_p = 2^7(16(2 + 1/p))^{1/p} \text{ in the real case}$$

and
$$c_p = 2^7(32(2 + 1/p))^{1/p} \text{ in the complex case.}$$

Proof. We first consider natural numbers
$$n = 2^N, \quad N = 1, 2, \ldots$$

According to the definition of the approximation numbers $a_k(T)$ we can determine operators $A_j \in L(E, F)$ with rank $(A_j) < 2^j$ and
$$\| T - A_j \| \leqslant 2a_{2^j}(T) \quad \text{for } j = 0, 1, 2, \ldots, N, \tag{3.1.2}$$

where $A_0 = 0$. If we take the differences $A_j - A_{j-1}$, we have
$$\text{rank}\,(A_j - A_{j-1}) < 2^{j+1} \quad \text{for } j = 1, 2, \ldots, N.$$

Moreover, these differences allow a representation
$$T = \sum_{j=1}^{N} (A_j - A_{j-1}) + (T - A_N),$$

which can be employed to obtain the desired estimate (3.1.1). Indeed, using the additivity of the dyadic entropy numbers (DE2) we may conclude that
$$e_{n_1 + \cdots + n_N - (N-1)}(T) \leqslant \sum_{j=1}^{N} e_{n_j}(A_j - A_{j-1}) + \| T - A_N \| \tag{3.1.3}$$

for natural numbers n_j to be chosen later. Hence it remains to estimate the values $e_{n_j}(A_j - A_{j-1})$. Since rank $(A_j - A_{j-1}) < 2^{j+1}$ we have
$$e_{n_j}(A_j - A_{j-1}) \leqslant 4 \cdot 2^{-(n_j-1)/2^{j+1}} \| A_j - A_{j-1} \|, \quad j = 1, \ldots, N,$$

(for the real version, see (1.3.36)). On the other hand, $\| A_j - A_{j-1} \|$ can be estimated by
$$\| A_j - A_{j-1} \| \leqslant \| A_j - T \| + \| T - A_{j-1} \| \leqslant 4a_{2^{j-1}}(T),$$

$j = 1, 2, \ldots, N$. Thus it follows that
$$e_{n_j}(A_j - A_{j-1}) \leqslant 2^4 2^{-(n_j-1)/2^{j+1}} a_{2^{j-1}}(T), \quad j = 1, \ldots, N. \tag{3.1.4}$$

If we now replace the terms $e_{n_j}(A_j - A_{j-1})$ and $\| T - A_N \|$ on the right-hand side of (3.1.3) by the upper bounds for $e_{n_j}(A_j - A_{j-1})$ and $\| T - A_N \|$ on the right-hand sides of (3.1.4) and (3.1.2), respectively, we arrive at
$$e_{n_1 + \cdots + n_N - (N-1)}(T) \leqslant 2^4 \sum_{j=1}^{N} 2^{-(n_j-1)/2^{j+1}} a_{2^{j-1}}(T) + 2a_{2^N}(T).$$

We recognize that
$$\sum_{j=1}^{N} 2^{-(n_j-1)/2^{j+1}} a_{2^{j-1}}(T) \leqslant \left(\sum_{j=1}^{N} 2^{-(n_j-1)/2^{j+1} - (j-1)/p} \right) \cdot \sup_{1 \leqslant j \leqslant N} 2^{(j-1)/p} a_{2^{j-1}}(T)$$
$$\leqslant \left(\sum_{j=1}^{N} 2^{-(n_j-1)/2^{j+1} - (j-1)/p} \right) \cdot \sup_{1 \leqslant j \leqslant 2^N} j^{1/p} a_j(T).$$

Moreover, $a_{2^N}(T)$ can obviously be given an upper bound by

$$a_{2^N}(T) \leqslant 2^{-N/p} \sup_{1 \leqslant j \leqslant 2^N} j^{1/p} a_j(T).$$

Combining these estimates we obtain

$$e_{n_1 + \cdots + n_N - (N-1)}(T) \leqslant \left(2^4 \sum_{j=1}^N 2^{-(n_j-1)/2^{j+1} - (j-1)/p} + 2 \cdot 2^{-N/p} \right)$$

$$\sup_{1 \leqslant j \leqslant 2^N} j^{1/p} a_j(T) \qquad (3.1.5)$$

Now we choose a natural number K with

$$1 + \frac{1}{p} \leqslant K \leqslant 2 + \frac{1}{p}$$

and put

$$n_j = 1 + K(N-j)2^{j+1}, \quad j = 1, \ldots, N. \qquad (3.1.6)$$

We then have

$$2^{-(n_j-1)/2^{j+1}} = 2^{-K(N-j)}.$$

Furthermore, the summation formula

$$\sum_{j=1}^N 2^{-K(N-j)-(j-1)/p} = 2^{1/p} 2^{-KN} \sum_{j=1}^N 2^{(K-1/p)j}$$

$$= 2^{1/p} 2^{-KN} 2^{(K-1/p)} \frac{2^{(K-1/p)N} - 1}{2^{(K-1/p)} - 1}$$

leads us to the conclusion that

$$\sum_{j=1}^N 2^{-(n_j-1)/2^{j+1} - (j-1)/p} = 2^{-KN} 2^K \frac{2^{(K-1/p)N} - 1}{2^{(K-1/p)} - 1}$$

$$\leqslant 2^2 2^{1/p} 2^{-N/p}$$

because $1 + 1/p \leqslant K \leqslant 2 + 1/p$. Hence the estimate (3.1.5) turns into

$$e_{n_1 + \cdots + n_N - (N-1)}(T) \leqslant 2^7 2^{1/p} 2^{-N/p} \sup_{1 \leqslant j \leqslant 2^N} j^{1/p} a_j(T).$$

It remains to limit the subscript $n_1 + \cdots + n_N - (N-1)$. From

$$n_1 + \cdots + n_N - N = K \sum_{j=1}^N (N-j) 2^{j+1}$$

(see (3.1.6)) and the summation formula

$$\sum_{j=1}^N (N-j) 2^{j+1} = 4(2^N - (N+1))$$

we have

$$n_1 + \cdots + n_N - (N-1) \leqslant 4K 2^N.$$

By the monotonicity of the entropy numbers we obtain the estimate

$$e_{4K2^N}(T) \leqslant 2^7 2^{1/p} 2^{-N/p} \sup_{1 \leqslant j \leqslant 2^N} j^{1/p} a_j(T), \tag{3.1.7}$$

for $N = 1, 2, \ldots$, where $1 + 1/p \leqslant K \leqslant 2 + 1/p$.

Finally, we estimate $e_n(T)$ for an arbitrary natural number n. First let us assume $n \geqslant 8K$. We may then choose a natural number $N \geqslant 1$ such that

$$8K\, 2^{N-1} \leqslant n \leqslant 8K\, 2^N.$$

Hence we get

$$2^{-N/p} \sup_{1 \leqslant j \leqslant 2^N} j^{1/p} a_j(T) \leqslant (8K)^{1/p} n^{-1/p} \sup_{1 \leqslant j \leqslant n} j^{1/p} a_j(T).$$

Moreover, the inequality (3.1.7) guarantees that

$$e_n(T) \leqslant e_{4K2^N}(T) \leqslant 2^7 2^{1/p} (8K)^{1/p} n^{-1/p} \sup_{1 \leqslant j \leqslant n} j^{1/p} a_j(T),$$

since $n \geqslant 4K2^N$. Thus we have finally

$$n^{1/p} e_n(T) \leqslant 2^7 (16K)^{1/p} \sup_{1 \leqslant j \leqslant n} j^{1/p} a_j(T).$$

This estimate also remains valid for $1 \leqslant n \leqslant 8K$. Indeed, we then have simply

$$n^{1/p} e_n(T) \leqslant (8K)^{1/p} \| T \| \leqslant 2^7 (16K)^{1/p} \sup_{1 \leqslant j \leqslant n} j^{1/p} a_j(T).$$

Since $K \leqslant 2 + 1/p$ we now obtain

$$n^{1/p} e_n(T) \leqslant 2^7 (16(2 + 1/p))^{1/p} \sup_{1 \leqslant j \leqslant n} j^{1/p} a_j(T)$$

for all $n = 1, 2, 3, \ldots$. Taking the supremum with respect to $n \leqslant m$ on both sides of this inequality we arrive at the desired estimate

$$\sup_{1 \leqslant n \leqslant m} n^{1/p} e_n(T) \leqslant c_p \sup_{1 \leqslant n \leqslant m} n^{1/p} a_n(T), \quad m = 1, 2, 3, \ldots,$$

with $c_p = 2^7 (16(2 + 1/p))^{1/p}$ (real case). In the complex case one can make use of

$$e_{n_j}(A_j - A_{j-1}) \leqslant 4 \cdot 2^{(n_j - 1)/2^{j+2}} \| A_j - A_{j-1} \|, \quad j = 1, \ldots, N,$$

(cf. (1.3.36)′) and choose

$$n_j = 1 + K(N - j)2^{j+2}, \quad j = 1, \ldots, N. \tag{3.1.6′}$$

The result is the assertion (3.1.1) with $c_p = 2^7 (32(2 + 1/p))^{1/p}$. ∎

The estimate (3.1.1) implies an estimate for the individual entropy number $e_m(T)$ by the arithmetic mean of order p of the approximation numbers from above. Indeed, we have

$$e_m(T) \leqslant m^{-1/p} \sup_{1 \leqslant k \leqslant m} k^{1/p} e_k(T) \leqslant c_p \cdot m^{-1/p} \sup_{1 \leqslant k \leqslant m} k^{1/p} a_k(T) \quad (3.1.8)$$

as an immediate consequence of (3.1.1). Furthermore it is

$$k^{1/p} a_k(T) \leqslant \left(\sum_{i=1}^{k} a_i(T)^p \right)^{1/p}$$

because of the monotonicity of the $a_i(T)$ and hence also

$$\sup_{1 \leqslant k \leqslant m} k^{1/p} a_k(T) \leqslant \left(\sum_{i=1}^{m} a_i(T)^p \right)^{1/p}.$$

Therefore the estimate (3.1.8) can be extended to

$$e_m(T) \leqslant c_p \left(\frac{\sum_{i=1}^{m} a_i(T)^p}{m} \right)^{1/p}, \quad m = 1, 2, \dots \quad (3.1.9)$$

As a counterpart to (3.1.1) we can now prove an estimate for $(\sum_{k=1}^{m} k^{t/s-1} e_k(T)^t)^{1/t}$ from above by $(\sum_{k=1}^{m} k^{t/s-1} a_k(T)^t)^{1/t}$.

Theorem 3.1.2. *Let $0 < s \leqslant \infty, 0 < t < \infty$, and let $T \in L(E, F)$ be an operator between arbitrary Banach spaces E and F. Then*

$$\left(\sum_{k=1}^{m} k^{t/s-1} e_k(T)^t \right)^{1/t} \leqslant C_{s,t} \left(\sum_{k=1}^{m} k^{t/s-1} a_k(T)^t \right)^{1/t} \quad (3.1.10)$$

for $m = 1, 2, 3, \dots$ with a constant $C_{s,t}$ depending on s and t only.

Proof. We choose $p = \frac{1}{2} \min(s, t)$ and estimate $e_k(T)$ from above according to (3.1.9), so that we have

$$\left(\sum_{k=1}^{m} k^{t/s-1} e_k(T)^t \right)^{1/t} \leqslant c_p \left(\sum_{k=1}^{m} k^{t/s-1} \left(\frac{\sum_{i=1}^{k} a_i(T)^p}{k} \right)^{t/p} \right)^{1/t}. \quad (3.1.11)$$

Since $p < \min(s, t)$ we can apply Hardy's inequality (see Lemma 1.5.3) to the right-hand side of (3.1.11), giving in detail

$$\left(\sum_{k=1}^{m} k^{t/s-1} \left(\frac{\sum_{i=1}^{k} a_i(T)^p}{k} \right)^{t/p} \right)^{1/t} \leqslant \left(1 + \frac{s}{s-p} \right)^{1/p} \left(\sum_{k=1}^{m} k^{t/s-1} a_k(T)^t \right)^{1/t}$$

for $0 < s, t < \infty$ and

$$\left(\sum_{k=1}^{m} k^{-1} \left(\frac{\sum_{i=1}^{k} a_i(T)^p}{k} \right)^{t/p} \right)^{1/t} \leqslant 2^{1/p} \left(\sum_{k=1}^{m} k^{-1} a_k(T)^t \right)^{1/t}$$

for $s = \infty, 0 < t < \infty$ (see (1.5.12) and (1.5.13), respectively). This proves (3.1.10) for $0 < s \leqslant \infty$ and $0 < t < \infty$. ∎

The two Theorems 3.1.1 and 3.1.2 are the basis for the main theorem of Bernstein type concerning the relation between approximation and entropy ideals.

Theorem 3.1.3. *The approximation ideal $L_{p,q}^{(a)}$ is contained in the entropy ideal $L_{p,q}^{(e)}$, that is*

$$L_{p,q}^{(a)} \subseteq L_{p,q}^{(e)} \quad \text{for } 0 < p, q \leqslant \infty. \tag{3.1.12}$$

Proof. Given $T \in L_{p,q}^{(a)}(E, F)$ we have $\sup_{1 \leqslant k < \infty} k^{1/p} a_k(T) < \infty$ when $q = \infty$ and $\sum_{k=1}^{\infty} k^{q/p-1} a_k(T)^q < \infty$ when $0 < q < \infty$. By (3.1.1) and (3.1.10) it follows that $\sup_{1 \leqslant k < \infty} k^{1/p} e_k(T) < \infty$ and $\sum_{k=1}^{\infty} k^{q/p-1} e_k(T)^q < \infty$, respectively, which proves the assertion. ∎

Supplement. Since the outer entropy numbers $e_n(T)$ are surjective and injective up to a factor of 2 (see section 1.3) we may replace T by $J_F T Q_E$, insert

$$t_n(T) = a_n(J_F T Q_E), \tag{2.6.1}$$

and thus conclude that

$$\sup_{1 \leqslant k \leqslant m} k^{1/p} e_k(T) \leqslant 2c_p \sup_{1 \leqslant k \leqslant m} k^{1/p} t_k(T) \quad \text{for } m = 1, 2, 3, \ldots \tag{3.1.13}$$

and

$$\left(\sum_{k=1}^{m} k^{q/p-1} e_k(T)^q \right)^{1/q} \leqslant 2C_{p,q} \left(\sum_{k=1}^{m} k^{q/p-1} t_k(T)^q \right)^{1/q} \quad \text{for } m = 1, 2, 3, \ldots \tag{3.1.14}$$

The ideal-theoretic version of this result is

$$L_{p,q}^{(t)} \subseteq L_{p,q}^{(e)} \tag{3.1.15}$$

which implies

$$L_{p,q}^{(d)} \subseteq L_{p,q}^{(e)} \tag{3.1.16}$$

and

$$L_{p,q}^{(c)} \subseteq L_{p,q}^{(e)}. \tag{3.1.17}$$

3.2. Inequalities of Jackson type

The inequalities of Jackson type to be derived in this section consist of estimates for the geometric means of the Kolmogorov and the Gelfand numbers from above by the entropy moduli. We start with the Kolmogorov numbers and first prove an auxiliary lemma concerning the relation of the Kolmogorov numbers $d_k(T)$ of an arbitrary operator $T : E \to F$ to the diagonal elements of a certain lower triangular matrix connected with T.

Lemma 3.2.1. (see Carl and Pietsch 1978; Pietsch 1978). *Let E and F be arbitrary Banach spaces and $T \in L(E, F)$. Given $\delta > 0$ there exist elements*

$x_1, x_2, \ldots \in E$ *and functionals* $b_1, b_2, \ldots \in F'$ *with* $\| x_i \| \leqslant 1$, $\| b_k \| \leqslant 1$, *such that*

$$\langle Tx_i, b_k \rangle = 0 \quad \text{for } i < k, \tag{3.2.1}$$

and

$$d_k(T) \leqslant (1 + \delta) |\langle Tx_k, b_k \rangle| \quad \text{for } k = 1, 2, 3, \ldots \tag{3.2.2}$$

Proof. We proceed by induction. To begin with we recall that $d_1(T) = \| T \| = \sup_{\| x \| \leqslant 1} \| Tx \|$ and choose $x_1 \in E$ with $\| x_1 \| \leqslant 1$ and $d_1(T) \leqslant (1 + \delta) \| Tx_1 \|$. By the Hahn–Banach theorem we can express the norm of the element $Tx_1 \in F$ as

$$\| Tx_1 \| = |\langle Tx_1, b_1 \rangle|$$

with an appropriate functional $b_1 \in F'$ where $\| b_1 \| \leqslant 1$. Thus (3.2.2) holds for $k = 1$. So let us assume that $x_1, x_2, \ldots, x_{n-1} \in E$ and $b_1, b_2, \ldots, b_{n-1} \in F'$ with the properties stated above have already been determined. Then we consider the subspace

$$N = \text{span}\{ Tx_1, Tx_2, \ldots, Tx_{n-1} \}$$

of F. Since $\dim(N) < n$ we have

$$d_n(T) \leqslant \| Q_N^F T \|$$

by Proposition 2.2.2. By the same method as for $n = 1$ we obtain an element $x_n \in E$ with $\| x_n \| \leqslant 1$ and a functional $\bar{b}_n \in (F/N)'$ with $\| \bar{b}_n \| \leqslant 1$ such that

$$\| Q_N^F T \| \leqslant (1 + \delta) |\langle Q_N^F T x_n, \bar{b}_n \rangle|.$$

The functional b_n over F defined by $b_n = (Q_N^F)' \bar{b}_n$ obviously satisfies the required condition

$$\| b_n \| \leqslant 1, \quad d_n(T) \leqslant (1 + \delta) |\langle Tx_n, b_n \rangle|.$$

Moreover we have

$$\langle Tx_i, b_n \rangle = 0 \quad \text{for } 1 \leqslant i \leqslant n - 1, \tag{3.2.1}'$$

since $Q_N^F Tx_i = 0$ for $1 \leqslant i \leqslant n - 1$. Hence the assertion is proved. ∎

Theorem 3.2.1. (see Carl 1982; Carl 1985). *Let* $T \in L(E, F)$ *be an operator between arbitrary Banach spaces* E *and* F. *Then*

$$\left(\prod_{k=1}^{n} d_k(T) \right)^{1/n} \leqslant n g_n(T) \quad \text{for } n = 1, 2, 3, \ldots \tag{3.2.3}$$

Proof. From Lemma 3.2.1 we can make use of elements $x_i \in E$ with $\| x_i \| \leqslant 1$ and functionals $b_i \in F'$ with $\| b_i \| \leqslant 1$, $1 \leqslant i \leqslant n$, so that $(\langle Tx_i, b_k \rangle)_{1 \leqslant i, k \leqslant n}$ takes the form of a triangular matrix. The corresponding determinant $\det(\langle Tx_i, b_k \rangle)$ therefore reduces to the product of the diagonal elements, that is to say

$$\det\left(\langle Tx_i, b_k\rangle\right) = \prod_{k=1}^{n} \langle Tx_k, b_k\rangle.$$

Consequently we have

$$\prod_{k=1}^{n} d_k(T) \leqslant (1+\delta)^n \prod_{k=1}^{n} |\langle Tx_k, b_k\rangle| = (1+\delta)^n |\det(\langle Tx_i, b_k\rangle)|$$

by (3.2.2). On the other hand, the absolute value of the determinant $\det(\langle Tx_i, b_k\rangle)$ represents the enlargement of the volume $\mathrm{vol}_n(U_2^n)$ of the n-dimensional euclidean unit ball U_2^n under the linear transformation $S: l_2^n \to l_2^n$ defined by

$$\eta_k = \sum_{i=1}^{n} \langle Tx_i, b_k\rangle \xi_i, \quad 1 \leqslant k \leqslant n, \tag{3.2.4}$$

for the n-tuples $(\xi_1, \xi_2, \ldots, \xi_n)$ and $(\eta_1, \eta_2, \ldots, \eta_n)$ of l_2^n, namely

$$|\det(\langle Tx_i, b_k\rangle)| = \frac{\mathrm{vol}_n S(U_2^n)}{\mathrm{vol}_n(U_2^n)}. \tag{3.2.5}$$

In view of (3.2.5) we remind the reader of the consideration of volume ratios in section 1.2. As a consequence of (1.2.3) and (1.2.4) we obtain

$$|\det(\langle Tx_i, b_k\rangle)|^{1/n} \leqslant g_n(S(U_2^n)) = g_n(S)$$

and therefore

$$\left(\prod_{k=1}^{n} d_k(T)\right)^{1/n} \leqslant (1+\delta)g_n(S).$$

It remains to estimate $g_n(S)$ from above by $g_n(T)$. For this purpose we construct a factorization of S with T as a factor. We introduce the operators

$$Az = \sum_{i=1}^{n} \langle z, e_i\rangle x_i$$

from l_2^n into E and

$$By = \sum_{i=1}^{n} \langle y, b_i\rangle e_i$$

from F into l_2^n and check that

$$BTAz = \sum_{k=1}^{n} \left(\sum_{i=1}^{n} \langle Tx_i, b_k\rangle \langle z, e_i\rangle\right) e_k \quad \text{for } z \in l_2^n$$

which amounts to

$$S = BTA$$

(see (3.2.4)). Now the simplified versions of the multiplicativity (M2)(a) and

(M2)(b) of the entropy moduli yield

$$g_n(S) \leqslant \|B\| g_n(T) \|A\|.$$

The calculations

$$\|A\| = \sup\left\{ \left\| \sum_{i=1}^{n} \xi_i x_i \right\| : \left(\sum_{i=1}^{n} |\xi_i|^2 \right)^{1/2} \leqslant 1 \right\}$$

$$\leqslant \sup\left\{ \left(\sum_{i=1}^{n} |\xi_i|^2 \right)^{1/2} \cdot \left(\sum_{i=1}^{n} \|x_i\|^2 \right)^{1/2} : \left(\sum_{i=1}^{n} |\xi_i|^2 \right)^{1/2} \leqslant 1 \right\}$$

$$\leqslant \sqrt{n}$$

and

$$\|B\| = \sup\left\{ \left\| \sum_{i=1}^{n} \langle y, b_i \rangle e_i \right\| : \|y\| \leqslant 1 \right\}$$

$$= \sup\left\{ \left(\sum_{i=1}^{n} |\langle y, b_i \rangle|^2 \right)^{1/2} : \|y\| \leqslant 1 \right\} \leqslant \left(\sum_{i=1}^{n} \|b_i\|^2 \right)^{1/2} \leqslant \sqrt{n}$$

finally enable us to conclude that

$$\left(\prod_{k=1}^{n} d_k(T) \right)^{1/n} \leqslant (1+\delta) g_n(S) \leqslant (1+\delta)\|B\| g_n(T) \|A\| \leqslant (1+\delta) n g_n(T).$$

Since $\delta > 0$ can be chosen arbitrarily small we get the desired result (3.2.3).
∎

Theorem 3.2.1 remains true if the Kolmogorov numbers $d_k(T)$ are replaced by the Gelfand numbers $c_k(T)$. This is due to a lemma which, in analogy to Lemma 3.2.1, gives an estimate for the Gelfand numbers $c_k(T)$ of an arbitrary operator $T: E \to F$ by the diagonal elements of a triangular matrix derived from T, in this situation an upper triangular matrix.

Lemma 3.2.2. (see Carl and Pietsch 1978; Pietsch 1987). *Let $T \in L(E,F)$ be an operator between arbitrary Banach spaces E and F. Given $\delta > 0$ there exist elements $x_1, x_2, \ldots \in E$ and functionals $b_1, b_2, \ldots \in F'$ such that $\|x_i\| \leqslant 1$, $\|b_k\| \leqslant 1$,*

$$\langle T x_i, b_k \rangle = 0 \quad \text{for } i > k, \tag{3.2.6}$$

and

$$c_k(T) \leqslant (1+\delta)|\langle T x_k, b_k \rangle| \quad \text{for } k = 1, 2, 3, \ldots \tag{3.2.7}$$

Proof. The fact that $c_1(T) = \|T\| = \sup_{\|x\| \leqslant 1} \|Tx\|$ again serves as the beginning of an inductive proof. We continue by assuming that we have already determined $x_1, x_2, \ldots, x_{n-1} \in E$ and $b_1, b_2, \ldots, b_{n-1} \in F'$ with the properties (3.2.6) and (3.2.7) and consider the subspace

$$M = \{ x \in E : \langle Tx, b_k \rangle = 0 \text{ for } 1 \leqslant k \leqslant n-1 \}$$

of E. Since codim $(M) < n$ we have

$$c_n(T) \leqslant \| TI_M^E \|$$

by Proposition 2.3.2 and, accordingly, can find an element $x_n \in M$ with $\|x_n\| \leqslant 1$ and a functional $b_n \in F'$ with $\|b_n\| \leqslant 1$ such that

$$c_n(T) \leqslant (1 + \delta)\| Tx_n \| = (1 + \delta)|\langle Tx_n, b_n \rangle|.$$

Owing to $x_n \in M$ the equations

$$\langle Tx_n, b_k \rangle = 0 \qquad (3.2.6)'$$

are satisfied for $k = 1, 2, \ldots, n - 1$. This finishes the proof of the assertion. ∎

Theorem 3.2.2. (see Carl 1982; Carl 1985). *Let $T \in L(E, F)$ be an operator between arbitrary Banach spaces E and F. Then*

$$\left(\prod_{k=1}^{n} c_k(T) \right)^{1/n} \leqslant n g_n(T) \quad \text{for } n = 1, 2, 3, \ldots \qquad (3.2.8)$$

Proof. Employing Lemma 3.2.2 we can estimate the product $\prod_{k=1}^{n} c_k(T)$ from above by

$$\prod_{k=1}^{n} c_k(T) \leqslant (1 + \delta)^n \prod_{k=1}^{n} |\langle Tx_k, b_k \rangle| = (1 + \delta)^n |\det \langle Tx_i, b_i \rangle|$$

and then proceed in the same way as for the proof of (3.2.3). ∎

So far the operator T has been tacitly assumed to act between real Banach spaces. The final results (3.2.3) and (3.2.8) are not, however, affected by the change from the real to the complex case, if we take note of the modified definition of the entropy moduli $g_n(T)$ in the complex case (see $(1.4.1)'$).

For the geometric means of the approximation numbers, the corresponding universal estimate from above by the entropy moduli is weaker, namely

$$\left(\prod_{k=1}^{n} a_k(T) \right)^{1/n} \leqslant \sqrt{2} \cdot n^{3/2} g_n(T) \quad \text{for } n = 1, 2, 3, \ldots \qquad (3.2.9)$$

This is due to the factor $(2n)^{1/2}$ in the estimates (2.4.14) or (2.4.32) for $a_n(T)$ by the Gelfand or the Kolmogorov numbers, respectively.

3.3. Geometrical parameters

The inequalities of Bernstein type derived in section 3.1, in a similar way to the inequalities of Jackson type derived in section 3.2, have a universal character in so far as they reflect general relations between certain approximation and entropy quantities irrespective of the particular geometrical properties of the underlying Banach spaces. They have been adapted to the classification scheme of the Lorentz sequence spaces in Theorem 3.1.3. But in the case where the approximation numbers $a_n(T)$ form a rapidly decreasing sequence, which means $(a_n(T)) \in l_{p,q}$ for all p and q, we only obtain the information that the sequence $e_n(T)$ of the dyadic entropy numbers is also rapidly decreasing. The asymptotic behaviour of the sequence $e_n(T)$ may, however, differ in detail from the asymptotic behaviour of the sequence $a_n(T)$ as we demonstrated in section 1.5 when discussing the entropy behaviour of the diagonal operator $D:l_p \to l_p$ generated by the sequence $\sigma_n = 2^{-n}$. We recall that in this situation

$$a_n(D) = 2^{-n}$$

(see (2.1.8)), while

$$\tfrac{1}{2} \cdot 2^{-\sqrt{2n}} \leqslant e_{n+1}(D) \leqslant 3 \cdot \sqrt{2 \cdot 2^{-\sqrt{2n}}}. \tag{1.5.21}$$

The result (1.5.21) can be regained, at least approximately, by the general methods to be developed in this section. The inclusion of the Lorentz approximation ideals in the Lorentz entropy ideals (see Theorem 3.1.3) may be reproduced in the weaker form

$$L_{s,\infty}^{(a)} \subseteq L_{r,\infty}^{(e)} \quad \text{for } r > s \tag{3.3.1}$$

within the new framework as well. The main point of the new methods, however, is that they take account of certain geometrical parameters of the underlying Banach spaces as mentioned in the title of this section. For operators T acting between Hilbert spaces the corresponding estimates will enable us to describe the asymptotic behaviour of the entropy moduli $g_k(T)$ and of the entropy numbers $\varepsilon_n(T)$ by using the geometric means $(\prod_{i=1}^{k} a_i(T))^{1/k}$ of the approximation numbers $a_i(T)$ (see section 3.4).

The definition of the geometrical parameters in question is based on *the Banach–Mazur distance*

$$d(E,F) = \inf \{ \| T \| \, \| T^{-1} \| : T \in L(E,F) \text{ is an isomorphism} \}$$

of two isomorphic Banach spaces E and F. We put

$$\delta_n(E) = \sup \{ d(M, l_2^m) : M \subseteq E, m = \dim(M) \leqslant n \} \tag{3.3.2}$$

and

$$\delta^{(n)}(E) = \sup \{ d(E/M, l_2^m) : M \subseteq E, m = \operatorname{codim}(M) \leqslant n \}, \tag{3.3.3}$$

calling $\delta_n(E)$ *the nth local injective distance of E to l_2* and $\delta^{(n)}(E)$ *the nth local surjective distance of E to l_2.* If $E = H$ itself is a Hilbert space, we obviously have

$$\delta_n(H) = 1 \quad \text{and} \quad \delta^{(n)}(H) = 1 \quad \text{for } n = 1, 2, 3, \dots \tag{3.3.4}$$

In the case of an arbitrary Banach space E we may conclude that

$$\delta_n(E) \leqslant \sqrt{n} \text{ and } \delta^{(n)}(E) \leqslant \sqrt{n} \quad \text{for } n = 1, 2, 3, \ldots \tag{3.3.5}$$

by Lemma 2.4.1.

We first consider finite rank operators.

Lemma 3.3.1. *Let E and F be arbitrary Banach spaces and $A \in L(E, F)$ an operator with $\operatorname{rank}(A) \leqslant m$. Then*

$$\varepsilon_n(A) \leqslant 6\delta_m(F)\delta^{(m)}(E) \cdot \sup_{1 \leqslant k \leqslant m} n^{-1/k} \left(\prod_{i=1}^{k} a_i(A) \right)^{1/k} \tag{3.3.6}$$

for $n = 1, 2, 3, \ldots$ (real version).

Proof. We start with the canonical factorization

$$A = I A_0 Q \tag{3.3.7}$$

of A over the quotient space E/N of E with respect to the null space $N = N(A)$ of A and the range $M = R(A)$ of A. Because $\operatorname{rank}(A) \leqslant m$ we have $\dim(E/N) = \dim(M) = r \leqslant m$. Given $\varepsilon > 0$ let X and Y be isomorphic mappings from E/N and M, respectively, onto l_2^r so that

$$\|X^{-1}\| \|X\| \leqslant d(E/N, l_2^r) + \varepsilon \tag{3.3.8}$$

and

$$\|Y^{-1}\| \|Y\| \leqslant d(M, l_2^r) + \varepsilon. \tag{3.3.9}$$

Writing

$$A_0 = Y^{-1}(Y A_0 X^{-1})X = Y^{-1}SX \tag{3.3.10}$$

with

$$S = Y A_0 X^{-1} \tag{3.3.11}$$

we may reduce the problem of estimating $\varepsilon_n(A)$ to the problem of estimating the nth entropy number $\varepsilon_n(S)$ of an operator $S: l_2^r \to l_2^r$ acting in an r-dimensional euclidean space. To this operator we may apply Schmidt's representation formula (2.1.9) and the corresponding product formulas

$$S = UDV^* \tag{2.1.10'}$$

and

$$D = U^*SV. \tag{2.1.11'}$$

Here D is a diagonal operator in l_2^r generated by a non-increasing sequence $\sigma_1 \geqslant \sigma_2 \geqslant \cdots \geqslant \sigma_r > 0$ of positive numbers, and U, U^*, V, and V^* are isometric mappings in l_2^r. Therefore we have

$$\varepsilon_n(S) = \varepsilon_n(D). \tag{3.3.12}$$

An estimate for $\varepsilon_n(D)$, however, is given by Proposition 1.3.2. Indeed, formula (1.3.16) tells us that

$$\varepsilon_n(D) \leqslant 6 \cdot \sup_{1 \leqslant k \leqslant m} n^{-1/k} \left(\prod_{i=1}^{k} \sigma_i \right)^{1/k} \tag{3.3.13}$$

since $\sigma_i = 0$ for $i > m \geqslant r$. But as we have seen in chapter 2 the various approximation quantities of a diagonal operator $D: l_p \to l_p$ generated by a non-increasing sequence $\sigma_i \geqslant 0$ coincide with the coefficients σ_i, in particular

$$t_i(D) = a_i(D) = \sigma_i. \tag{2.6.5}$$

Furthermore, referring to $(2.1.10)'$ and $(2.1.11)'$ we can state that

$$t_i(S) = t_i(D) \tag{3.3.14}$$

analogously to (3.3.12). By using these results (3.3.13) may be written as

$$\varepsilon_n(S) \leqslant 6 \cdot \sup_{1 \leqslant k \leqslant m} n^{-1/k} \left(\prod_{i=1}^{k} t_i(S) \right)^{1/k}. \tag{3.3.15}$$

Now using (3.3.10) and (3.3.11) we can change (3.3.15) into a corresponding estimate for $\varepsilon_n(A_0)$, that is

$$\varepsilon_n(A_0) \leqslant 6 \| Y^{-1} \| \, \| Y \| \, \| X \| \, \| X^{-1} \| \sup_{1 \leqslant k \leqslant m} n^{-1/k} \left(\prod_{i=1}^{k} t_i(A_0) \right)^{1/k}. \tag{3.3.16}$$

Finally we change from A_0 to A which is possible because of the relation (3.3.7) between A and A_0. Indeed, (3.3.7) implies that

$$\varepsilon_n(A) \leqslant \varepsilon_n(A_0),$$

and

$$t_i(A) = t_i(A_0)$$

for the symmetrized approximation numbers are surjective and injective. Thus we obtain

$$\varepsilon_n(A) \leqslant 6 \| Y^{-1} \| \, \| Y \| \, \| X \| \, \| X^{-1} \| \sup_{1 \leqslant k \leqslant m} n^{-1/k} \left(\prod_{i=1}^{k} t_i(A) \right)^{1/k}$$

from (3.3.16). Estimating the products $\| X \| \, \| X^{-1} \|$ and $\| Y^{-1} \| \, \| Y \|$ by using (3.3.8) and (3.3.9), and taking account of the definitions (3.3.2) and (3.3.3) for $\delta_n(E)$ and $\delta^{(n)}(E)$, respectively, we arrive at

$$\varepsilon_n(A) \leqslant 6(\delta_m(F) + \varepsilon)(\delta^{(m)}(E) + \varepsilon) \sup_{1 \leqslant k \leqslant m} n^{-1/k} \left(\prod_{i=1}^{k} t_i(A) \right)^{1/k}. \tag{3.3.17}$$

Since $t_i(A) \leqslant a_i(A)$ (see (2.6.1)) we can replace the symmetrized approximation numbers by the usual approximation numbers on the right-hand side of (3.3.17). The proof of (3.3.6) is completed by noting that $\varepsilon > 0$ can be chosen arbitrarily small. ∎

In the case of an arbitrary operator $T:E \to F$ we take an operator $A:E \to F$ with rank $(A) \leqslant m$ and then employ the estimate (3.3.6) for $\varepsilon_n(A)$. The corresponding estimate for $\varepsilon_n(T)$ obtained in this way then of course contains the $(m+1)$th approximation number $a_{m+1}(T)$ of T as an additive term on the right-hand side. Besides that, the factor 6 increases.

Theorem 3.3.1. *Let $T \in L(E, F)$ be an operator between arbitrary Banach spaces E and F. Then*

$$\varepsilon_n(T) \leqslant a_{m+1}(T) + 12\delta_m(F)\delta^{(m)}(E) \cdot \sup_{1 \leqslant k \leqslant m} n^{-1/k} \left(\prod_{i=1}^{k} a_i(T) \right)^{1/k} \qquad (3.3.18)$$

for $n, m = 1, 2, 3, \ldots$ (real version).

Proof. Given $\varepsilon > 0$ we choose an operator $A \in L(E, F)$ with rank $(A) \leqslant m$ such that

$$\| T - A \| \leqslant (1 + \varepsilon)a_{m+1}(T).$$

Then we estimate the entropy number $\varepsilon_n(T)$ of T from above by the entropy number $\varepsilon_n(A)$ of A and the approximation number $a_{m+1}(T)$ of T, namely

$$\varepsilon_n(T) \leqslant \| T - A \| + \varepsilon_n(A) \leqslant (1 + \varepsilon)a_{m+1}(T) + \varepsilon_n(A).$$

Now we can apply the inequality (3.3.6) to $\varepsilon_n(A)$. However, instead of the geometric mean $(\prod_{i=1}^{k} a_i(A))^{1/k}$ of the $a_i(A)$ we would like to have the geometric mean of the $a_i(T)$. For the pupose of eliminating the $a_i(A)$ we observe that

$$a_i(A) \leqslant \| A - T \| + a_i(T) \leqslant (1 + \varepsilon)a_{m+1}(T) + a_i(T)$$

$$\leqslant (2 + \varepsilon)a_i(T) \quad \text{for } i \leqslant k \leqslant m. \qquad (3.3.19)$$

Hence it follows that

$$\varepsilon_n(T) \leqslant (1 + \varepsilon)a_{m+1}(T) + 6(2 + \varepsilon)\delta_m(F)\delta^{(m)}(E) \cdot \sup_{1 \leqslant k \leqslant m} n^{-1/k} \left(\prod_{i=1}^{k} a_i(T) \right)^{1/k}$$

The assertion (3.3.18) results from letting $\varepsilon \to 0$. ∎

If we want to compare the nth entropy number $\varepsilon_n(T)$ with the first n approximation numbers $a_1(T), a_2(T), \ldots, a_n(T)$ we have to put $m = n$ in formula (3.3.18). Then the corresponding estimate can be given a more comprehensive form since

$$a_{n+1}(T) \leqslant n^{1/n} \cdot n^{-1/n} \left(\prod_{i=1}^{n} a_i(T) \right)^{1/n} \leqslant 2 \cdot \sup_{1 \leqslant k \leqslant n} n^{-1/k} \left(\prod_{i=1}^{k} a_i(T) \right)^{1/k}.$$

Because

$$\delta_n(F) \geqslant 1 \quad \text{and} \quad \delta^{(n)}(E) \geqslant 1 \qquad (3.3.20)$$

is always true the right-hand side of (3.3.18) can be summarized to

$$\varepsilon_n(T) \leqslant 14 \delta_n(F) \delta^{(n)}(E) \cdot \sup_{1 \leqslant k \leqslant n} n^{-1/k} \left(\prod_{i=1}^{k} a_i(T) \right)^{1/k}. \qquad (3.3.21)$$

If we disregard the particular geometrical properties of the Banach spaces E and F by employing (3.3.5), the inequality (3.3.18) changes into the *universal estimate*

$$\varepsilon_n(T) \leqslant a_{m+1}(T) + 12m \cdot \sup_{1 \leqslant k \leqslant m} n^{-1/k} \left(\prod_{i=1}^{k} a_i(T) \right)^{1/k} \qquad \text{for } n, m = 1, 2, 3, \ldots$$

$$(3.3.22)$$

With the dyadic entropy numbers $\varepsilon_{2^n}(T) = e_{n+1}(T)$ instead of the original ones this means

$$e_{n+1}(T) \leqslant a_{m+1}(T) + 12m \cdot \sup_{1 \leqslant k \leqslant m} 2^{-n/k} \left(\prod_{i=1}^{k} a_i(T) \right)^{1/k} \qquad \text{for } n, m = 1, 2, 3, \ldots$$

$$(3.3.23)$$

Though (3.3.21) at first glance seems to be clearer than (3.3.18), (3.3.22) and (3.3.23), these other versions provide more precise information in special circumstances. To demonstrate this, we derive the two approximative results mentioned at the beginning of this section.

First we consider an operator T between arbitrary Banach spaces E and F such that

$$a_n(T) \leqslant C \cdot 2^{-n} \quad \text{for } n = 1, 2, 3, \ldots$$

The estimate (3.3.23) then yields

$$e_{n+1}(T) \leqslant C \cdot 2^{-(m+1)} + 12C \cdot m \sup_{1 \leqslant k \leqslant m} 2^{-n/k} \left(\prod_{i=1}^{k} 2^{-i} \right)^{1/k}$$

for $n, m = 1, 2, 3, \ldots$. The expression

$$\tau_{n+1} = \sup_{1 \leqslant k < \infty} 2^{-n/k} \left(\prod_{i=1}^{k} 2^{-i} \right)^{1/k} = \frac{1}{\sqrt{2}} \sup_{1 \leqslant k < \infty} 2^{-(n/k + k/2)}$$

has already been estimated in section 1.5, namely by

$$\tau_{n+1} \leqslant \frac{1}{\sqrt{2}} 2^{-\sqrt{2n}}.$$

If we now choose $m = [\sqrt{2n}]$ so that $m \leqslant \sqrt{2n} < m+1$ we get

$$e_{n+1}(T) \leqslant C \cdot 2^{-\sqrt{2n}} + 12C \sqrt{2n} 2^{-\sqrt{2n}} \leqslant 13C \cdot \sqrt{2n} 2^{-\sqrt{2n}}.$$

This estimate is a little bit worse than the right-hand part of the original estimate (1.5.21) derived for the diagonal operator $D: l_p \to l_p$ with $\sigma_n = 2^{-n}$ as generating sequence. But on the basis of the inequality

$$\varepsilon \sqrt{2n} \log 2 \leqslant 2^{\varepsilon \sqrt{2n}}.$$

the annoying factor $\sqrt{2n}$ can be removed in favour of a slightly modified exponential term, giving

$$e_{n+1}(T) \leqslant \frac{13C}{\varepsilon \log 2} 2^{-(1-\varepsilon)\sqrt{2n}}, \quad n = 1, 2, 3, \ldots,$$

for fixed ε with $0 < \varepsilon < 1$. In contrast with this, the inequality (3.3.21) does not even provide the limit relation $\lim_{n \to \infty} e_n(T) = 0$ in the most unfavourable case

$$\delta_{2^n}(F) = 2^{n/2} \text{ and } \delta^{(2^n)}(E) = 2^{n/2}.$$

Now we turn to an operator $T : E \to F$ with the property

$$a_n(T) \leqslant C \cdot n^{-1/s}, \quad n = 1, 2, 3, \ldots, \tag{3.3.24}$$

for some $s \geqslant 0$. In order to show that T belongs to $L_{r,\infty}^{(e)}$ for any $r > s$ we again make use of (3.3.23) replacing the supremum $\sup_{1 \leqslant k \leqslant m} 2^{-n/k} (\prod_{i=1}^{k} a_i(T))^{1/k}$ by the upper bound $2^{-n/m} C$, so that we get

$$e_{n+1}(T) \leqslant C \cdot m^{-1/s} + 12C \cdot m \cdot 2^{-n/m}. \tag{3.3.25}$$

We put

$$\alpha = s/r$$

and fix β with

$$\alpha < \beta < 1.$$

For sufficiently large $n \geqslant n(\alpha, \beta)$ we can then determine a natural number m such that

$$n^\alpha \leqslant m \leqslant n^\beta.$$

If we replace m by the corresponding powers of n on the right-hand side of (3.3.25) we obtain

$$e_{n+1}(T) \leqslant C \cdot (n^{-1/r} + 12n^\beta 2^{-n^{1-\beta}}).$$

Since the sequence $\xi_n = n^\beta 2^{-n^{1-\beta}}$ is rapidly decreasing we may conclude that $T \in L_{r,\infty}^{(e)}(E, F)$.

For the sake of completeness we remark that for an operator T acting between complex Banach spaces E and F the supremum on the right-hand side of (3.3.6), and hence also the one on the right-hand sides of (3.3.18), (3.3.21), (3.3.22) and (3.3.23), have to be replaced by

$$\sup_{1 \leqslant k \leqslant m} n^{-1/2k} \left(\prod_{i=1}^{k} a_i(A) \right)^{1/k} \text{ and } \sup_{1 \leqslant k \leqslant m} 2^{-n/2k} \left(\prod_{i=1}^{k} a_i(T) \right)^{1/k},$$

respectively. This is due to the corresponding estimate (1.3.16)' for the entropy numbers $\varepsilon_n(D)$ of a diagonal operator D in a complex l_p space. As we have already observed, the estimate (1.4.7) for the entropy moduli

$g_k(D)$ of a diagonal operator $D:l_p \to l_p$, like all other estimates for entropy moduli proved so far, is not affected by the change from the real to the complex case. In particular, the following estimation of the entropy moduli $g_k(T)$ from above by the geometric means $(\prod_{i=1}^{k} a_i(T))^{1/k}$ of the approximation numbers has the same form in the real and complex cases. It can be regarded as an immediate generalization of the inequality (1.4.7) proved in Proposition 1.4.1 for diagonal operators $D:l_p \to l_p$.

Theorem 3.3.2. *Let* $T \in L(E, F)$ *be an operator between arbitrary Banach spaces* E *and* F. *Then*

$$g_k(T) \leqslant 10\delta_k(F)\delta^{(k)}(E)\left(\prod_{i=1}^{k} a_i(T) \right)^{1/k} \quad for \ k = 1, 2, 3, \ldots \quad (3.3.26)$$

Proof. In a similar way to the proof of Theorem 3.3.1 we fix $\varepsilon > 0$ and first choose an operator $A \in L(E, F)$ with rank $(A) = r < k$ such that

$$\| T - A \| \leqslant (1 + \varepsilon)a_k(T)$$

and, correspondingly,

$$\varepsilon_n(T) \leqslant (1 + \varepsilon)a_k(T) + \varepsilon_n(A). \quad (3.3.27)$$

To estimate $\varepsilon_n(A)$ from above we again use the canonical factorization

$$A = IA_0Q \quad (3.3.7)$$

of A which leads us to

$$\varepsilon_n(A) \leqslant \varepsilon_n(A_0),$$

as well as a factorization

$$A_0 = Y^{-1}SX \quad (3.3.10)$$

of A_0 over l_2^r with

$$S = YA_0X^{-1} \quad (3.3.11)$$

under the additional conditions (3.3.8) and (3.3.9). This yields

$$\varepsilon_n(A) \leqslant \| Y^{-1} \| \varepsilon_n(S) \| X \|. \quad (3.3.28)$$

According to (3.3.12) the nth entropy number of the operator $S:l_2^r \to l_2^r$ coincides with the nth entropy number of the diagonal operator $D:l_2^r \to l_2^r$ related to S by the formulas (2.1.10)' and (2.1.11)' in this section. Let us recall that the generating sequence $\sigma_1 \geqslant \sigma_2 \geqslant \cdots \geqslant \sigma_r > 0$ of D is given by

$$\sigma_i = a_i(D) = t_i(D), \quad 1 \leqslant i \leqslant r, \quad (2.6.5)$$

which implies

$$\sigma_i = a_i(S) = t_i(S), \quad 1 \leqslant i \leqslant r,$$

(see (3.3.14)). Since the symmetrized approximation numbers are both surjective and injective we obtain

$$t_i(S) \leqslant \| Y \| t_i(A_0) \| X^{-1} \| = \| Y \| t_i(A) \| X^{-1} \|$$

and thus

$$\sigma_i \leqslant \| Y \| a_i(A) \| X^{-1} \|.$$

Finally, the operator A can be eliminated in favour of T by using $a_i(A) \leqslant (2 + \varepsilon)a_i(T)$ (see (3.3.19)). Having obtained the inequalities

$$\sigma_i \leqslant (2 + \varepsilon) \| Y \| \ \| X^{-1} \| a_i(T), \quad 1 \leqslant i \leqslant r, \tag{3.3.29}$$

we change from the diagonal operator $D: l_2^r \to l_2^r$ generated by the sequence $\sigma_1 \geqslant \sigma_2 \geqslant \cdots \geqslant \sigma_r > 0$ to the diagonal operator $\tilde{D}: l_2^r \to l_2^r$ generated by the sequence

$$\tilde{\sigma}_i = (2 + \varepsilon) \| Y \| \ \| X^{-1} \| a_i(T), \quad 1 \leqslant i \leqslant r. \tag{3.3.30}$$

From (3.3.29) we have

$$D(U_2^r) \subseteq \tilde{D}(U_2^r),$$

which implies

$$\varepsilon_n(D) \leqslant \varepsilon_n(\tilde{D}). \tag{3.3.31}$$

Combining (3.3.28), (3.3.12) and (3.3.31) we get

$$\varepsilon_n(A) \leqslant \| Y^{-1} \| \ \| X \| \varepsilon_n(\tilde{D}). \tag{3.3.32}$$

We are now in a position to continue the estimate (3.3.27) for $\varepsilon_n(T)$. Without loss of generality we may assume $a_k(T) > 0$, since the assertion is trivial when $a_k(T) = 0$ (see section 2.1, (A4), and section 1.4, (M3)). We put

$$\rho = (2 + \varepsilon) \| Y \| \ \| X^{-1} \| a_k(T), \tag{3.3.33}$$

so that

$$\tilde{\sigma}_1 \geqslant \tilde{\sigma}_2 \geqslant \cdots \geqslant \tilde{\sigma}_r \geqslant \rho > 0 \tag{3.3.34}$$

is guaranteed. According to what has been carried out for the proof of Proposition 1.4.1 there is a system of elements $y_1, y_2, \ldots, y_N$ in $\tilde{D}(U_2^r)$ such that the balls $\{y_j + \rho U_2^r\}$ are pairwise disjoint for $1 \leqslant j \leqslant N$, and

$$\tilde{D}(U_2^r) \subseteq \bigcup_{j=1}^{N} \{y_j + 2\rho U_2^r\}. \tag{1.4.8'}$$

From (1.4.8)' it follows that

$$\varepsilon_N(\tilde{D}) \leqslant 2\rho. \tag{3.3.35}$$

On the other hand, the mutual disjointness of the balls $\{y_j + \rho U_2^r\}$ enables us to use the inclusion

$$\bigcup_{j=1}^{N} \{y_j + \rho U_2^r\} \subseteq \tilde{D}(U_2^r) + \rho U_2^r \tag{1.4.9'}$$

for a comparison of volumes in the r-dimensional euclidean space. The r-dimensional volume of the union $\bigcup_{j=1}^{N}\{y_j + \rho U_2^r\}$ is given by $N \cdot \rho^r \cdot \mathrm{vol}_r(U_2^r)$. Since

$$\tilde{D}(U_2^r) + \rho U_2^r \subseteq \tilde{D}(U_2^r) + \tilde{D}(U_2^r) = 2\tilde{D}(U_2^r)$$

by (3.3.34), the r-dimensional volume of the set $\tilde{D}(U_2') + \rho U_2'$ can be estimated from above by $2^r \cdot \tilde{\sigma}_1 \tilde{\sigma}_2 \cdots \tilde{\sigma}_r \cdot \mathrm{vol}_r(U_2')$. Hence we obtain

$$N \leqslant 2^r \prod_{i=1}^r \frac{\tilde{\sigma}_i}{\rho}. \tag{1.4.10}'$$

Because

$$\frac{\tilde{\sigma}_i}{\rho} = \frac{a_i(T)}{a_k(T)} \geqslant 1 \quad \text{for } 1 \leqslant i \leqslant k,$$

we have

$$N \leqslant 2^r \prod_{i=1}^r \frac{a_i(T)}{a_k(T)} \leqslant \left(\frac{2}{a_k(T)} \right)^k \prod_{i=1}^k a_i(T)$$

and thus we may conclude that

$$N^{1/k} \leqslant \frac{2}{a_k(T)} \left(\prod_{i=1}^k a_i(T) \right)^{1/k}. \tag{3.3.36}$$

Next we consider inequality (3.3.27) with N instead of n and obtain

$$\varepsilon_N(T) \leqslant (1 + \varepsilon) a_k(T) + 2 \| Y^{-1} \| \, \| X \| \rho \tag{3.3.37}$$

in view of (3.3.32) and (3.3.35). The two estimates (3.3.36) and (3.3.37) are the key to the desired estimate (3.3.26) of the entropy modulus $g_k(T)$, namely

$$g_k(T) \leqslant N^{1/k} \varepsilon_N(T)$$

$$\leqslant 2(1 + \varepsilon) \left(\prod_{i=1}^k a_i(T) \right)^{1/k} + 4 \| Y^{-1} \| \, \| X \| \left(\prod_{i=1}^k a_i(T) \right)^{1/k} \frac{\rho}{a_k(T)}.$$

Using the right-hand side of (3.3.33) as a substitute for ρ we arrive at

$$g_k(T) \leqslant 2(1 + \varepsilon) \left(\prod_{i=1}^k a_i(T) \right)^{1/k}$$

$$+ 4(2 + \varepsilon) \| Y^{-1} \| \, \| Y \| \cdot \| X \| \, \| X^{-1} \| \left(\prod_{i=1}^k a_i(T) \right)^{1/k}.$$

As stated above the isomorphisms $X : E/N(A) \to l_2^r$ and $Y : R(A) \to l_2^r$ are supposed to satisfy the conditions (3.3.8) and (3.3.9), respectively. Therefore the product $\| Y^{-1} \| \, \| Y \| \cdot \| X \| \, \| X^{-1} \|$ can again be estimated by

$$\| Y^{-1} \| \, \| Y \| \cdot \| X \| \, \| X^{-1} \| \leqslant (\delta_k(F) + \varepsilon)(\delta^{(k)}(E) + \varepsilon).$$

We can then take the limit $\varepsilon = 0$ on the right-hand side of the corresponding estimate for $g_k(T)$ which yields

$$g_k(T) \leqslant 2 \left(\prod_{i=1}^k a_i(T) \right)^{1/k} + 8\delta_k(F)\delta^{(k)}(E) \left(\prod_{i=1}^k a_i(T) \right)^{1/k}.$$

To complete the proof of (3.3.26) we replace $2(\prod_{i=1}^k a_i(T))^{1/k}$ by $2\delta_k(F)\delta^{(k)}(E)(\prod_{i=1}^k a_i(T))^{1/k}$ (see (3.3.20)) and then sum the two items on the right-hand side. ∎

In analogy to (3.3.23) we write down the universal estimate

$$g_k(T) \leqslant 10k \left(\prod_{i=1}^{k} a_i(T) \right)^{1/k}, \quad k = 1, 2, 3, \ldots, \tag{3.3.38}$$

which results from (3.3.26) if we disregard the particular geometrical properties of the Banach spaces E and F (see (3.3.5)).

3.4. The Hilbert space setting

For operators T acting between Hilbert spaces H and K the asymptotic behaviour of the entropy moduli $g_k(T)$ is reflected exactly by the asymptotic behaviour of the geometric means $(\prod_{i=1}^{k} a_i(T))^{1/k}$ of the approximation numbers $a_i(T)$. The essential part of the proof consists in an iterated application of the universal estimate

$$\left(\prod_{k=1}^{n} d_k(T) \right)^{1/n} \leqslant n g_n(T) \tag{3.2.3}$$

and a limit process which removes the factor n on the right-hand side. This procedure becomes possible by specific properties of operators in Hilbert spaces, particularly properties connected with the notion of the adjoint and the absolute value of an operator $T : H \to K$.

The *adjoint* $T^* : K \to H$ *of an operator* $T : H \to K$ is defined by

$$(Tx, y) = (x, T^*y) \quad \text{for } x \in H \text{ and } y \in K,$$

where $(\cdot, \cdot)$ stands for the scalar product both in H and K. In the case of real Hilbert spaces H and K the adjoint operator T^* coincides with the dual T' of T (see section 2.5). In the complex case however, we have to take account of the fact that the isometry C_H between H and its dual H' is conjugate-linear in the sense that

$$C_H(\lambda x) = \bar{\lambda} C_H(x) \quad \text{for all } x \in H.$$

Hence, the adjoint operator $T^* : K \to H$ of $T : H \to K$ is related to the dual $T' : K' \to H'$ by

$$T^* = C_H^{-1} T' C_K$$

(cf. Taylor and Lay 1980). To be more concrete, let us refer to a matrix operator $T = (\tau_{ik})$ of the complex l_2 space into itself. The adjoint T^* of T is the matrix operator $T^* = (\tau_{ik}{}^*)$ with the elements

$$\tau_{ik}{}^* = \overline{\tau_{ki}},$$

whereas the matrix elements τ'_{ik} of the dual T' are determined by

$$\tau'_{ik} = \tau_{ki}.$$

The operator $T^{**} : H \to K$, in the same way as the operator $T'' : H \to K$, coincides with the original operator $T : H \to K$ in the real as well as in the complex situation, as can easily be verified.

If we consider operators T mapping a Hilbert space H into itself we have the notions of a *self-adjoint operator* and a *positive operator*. An operator $T \in L(H, H)$ is called *self-adjoint*, if $T^* = T$ and *positive*, if $(Tx, x) \geqslant 0$ for all $x \in H$. In the case of a complex Hilbert space every positive operator is automatically self-adjoint. With the help of matrix operators in the real l_2 space, or simply in the real euclidean space l_2^n, one can demonstrate that this fails to be true in the real case.

The composition T^*T of $T: H \to K$ and its adjoint $T^*: K \to H$ is a positive and self-adjoint operator in H because $T^{**} = T$ and therefore

$$(T^*Tx, y) = (Tx, Ty) = (x, T^*Ty) \quad \text{for all } x, y \in H. \tag{3.4.1}$$

In particular (3.4.1) implies

$$(T^*Tx, x) = \|Tx\|^2 \geqslant 0 \quad \text{for all } x \in H. \tag{3.4.2}$$

On the other hand, the Cauchy–Schwarz inequality states that

$$|(T^*Tx, x)| \leqslant \|T^*Tx\| \, \|x\| \leqslant \|T^*T\| \, \|x\|^2. \tag{3.4.3}$$

Combining (3.4.2) and (3.4.3) we may conclude that

$$\|T\|^2 = \sup_{\|x\| \leqslant 1} \|Tx\|^2 \leqslant \|T^*T\| \leqslant \|T^*\| \, \|T\|$$

and thus get

$$\|T\| \leqslant \|T^*\|.$$

The converse inequality now follows from

$$\|T^*\| \leqslant \|T^{**}\| = \|T\|.$$

The final result is

$$\|T^*\| = \|T\|. \tag{3.4.4}$$

As a byproduct we have

$$\|T\|^2 = \|T^*T\| = \|TT^*\| \tag{3.4.5}$$

(cf. Taylor and Lay 1980).

Given a positive operator $T \in L(H, H)$ there always exists a positive operator $R \in L(H, H)$ such that $R^2 = T$. The operator R is uniquely determined and called *the square root of the operator T* (see Kantorowitsch and Akilow 1978; Riesz and Nagy 1982), the symbol $T^{1/2}$ being the usual notation for the square root. If T is self-adjoint, the square root $T^{1/2}$ is also self-adjoint. Since the composition T^*T of an arbitrary operator $T: H \to K$ and its adjoint $T^*: K \to H$ is always a positive and self-adjoint operator, the square root $(T^*T)^{1/2}$ makes sense and is a positive and self-adjoint operator. We write

$$|T| = (T^*T)^{1/2} \tag{3.4.6}$$

calling $|T|: H \to H$ *the absolute value of T*. In analogy to the polar representation of a complex number every operator $T: H \to K$ allows a *polar decomposition*

$$T = U|T| \tag{3.4.7}$$

made up of the absolute value $|T|:H \to H$ of T and an operator $U:H \to K$ which vanishes on the null space $N(T)$ of T and is an isometry on the orthogonal complement $N(T)^{\perp}$ (see Halmos 1967; Riesz and Nagy 1982).

A non-vanishing operator $U:H \to K$ such that the restriction to the orthogonal complement $N(U)^{\perp}$ of the null space $N(U)$ is an isometry is called a *partial isometry*. Obviously, any partial isometry U is of norm 1,

$$\|U\| = 1.$$

If $U:H \to K$ is a partial isometry, then the adjoint operator $U^*:K \to H$ is a partial isometry as well. The composition $P = U^*U$ of a partial isometry U and its adjoint U^* is a projection of H onto the orthogonal complement $N(U)^{\perp}$ of the null space $N(U)$ of U with

$$\|P\| = \|U^*\| \, \|U\| = 1$$

(cf. Halmos 1967).

The partial isometry U in the polar decomposition (3.4.7) of the operator $T:H \to K$ has the null space

$$N(U) = N(T).$$

Furthermore, since

$$|T|^2 = T^*T$$

and the operator $|T|$ is self-adjoint, equation (3.4.2) can be expressed as

$$(|T|^2 x, x) = \| \, |T|x \|^2 = \|Tx\|^2$$

and hence ensures that

$$N(T) = N(|T|).$$

A self-adjoint operator S, however, quite generally possesses the property that its null space $N(S)$ coincides with the orthogonal complement $R(S)^{\perp}$ of its range $R(S)$, that is

$$N(S) = R(S)^{\perp}.$$

Thus we arrive at

$$N(U) = R(|T|)^{\perp}$$

and obtain

$$N(U)^{\perp} = R(|T|)^{\perp\perp}.$$

As stated above, the subspace $N(U)^{\perp} \subseteq H$ is the range of the projection $P = U^*U$ so that, in particular, U^*U is the identity map on $R(|T|) \subseteq R(|T|)^{\perp\perp}$. Hence, if we multiply (3.4.7) from the left by U^* we get

$$|T| = U^*T \tag{3.4.8}$$

as a counterpart to (3.4.7) (cf. Halmos 1967).

Now we are prepared to derive the estimates mentioned above for the entropy moduli $g_k(T)$ of arbitrary operators T acting between Hilbert spaces.

Theorem 3.4.1. *Let H and K be Hilbert spaces and $T \in L(H, K)$. Then*

$$\left(\prod_{i=1}^{k} a_i(T) \right)^{1/k} \leqslant g_k(T) \leqslant 10 \left(\prod_{i=1}^{k} a_i(T) \right)^{1/k} \quad \text{for } k = 1, 2, 3, \ldots \quad (3.4.9)$$

Proof. The right-hand inequality is an immediate consequence of Theorem 3.3.2 and of $\delta_k(K) = \delta^{(k)}(H) = 1$ for $k = 1, 2, 3, \ldots$ (see (3.3.4)). To prove the left-hand inequality we make use of the polar decomposition (3.4.7) of T and the converse formula (3.4.8) for $|T|$. Applying the weak form of multiplicativity (A3)(a) and (M2)(a) of the approximation numbers and the entropy moduli, respectively, to (3.4.7) and (3.4.8) we see that

$$a_n(T) = a_n(|T|) \quad (3.4.10)$$

and

$$g_n(T) = g_n(|T|). \quad (3.4.11)$$

Next we show that

$$a_n^2(T) \leqslant a_n(T^*T). \quad (3.4.12)$$

For this purpose we fix $\varepsilon > 0$ and then, by means of Proposition 2.4.2, determine an orthogonal projection $P \in L(H, H)$ with $\operatorname{rank}(P) < n$ such that

$$\| T^*T - T^*TP \| \leqslant (1 + \varepsilon) a_n(T^*T).$$

Since

$$a_n(T) \leqslant \| T - TP \| = \| T(I - P) \|$$

is always true, we may conclude that

$$a_n^2(T) \leqslant \| T(I - P) \|^2 = \| (I - P)T^*T(I - P) \|$$
$$\leqslant \| T^*T - T^*TP \| \leqslant (1 + \varepsilon) a_n(T^*T)$$

by using (3.4.5) with $S = T(I - P)$ in place of T. This proves (3.4.12) since $\varepsilon > 0$ can be chosen arbitrarily small. With $|T|$ instead of T we obtain

$$a_n^2(|T|) \leqslant a_n(|T|^2) \quad (3.4.13)$$

from (3.4.12) because $|T| : H \to H$ is a self-adjoint operator. Furthermore, we have

$$a_n(|T|^2) = d_n(|T|^2)$$

(see section 2.4, (2.2.14)). Therefore Theorem 3.2.1 yields

$$\left(\prod_{i=1}^{k} a_i(|T|^2) \right)^{1/k} \leqslant k g_k(|T|^2) \quad (3.4.14)$$

with $|T|^2$ instead of T. The left-hand side of the inequality (3.4.14) can be estimated from below by means of (3.4.13), while the right-hand side, owing to the multiplicativity (M2) of the entropy moduli, can be estimated by

$$g_k(|T|^2) \leqslant g_k^2(|T|)$$

from above. The result is

$$\left(\prod_{i=1}^{k} a_i(|T|) \right)^{2/k} \leqslant k g_k^2(|T|)$$

which, in view of (3.4.10) and (3.4.11), means

$$\left(\prod_{i=1}^{k} a_i(T) \right)^{1/k} \leqslant k^{1/2} g_k(T).$$

Iterating this procedure m times we arrive at

$$\left(\prod_{i=1}^{k} a_i(T) \right)^{1/k} \leqslant k^{2^{-m}} g_k(T).$$

The desired estimate follows by letting $m \to \infty$. ∎

On the basis of Theorem 3.4.1 we can also give an exact description of the asymptotic behaviour of the entropy numbers $\varepsilon_n(T)$ of a Hilbert space operator $T : H \to K$.

Theorem 3.4.2. *Let H and K be (real) Hilbert spaces and $T \in L(H, K)$. Then*

$$\sup_{1 \leqslant k < \infty} n^{-1/k} \left(\prod_{i=1}^{k} a_i(T) \right)^{1/k} \leqslant \varepsilon_n(T) \leqslant 14 \cdot \sup_{1 \leqslant k < \infty} n^{-1/k} \left(\prod_{i=1}^{k} a_i(T) \right)^{1/k}$$

(3.4.15)

for $n = 1, 2, 3, \ldots$

Proof. The right-hand inequality follows from (3.3.21) and $\delta_n(K) = \delta^{(n)}(H) = 1$ for $n = 1, 2, 3, \ldots$ To prove the left-hand inequality we refer to the definition

$$g_k(T) = \inf_{1 \leqslant n < \infty} n^{1/k} \varepsilon_n(T) \qquad (1.4.1)$$

of the kth entropy modulus $g_k(T)$ which implies

$$\sup_{1 \leqslant k < \infty} n^{-1/k} g_k(T) \leqslant \varepsilon_n(T).$$

Estimating $g_k(T)$ from below by using (3.4.9) we obtain the desired estimate for $\varepsilon_n(T)$ from below. ∎

In the case of an operator T acting between complex Hilbert spaces H

and K we have

$$\sup_{1\leqslant k<\infty} n^{-1/2k}\left(\prod_{i=1}^{k} a_i(T)\right)^{1/k} \leqslant \varepsilon_n(T) \leqslant 14\cdot \sup_{1\leqslant k<\infty} n^{-1/2k}\left(\prod_{i=1}^{k} a_i(T)\right)^{1/k}$$

$$(3.4.15)'$$

for $n=1,2,3,\ldots$ instead of (3.4.15), while (3.4.9) remains unchanged if we move from the real to the complex case.

The decisive point in the Hilbert space setting is obviously the estimate (3.4.9) of the $g_k(T)$ from below by the geometric means $(\prod_{i=1}^{k} a_i(T))^{1/k}$ of the approximation numbers. It implies

$$a_n(T) \leqslant g_n(T), \quad n=1,2,3,\ldots$$

On the other hand we have

$$g_n(T) \leqslant 2^{(n-1)/n}\varepsilon_{2^{n-1}}(T) \leqslant 2e_n(T), \quad n=1,2,3,\ldots,$$

by the definition of $g_n(T)$, and thus

$$a_n(T) \leqslant 2e_n(T) \tag{3.0.9}$$

as already mentioned in the introduction to this chapter. Hence the class $L_{p,q}^{(g)}(H,K)$ of operators $T:H\to K$ acting between Hilbert spaces H and K with $(g_n(T))\in l_{p,q}$ is included between the entropy class $L_{p,q}^{(e)}(H,K)$ and the approximation class $L_{p,q}^{(a)}(H,K)$, that is

$$L_{p,q}^{(e)}(H,K) \subseteq L_{p,q}^{(g)}(H,K) \subseteq L_{p,q}^{(a)}(H,K).$$

By Theorem 3.1.3, however, the approximation ideal $L_{p,q}^{(a)}$ is contained in the entropy ideal $L_{p,q}^{(e)}$. Thus we arrive at

$$L_{p,q}^{(e)}(H,K) = L_{p,q}^{(g)}(H,K) = L_{p,q}^{(a)}(H,K) \tag{3.4.16}$$

(see Carl 1981). The coincidence of the class of operators $L_{p,q}^{(g)}(H,K)$ with $L_{p,q}^{(e)}(H,K)$ or $L_{p,q}^{(a)}(H,K)$ in particular makes it clear that $L_{p,q}^{(g)}(H,K)$ is a linear space for Hilbert spaces H and K. Since the entropy moduli $g_n(T)$ are not additive (see section 1.4) we do not know if the class of operators $L_{p,q}^{(g)}(E,F)$ is a linear space for an arbitrary pair of Banach spaces E,F.

3.5. Powers of operators

When studying spectral properties of operators T mapping a Banach space E into itself one is presented with the powers T^k of T as is well known from the so-called Riesz theory (see chapter 4). However, it turns out that the powers T^k of an operator $T\in L(E,E)$ also reflect basic relations between approximation and entropy quantities which can be proved without reference to the spectral properties of T. On the contrary, in the next chapter we shall employ these relations for disclosing the dependence of the eigenvalues $\lambda_n(T)$ of an operator $T:E\to E$ on its entropy behaviour.

Proposition 3.5.1. *Let E be an arbitrary Banach space and $T \in L(E, E)$. Then the limits $\lim_{k \to \infty} g_n^{1/k}(T^k)$ and $\lim_{k \to \infty} a_n^{1/k}(T^k)$ exist for every $n = 1, 2, 3, \ldots$, and*

$$\lim_{k \to \infty} a_n^{1/k}(T^k) = \frac{\lim_{k \to \infty} g_n^{n/k}(T^k)}{\lim_{k \to \infty} g_{n-1}^{(n-1)/k}(T^k)} \quad \text{for } n = 2, 3, \ldots \quad (3.5.1)$$

if $\lim_{k \to \infty} g_{n-1}^{(n-1)k}(T^k) \neq 0$, while

$$\lim_{k \to \infty} a_n^{1/k}(T^k) = 0 \quad \text{if } \lim_{k \to \infty} g_{n-1}^{(n-1)/k}(T^k) = 0. \quad (3.5.1)'$$

Furthermore,

$$\lim_{k \to \infty} g_n^{1/k}(T^k) = \left(\prod_{i=1}^{n} \lim_{k \to \infty} a_i^{1/k}(T^k) \right)^{1/n} \quad \text{for } n = 1, 2, 3, \ldots \quad (3.5.2)$$

Proof. We first prove the existence of $\lim_{k \to \infty} g_n^{1/k}(T^k)$ by showing that this limit coincides with

$$\rho_n(T) = \inf_{1 \leqslant k < \infty} g_n^{1/k}(T^k) \quad \text{for } n = 1, 2, 3, \ldots$$

So let $\varepsilon > 0$ be an arbitrary positive number and m be chosen such that

$$g_n(T^m) < (\rho_n(T) + \varepsilon)^m.$$

Then for each natural number k there are integers p_k and q_k with $p_k \geqslant 0$ and $0 \leqslant q_k < m$ such that

$$k = p_k m + q_k.$$

This decomposition of k enables us to estimate $g_n(T^k)$ by using the multiplicativity (M2) of the entropy moduli g_n, namely

$$g_n(T^k) = g_n((T^m)^{p_k} T^{q_k}) \leqslant g_n^{p_k}(T^m) g_n^{q_k}(T).$$

It follows that

$$\rho_n(T) \leqslant g_n^{1/k}(T^k) < (\rho_n(T) + \varepsilon)^{p_k m/k} g_n^{q_k/k}(T) = (\rho_n(T) + \varepsilon) \left(\frac{g_n(T)}{\rho_n(T) + \varepsilon} \right)^{q_k/k}. \quad (3.5.3)$$

Since $0 \leqslant q_k < m$ for any k we have $\lim_{k \to \infty} (q_k/k) = 0$ and, correspondingly,

$$\left(\frac{g_n(T)}{\rho_n(T) + \varepsilon} \right)^{q_k/k} < \frac{\rho_n(T) + 2\varepsilon}{\rho_n(T) + \varepsilon} \quad (3.5.4)$$

for sufficiently large k, say $k > k_\varepsilon$. If we combine (3.5.3) and (3.5.4) we get

$$\rho_n(T) \leqslant g_n^{1/k}(T^k) < \rho_n(T) + 2\varepsilon \quad \text{for } k > k_\varepsilon.$$

This proves the existence of $\lim_{k \to \infty} g_n^{1/k}(T^k)$, namely

$$\lim_{k \to \infty} g_n^{1/k}(T^k) = \inf_{1 \leqslant k < \infty} g_n^{1/k}(T^k) \quad \text{for } n = 1, 2, 3, \ldots \quad (3.5.5)$$

To prove (3.5.1) we start from the two universal estimates (3.2.9) and (3.3.38) for $g_n(T)$ from below and above, respectively, by the geometric means $(\prod_{i=1}^{n} a_i(T))^{1/n}$ of the approximation numbers,

$$C_n \left(\prod_{i=1}^{n} a_i(T) \right)^{1/n} \leqslant g_n(T) \leqslant D_n \left(\prod_{i=1}^{n} a_i(T) \right)^{1/n} \qquad (3.5.6)$$

where

$$C_n = \frac{1}{\sqrt{2}} \cdot n^{-3/2} \text{ and } D_n = 10n \quad \text{for } n = 1, 2, 3, \ldots$$

If we replace T by T^k in (3.5.6), take the kth roots, and then let k tend to infinity, we get

$$\lim_{k \to \infty} g_n^{1/k}(T^k) = \lim_{k \to \infty} \left(\prod_{i=1}^{n} a_i^{1/k}(T^k) \right)^{1/n}, \quad n = 1, 2, 3, \ldots \qquad (3.5.7)$$

If $\lim_{k \to \infty} g_{n-1}^{(n-1)/k}(T^k) \neq 0$ the relation (3.5.1) now immediately follows from

$$\frac{\lim_{k \to \infty} g_n^{n/k}(T^k)}{\lim_{k \to \infty} g_{n-1}^{(n-1)/k}(T^k)} = \lim_{k \to \infty} \frac{\prod_{i=1}^{n} a_i^{1/k}(T^k)}{\prod_{i=1}^{n-1} a_i^{1/k}(T^k)} = \lim_{k \to \infty} a_n^{1/k}(T^k).$$

If $\lim_{k \to \infty} g_{n-1}^{(n-1)/k}(T^k) = 0$ we obtain

$$\lim_{k \to \infty} \left(\prod_{i=1}^{n-1} a_i^{1/k}(T^k) \right)^{1/(n-1)} = 0$$

from (3.5.7). Since

$$0 \leqslant a_n^{1/k}(T^k) \leqslant a_{n-1}^{1/k}(T^k) \leqslant \left(\prod_{i=1}^{n-1} a_i^{1/k}(T^k) \right)^{1/(n-1)}$$

this implies

$$\lim_{k \to \infty} a_n^{1/k}(T^k) = 0,$$

as claimed in (3.5.1)′.

Altogether the above considerations guarantee the existence of $\lim_{k \to \infty} a_n^{1/k}(T^k)$ for $n = 2, 3, \ldots$ The existence of $\lim_{k \to \infty} a_1^{1/k}(T^k)$ is guaranteed by (3.5.5) since

$$g_1(T^k) = \| T^k \| = a_1(T^k).$$

Hence the expression on the right-hand side of (3.5.2) actually makes sense for $n = 1, 2, 3, \ldots$ and can be calculated by

$$\left(\prod_{i=1}^{n} \lim_{k \to \infty} a_i^{1/k}(T^k) \right)^{1/n} = \lim_{k \to \infty} \left(\prod_{i=1}^{n} a_i^{1/k}(T^k) \right)^{1/n}.$$

The equation (3.5.2) itself then results from (3.5.7). ∎

Supplement. In contrast to $\lim_{k \to \infty} g_n^{1/k}(T^k)$ the corresponding limit $\lim_{k \to \infty} \varepsilon_n^{1/k}(T^k)$ formed with the individual entropy numbers ε_n does not depend on n because

$$\| T^k \| = g_1(T^k) \leqslant n\varepsilon_n(T^k) \leqslant n \| T^k \|$$

and, consequently,

$$\lim_{k \to \infty} \varepsilon_n^{1/k}(T^k) = \lim_{k \to \infty} \| T^k \|^{1/k} \quad \text{for } n = 1, 2, 3, \ldots \qquad (3.5.8)$$

Let us consider the diagonal operator

$$D(\xi_1, \xi_2, \ldots, \xi_n, \ldots) = (\sigma_1\xi_1, \sigma_2\xi_2, \ldots, \sigma_n\xi_n, \ldots), \qquad (1.3.15)$$

where $\sigma_1 \geqslant \sigma_2 \geqslant \cdots \geqslant \sigma_n \geqslant \cdots \geqslant 0$, of l_p into itself. The power D^k is obviously the diagonal operator generated by the sequence $\sigma_1^k \geqslant \sigma_2^k \geqslant \cdots \geqslant \sigma_n^k \geqslant \cdots \geqslant 0$. The norm $\| D^k \|$ of D^k is determined by

$$\| D^k \| = \sigma_1^k$$

so that

$$\lim_{k \to \infty} \| D^k \|^{1/k} = \sigma_1. \qquad (3.5.9)$$

On the other hand we have

$$a_n(D) = \sigma_n \quad \text{and} \quad a_n(D^k) = \sigma_n^k$$

(see (2.1.8)) and, correspondingly,

$$\lim_{k \to \infty} a_n^{1/k}(D^k) = \sigma_n \quad \text{for } n = 1, 2, 3, \ldots$$

Formula (3.5.2) then yields the result

$$\lim_{k \to \infty} g_n^{1/k}(D^k) = \left(\prod_{i=1}^n \sigma_i \right)^{1/n}$$

which, of course, can also be derived from the estimate (1.4.7) for $g_n(D)$. At any rate, $\lim_{k \to \infty} g_n^{1/k}(D^k)$ will depend on n unless $\sigma_1 = \sigma_2 = \cdots = \sigma_n = \cdots = \sigma$.

To get rid of the index n in $\lim_{k \to \infty} g_n^{1/k}(T^k)$ we take $\lim_{n \to \infty} \lim_{k \to \infty} g_n^{1/k}(T^k)$ and show that this iterated limit has the same value as the corresponding iterated limit taken over the terms $a_n^{1/k}(T^k)$.

Proposition 3.5.2. *Let E be an arbitrary Banach space and $T \in L(E, E)$. Then*

$$\lim_{n \to \infty} \lim_{k \to \infty} g_n^{1/k}(T^k) = \lim_{n \to \infty} \lim_{k \to \infty} a_n^{1/k}(T^k). \qquad (3.5.10)$$

Proof. Since the sequence of approximation numbers is monotonously decreasing, the sequence $\lim_{k \to \infty} a_n^{1/k}(T^k)$, $n = 1, 2, 3, \ldots$, is also monotonously decreasing. Therefore the limit $\lim_{n \to \infty} \lim_{k \to \infty} a_n^{1/k}(T^k)$ exists and coincides with the limit of the sequence of the geometric means $\lim_{n \to \infty} (\prod_{i=1}^n \lim_{k \to \infty} a_i^{1/k}(T^k))^{1/n}$. But by (3.5.2) we have

$$\lim_{n \to \infty} \left(\prod_{i=1}^n \lim_{k \to \infty} a_i^{1/k}(T^k) \right)^{1/n} = \lim_{n \to \infty} \lim_{k \to \infty} g_n^{1/k}(T^k).$$

Hence it follows that

$$\lim_{n \to \infty} \lim_{k \to \infty} a_n^{1/k}(T^k) = \lim_{n \to \infty} \lim_{k \to \infty} g_n^{1/k}(T^k). \qquad (3.5.10)$$

∎

As far as the meaning of the iterated limits $\lim_{n \to \infty} \lim_{k \to \infty} g_n^{1/k}(T^k)$ and $\lim_{n \to \infty} \lim_{k \to \infty} a_n^{1/k}(T^k)$ is concerned, for the moment we adopt neither a geometrical nor an analytical interpretation. Such an interpretation – an analytical rather than a geometrical one – will be given in the next chapter. In preparation for this we first reduce the above iterated limits to simple limits by introducing the *measures of non-compactness*

$$g(T) = \lim_{n \to \infty} g_n(T), \qquad (3.5.11)$$

$$e(T) = \lim_{n \to \infty} \varepsilon_n(T), \qquad (2.5.20)$$

$$d(T) = \lim_{n \to \infty} d_n(T), \qquad (3.5.12)$$

and the *measure of non-approximability*

$$a(T) = \lim_{n \to \infty} a_n(T) \qquad (3.5.13)$$

for operators T acting between arbitrary Banach spaces E and F. In the end, however, it turns out that the three measures of non-compactness $g(T)$, $e(T)$, and $d(T)$ coincide.

Proposition 3.5.3. *Let E and F be arbitrary Banach spaces and $T \in L(E, F)$. Then*

$$e(T) \leqslant a(T) \qquad (3.5.14)$$

and

$$e(T) = d(T) \qquad (3.5.15)$$

as well as

$$e(T) = g(T). \qquad (3.5.16)$$

Proof. By Lemma 2.5.2 we have

$$e(T) \leqslant a_n(T) \quad \text{for } n = 1, 2, 3, \dots \qquad (2.5.27)$$

and hence also

$$e(T) \leqslant a(T). \qquad (3.5.14)$$

Replacing T by TQ_E and realizing that $\varepsilon_n(TQ_E) = \varepsilon_n(T)$ (see (ES), section 1.3) and $a_n(TQ_E) = d_n(T)$ (see Theorem 2.2.1) we can read (3.5.14) as

$$e(T) \leqslant d(T).$$

The converse inequality can be verified by analysing the relations between the individual entropy numbers $\varepsilon_n(T)$ and the individual Kolmogorov numbers $d_n(T)$. Given $\varepsilon > \varepsilon_n(T)$ there exist elements $y_1, y_2, \ldots, y_n \in F$ such that

$$T(U_E) \subseteq \bigcup_{i=1}^{n} \{y_i + \varepsilon U_F\}.$$

These very elements y_i can also be used to span a subspace $N \subseteq F$. The above inclusion implies

$$T(U_E) \subset N + \varepsilon U_F$$

and hence $d_{n+1}(T) \leqslant \varepsilon_n(T)$ since $\dim(N) < n + 1$. Letting $n \to \infty$ we obtain $d(T) \leqslant e(T)$.

The representation formula

$$g(T) = \inf_{1 \leqslant n < \infty} \inf_{1 \leqslant k < \infty} k^{1/n} \varepsilon_k(T)$$

for the quantity $g(T)$ in the real case, or

$$g(T) = \inf_{1 \leqslant n < \infty} \inf_{1 \leqslant k < \infty} k^{1/2n} \varepsilon_k(T)$$

in the complex case, immediately makes it clear that

$$g(T) = \inf_{1 \leqslant k < \infty} \varepsilon_k(T) = e(T). \qquad \blacksquare$$

The equality $e(T) = d(T)$ in particular justifies the assertion of section 1.3 that the factor 2 sufficient for the injectivity of the outer entropy numbers $\varepsilon_n(T)$ in the weak sense (see (1.3.6)) cannot be reduced. Indeed, in section 2.5 we proved that

$$a_n(I_0) = 1, \quad n = 1, 2, 3, \ldots, \tag{2.5.7}$$

and

$$a_n(I_\infty) = \tfrac{1}{2}, \quad n = 2, 3, 4, \ldots, \tag{2.5.9}$$

for the identity maps $I_0 : l_1 \to c_0$ and $I_\infty : l_1 \to l_\infty$, respectively. Since l_1 possesses the metric lifting property we also have

$$d_n(I_0) = 1 \quad n = 1, 2, 3, \ldots,$$

and

$$d_n(I_\infty) = \tfrac{1}{2}, \quad n = 2, 3, 4, \ldots,$$

and, correspondingly,

$$d(I_0) = 1 \quad \text{and} \quad d(I_\infty) = \tfrac{1}{2}.$$

The result $e(T) = d(T)$ now yields

$$e(I_0) = 1 \quad \text{and} \quad e(I_\infty) = \tfrac{1}{2} \tag{3.5.17}$$

and thus makes it clear that the individual entropy numbers $\varepsilon_n(I_0)$ and $\varepsilon_n(I_\infty)$ of the operators I_0 and I_∞, connected by

$$I_\infty = J_0 I_0, \tag{2.5.14}$$

where J_0 is the natural embedding of c_0 into l_∞, cannot be related to each other by an inequality

$$\varepsilon_n(I_0) \leqslant \rho \varepsilon_n(J_0 I_0)$$

with a factor $\rho < 2$ independently of n.

Let us denote by

$$\beta(T) = g(T) = e(T) = d(T) \qquad (3.5.18)$$

the common value of the measures of non-compactness $g(T)$, $e(T)$, *and* $d(T)$. Obviously the quantity $\beta(T)$ fulfils *a strong multiplication law*

$$\beta(RS) \leqslant \beta(R)\beta(S) \qquad (3.5.19)$$

for $S \in L(E, Z)$ and $R \in L(Z, F)$. The same applies to the measure of non-approximability $a(T)$, namely

$$a(RS) \leqslant a(R)a(S) \qquad (3.5.20)$$

for $S \in L(E, Z)$ and $R \in L(Z, F)$. These strong multiplication laws are responsible for the validity of the equalities

$$\lim_{k \to \infty} \beta^{1/k}(T^k) = \inf_{1 \leqslant k < \infty} \beta^{1/k}(T^k) \qquad (3.5.21)$$

and

$$\lim_{k \to \infty} a^{1/k}(T^k) = \inf_{1 \leqslant k < \infty} a^{1/k}(T^k), \qquad (3.5.22)$$

for operators $T \in L(E, E)$, analogous to (3.5.5), as can easily be seen by recapitulating the proof of (3.5.5). We shall make immediate use of (3.5.21) to derive a formula equating the iterated limit $\lim_{n \to \infty} \lim_{k \to \infty} g_n{}^{1/k}(T^k)$ with the simple limit $\lim_{k \to \infty} \beta^{1/k}(T^k)$.

Proposition 3.5.3. *Let E be an arbitrary Banach space and* $T \in L(E, E)$. *Then*

$$\lim_{n \to \infty} \lim_{k \to \infty} g_n{}^{1/k}(T^k) = \lim_{k \to \infty} \beta^{1/k}(T^k). \qquad (3.5.23)$$

Proof. By (3.5.5) and in view of the monotonous decrease of the sequence of entropy moduli g_n we have

$$\lim_{n \to \infty} \lim_{k \to \infty} g_n{}^{1/k}(T^k) = \inf_{1 \leqslant n < \infty} \inf_{1 \leqslant k < \infty} g_n{}^{1/k}(T^k)$$

which can also be written as

$$\lim_{n \to \infty} \lim_{k \to \infty} g_n{}^{1/k}(T^k) = \inf_{1 \leqslant k < \infty} \inf_{1 \leqslant n < \infty} g_n{}^{1/k}(T^k) = \inf_{1 \leqslant k < \infty} \beta^{1/k}(T^k).$$

Applying (3.5.21) then yields the desired result (3.5.23). ∎

Remark. Since

$$\beta^{1/k}(T^k) = \lim_{n \to \infty} g_n{}^{1/k}(T^k), \qquad (3.5.24)$$

from definition (3.5.11) of the measure of non-compactness $g(T) = \beta(T)$, (3.5.23) amounts to the fact that the limit procedures with respect to k and n can be interchanged.

Next we connect the measure of non-compactness $\beta(T)$ with the measure of non-approximability $a(T)$.

Proposition 3.5.4. *Let E be an arbitrary Banach space and $T \in L(E, E)$. Then*

$$\lim_{k \to \infty} \beta^{1/k}(T^k) = \lim_{k \to \infty} a^{1/k}(T^k). \tag{3.5.25}$$

Proof. The inequality $\lim_{k \to \infty} \beta^{1/k}(T^k) \leqslant \lim_{k \to \infty} a^{1/k}(T^k)$ is an immediate consequence of $\beta(T^k) = e(T^k)$ and (3.5.14). For the proof of the converse inequality we remark that

$$\lim_{k \to \infty} a^{1/k}(T^k) \leqslant \inf_{1 \leqslant n < \infty} \lim_{k \to \infty} a_n^{1/k}(T^k) \leqslant \lim_{n \to \infty} \lim_{k \to \infty} a_n^{1/k}(T^k)$$

and then refer to (3.5.10) and (3.5.23). ∎

Remark. In analogy to (3.5.24) we have

$$a^{1/k}(T^k) = \lim_{n \to \infty} a_n^{1/k}(T^k). \tag{3.5.26}$$

Hence, if we calculate the iterated limit $\lim_{n \to \infty} \lim_{k \to \infty} a_n^{1/k}(T^k)$ by means of (3.5.10), (3.5.23) and (3.5.25), we end up with

$$\lim_{n \to \infty} \lim_{k \to \infty} a_n^{1/k}(T^k) = \lim_{k \to \infty} \lim_{n \to \infty} a_n^{1/k}(T^k). \tag{3.5.27}$$

A corresponding equality concerning the limits with respect to k and n of the terms $g_n^{1/k}(T^k)$ was the subject of the remark after Proposition 3.5.3, namely

$$\lim_{n \to \infty} \lim_{k \to \infty} g_n^{1/k}(T^k) = \lim_{k \to \infty} \lim_{n \to \infty} g_n^{1/k}(T^k). \tag{3.5.28}$$

Let us emphasize that the entropy terms $\varepsilon_n^{1/k}(T^k)$ do not allow an interchange of the limit procedures $k \to \infty$ and $n \to \infty$ since

$$\lim_{k \to \infty} \varepsilon_n^{1/k}(T^k) = \lim_{k \to \infty} \| T^k \|^{1/k} \tag{3.5.8}$$

independently of n, while

$$\lim_{n \to \infty} \varepsilon_n^{1/k}(T^k) = \beta^{1/k}(T^k)$$

and therefore

$$\lim_{k \to \infty} \lim_{n \to \infty} \varepsilon_n^{1/k}(T^k) = \lim_{k \to \infty} \beta^{1/k}(T^k) = \lim_{k \to \infty} a^{1/k}(T^k)$$

by Proposition 3.5.4. In the case of the diagonal operator (1.3.15) we have

$\lim_{k \to \infty} \| D^k \|^{1/k} = \sigma_1$ and in contrast with that

$$\lim_{k \to \infty} a^{1/k}(D^k) = \lim_{n \to \infty} \sigma_n. \qquad (3.5.29)$$

In the next chapter we shall first derive an analytical interpretation for the common value of the iterated limits $\lim_{n \to \infty} \lim_{k \to \infty} a_n^{1/k}(T^k)$ and $\lim_{n \to \infty} \lim_{k \to \infty} g_n^{1/k}(T^k)$ (see Proposition 4.1.1) and then connect the simple limits $\lim_{k \to \infty} a_n^{1/k}(T^k)$ and $\lim_{k \to \infty} g_n^{1/k}(T^k)$ with the eigenvalues of the operator T (see Theorem 4.3.1).

4

A refined Riesz theory

What nowadays is referred to as 'Riesz theory' or 'Riesz–Schauder theory' takes in the theory of operators $S: E \to E$ in a complex Banach space E which can be represented by

$$S = \lambda I_E - T \qquad (4.0.1)$$

with $\lambda \neq 0$ and a compact operator $T \in L(E, E)$. The investigation of such operators $S \in L(E, E)$ in the end serves to disclose the properties of compact operators T mapping a complex Banach space E into itself. In his pioneering paper 'Uber lineare Funktionalgleichungen' (1918) F. Riesz presented the foundation of the theory of operators $S: E \to E$ of the form (4.0.1) although he confined himself to the consideration of integral equations. His results were completed by Schauder in 1930 who proved corresponding statements for the dual operators. A short survey on the Riesz–Schauder theory from a modern point of view can be found in Caradus, Pfaffenberger and Yood (1974); König (1986) and Pietsch (1987). A full representation is given in Heuser (1975), Kantorowitsch and Akilow (1978), and Taylor and Lay (1980). We shall not go into details in this book, but rather complement that part of Riesz theory which deals with the spectral properties of the underlying operator T. The *refined Riesz theory* we are going to develop concerns the influence of entropy and approximation properties of an operator T on its spectral properties, in particular on the distribution of its eigenvalues. The main results along these lines were obtained by König (1977, 1979, and 1980), Carl and Triebel (1980), Carl (1981), and Makai and Zemánek (1982) (see section 4.2). König established a formula for the absolute value $|\lambda_n(T)|$ of the nth eigenvalue $\lambda_n(T)$ by the limit $\lim_{k \to \infty} a_n^{1/k}(T^k)$ formed with the nth approximation numbers $a_n(T^k)$ of the powers T^k, a matrix version of this representation formula having already been achieved by Yamamoto (1967). In contrast with that Carl developed an estimate for $|\lambda_n(T)|$ from above by the nth entropy modulus $g_n(T)$. This result was improved by Carl and Triebel in so far as the absolute value of the individual eigenvalue $\lambda_n(T)$ was replaced by the geometric mean $(\prod_{i=1}^{n} |\lambda_i(T)|)^{1/n}$. Finally, Makai and Zemánek gave a representation for the geometric mean $(\prod_{i=1}^{n} |\lambda_i(T)|)^{1/n}$

by the limit $\lim_{k \to \infty} g_n^{1/k}(T^k)$. Though some of the proofs in their original versions were carried out for compact operators only, they in fact do not require any additional assumptions about $T \in L(E, E)$. The crucial point is the way of selecting, arranging, and counting the eigenvalues, provided eigenvalues do exist at all. The preliminaries necessary for the 'refined Riesz theory' will be reviewed in the first section of this chapter.

4.1. Main aspects of classical Riesz theory

From now on E stands for a complex Banach space and $T: E \to E$ for an operator mapping E into itself. The *resolvent set* $\rho(T)$ of T is the subset of the set $\mathbb{C}$ of complex numbers defined by

$$\rho(T) = \{\lambda \in \mathbb{C} : \lambda I_E - T \text{ is invertible in } L(E, E)\}.$$

Note that the invertibility of $S_\lambda = \lambda I_E - T$ in $L(E, E)$ is granted by the convergence of the Neumann series $\sum_{k=0}^{\infty} T^k/\lambda^{k+1}$; if the series converges the inverse S_λ^{-1} is given by

$$S_\lambda^{-1} = \sum_{k=0}^{\infty} \frac{T^k}{\lambda^{k+1}}.$$

Certainly, the Neumann series converges for $|\lambda| > \lim_{k \to \infty} \|T^k\|^{1/k}$ and diverges for $|\lambda| < \lim_{k \to \infty} \|T^k\|^{1/k}$. Hence, the set

$$\sigma(T) = \mathbb{C} \setminus \rho(T),$$

the so-called *spectrum of the operator T*, is completely contained in the circle of radius

$$r(T) = \lim_{k \to \infty} \|T^k\|^{1/k} \tag{4.1.1}$$

with $\lambda = 0$ as centre. Basic facts from the theory of holomorphic functions of a complex variable guarantee that $\sigma(T)$ is not empty and, moreover,

$$r(T) = \sup_{\lambda \in \sigma(T)} |\lambda|. \tag{4.1.2}$$

The formula (4.1.2) for the radius $r(T)$ justifies the terminology *spectral radius*. Which other properties of the spectrum $\lambda(T)$, apart from (4.1.2), are worth mentioning? The resolvent set $\rho(T)$, as can easily be checked, is open. Therefore the spectrum $\sigma(T)$ is closed. All in all we can state that $\sigma(T)$ is a compact subset of the complex plane.

Among the elements of $\sigma(T)$ the eigenvalues λ of T are of particular interest. A value λ is called an *eigenvalue* of the operator T if there exists an element $x \neq 0$ in E with

$$Tx = \lambda x. \tag{4.1.3}$$

For an eigenvalue λ the operator $S_\lambda = \lambda I_E - T$ is obviously not invertible so that the eigenvalues of T in fact belong to the spectrum $\sigma(T)$ of T. In the case of a compact operator all elements $\lambda \neq 0$ of the spectrum are eigenvalues and the set of eigenvalues is at most countable. If $T: E \to E$ is a compact operator acting in an infinite-dimensional Banach space E, the value $\lambda = 0$ itself belongs to the spectrum of T, but $\lambda = 0$ need not be an eigenvalue of T. However, $\lambda = 0$ is the only point of accumulation of the set of eigenvalues of a compact operator T if T possesses infinitely many eigenvalues.

An element $x \neq 0$ in E satisfying (4.1.3) is called *an eigenvector of the eigenvalue λ*. An eigenvalue $\lambda \neq 0$ of a compact operator T has only finitely many linearly independent eigenvectors, which means that the kernel $N(S_\lambda)$ of the operator $S_\lambda = \lambda I_E - T$ is of finite dimension for $\lambda \neq 0$. This number dim $N(S_\lambda)$ is called *the geometric multiplicity of the eigenvalue λ*.

More significant than the geometric multiplicity is the so-called *algebraic multiplicity of an eigenvalue $\lambda \neq 0$* which is introduced via the *iterated kernels* $N((\lambda I_E - T)^m)$. Since

$$(\lambda I_E - T)^m = \lambda^m I_E - m\lambda^{m-1} T + \cdots + (-1)^m T^m$$

the power $(\lambda I_E - T)^m$ can again be represented as

$$(\lambda I_E - T)^m = \lambda^m I_E - \tilde{T},$$

where $\tilde{T} \in L(E, E)$ is a compact operator, provided T is compact itself. In analogy to the kernel $N(\lambda I_E - T)$, the kernel $N((\lambda I_E - T)^m)$ is also finite-dimensional for arbitrary m. Moreover, there is a smallest natural number m such that

$$N((\lambda I_E - T)^m) = N((\lambda I_E - T)^{m+1}).$$

This very natural number $m = d(T; \lambda)$ is called *the ascent of the operator* $S_\lambda = \lambda I_E - T$. The kernels $N(S_\lambda^m)$ do not increase any further for $m \geqslant d(T; \lambda)$, so we have

$$N((\lambda I_E - T)^{d(T;\lambda)}) = \bigcup_{m=1}^{\infty} N((\lambda I_E - T)^m).$$

The dimension

$$\dim \left(\bigcup_{m=1}^{\infty} N((\lambda I_E - T)^m) \right) = m(T; \lambda)$$

is defined as *the algebraic multiplicity of the eigenvalue $\lambda \neq 0$*. If T is a matrix operator mapping a finite-dimensional Banach space E into itself and $\lambda = \lambda_0$ is an eigenvalue of T, possibly even $\lambda_0 = 0$, then λ_0 is a zero of the characteristic polynomial $p(T; \lambda) = \det(\lambda I_E - T)$ and the algebraic multiplicity of λ_0 coincides with the multiplicity of λ_0 as a zero of $p(T; \lambda)$. This explains the adjective 'algebraic'. Since

$$N(\lambda I_E - T) \subseteq N((\lambda I_E - T)^m) \quad \text{for } m \geqslant 1$$

the algebraic multiplicity is larger than or equal to the geometric multiplicity.

As a counterpart to the notion of an eigenvector we have the notion of a principal vector. A *principal vector* is understood to be an element $x \neq 0$ with

$$(\lambda I_E - T)^m x = 0$$

for some positive integer m. Thus the principal vectors span the space

$$N_\infty(\lambda I_E - T) = \bigcup_{m=1}^{\infty} N((\lambda I_E - T)^m)$$

and the number of linearly independent principal vectors is the algebraic multiplicity $m(T; \lambda)$ of the eigenvalue λ.

All the kernels $N((\lambda I_E - T)^m)$ are invariant under T because

$$(\lambda I_E - T)^m T = T(\lambda I_E - T)^m.$$

The same argument makes it clear that the ranges $R((\lambda I_E - T)^m) = (\lambda I_E - T)^m(E)$ are also invariant subspaces of T where

$$R((\lambda I_E - T)^{m+1}) \subseteq R((\lambda I_E - T)^m).$$

Furthermore, if $d(T; \lambda)$ is the ascent of the operator $S_\lambda = \lambda I_E - T$ then $d(T; \lambda)$ simultaneously appears as the smallest natural number m with

$$(\lambda I_E - T)^m(E) = (\lambda I_E - T)^{m+1}(E),$$

and the spaces $(\lambda I_E - T)^m(E)$ do not decrease any further for $m \geqslant d(T; \lambda)$ so that

$$(\lambda I_E - T)^{d(T;\lambda)}(E) = \bigcap_{m=1}^{\infty} (\lambda I_E - T)^m(E).$$

For $\lambda \neq 0$ the two spaces $N_\infty(\lambda I_E - T)$ and

$$R_\infty(\lambda I_E - T) = \bigcap_{m=1}^{\infty} (\lambda I_E - T)^m(E)$$

span the whole space E, that is to say

$$E = N_\infty(\lambda I_E - T) \oplus R_\infty(\lambda I_E - T). \tag{4.1.4}$$

The projection $P_\lambda(T)$ of E onto $N_\infty(\lambda I_E - T)$ with

$$N(P_\lambda(T)) = R_\infty(\lambda I_E - T)$$

is called *the spectral projection of the eigenvalue* λ. Because the subspaces $N_\infty(\lambda I_E - T)$ and $R_\infty(\lambda I_E - T)$ are invariant under T the spectral projection $P_\lambda(T)$ commutes with T.

If we consider a finite set of distinct eigenvalues $\mu_1, \mu_2, \ldots, \mu_r$ we recognize that principal vectors $x_1, x_2, \ldots, x_r$ associated with these eigenvalues are linearly independent. Moreover, under the assumption

$\mu_i \neq 0$ for $1 \leqslant i \leqslant r$ the Banach space E has a decomposition

$$E = N_\infty(\mu_1 I_E - T) \oplus \cdots \oplus N_\infty(\mu_r I_E - T)$$

$$\oplus \bigcap_{i=1}^{r} R_\infty(\mu_i I_E - T) \tag{4.1.5}$$

which generalizes the decomposition (4.1.4) connected with a single eigenvalue $\lambda \neq 0$. Since

$$N_\infty(\mu_j I_E - T) \subseteq R_\infty(\mu_i I_E - T) \quad \text{for } i \neq j$$

we have

$$P_{\mu_i}(T)P_{\mu_j}(T) = 0 \quad \text{for } i \neq j$$

so that the sum

$$P = \sum_{i=1}^{r} P_{\mu_i}(T) \tag{4.1.6}$$

represents a projection of E onto $N_\infty(\mu_1 I_E - T) \oplus \cdots \oplus N_\infty(\mu_r I_E - T)$ with

$$N(P) = \bigcap_{i=1}^{r} R_\infty(\mu_i I_E - T).$$

Because

$$P_{\mu_i}(T)T = TP_{\mu_i}(T) \quad \text{for } 1 \leqslant i \leqslant r$$

this projection P also commutes with T.

So far we have been considering a compact operator $T: E \to E$. Following Zemanek (1983) and Makai and Zemanek (1982) we can extend these well-known results of Riesz theory for compact operators to arbitrary operators $T \in L(E, E)$ if we confine ourselves to an appropriate part of the spectrum $\sigma(T)$ of T. This part of the spectrum of T consists of those values λ for which $S_\lambda = \lambda I_E - T$ is 'almost invertible' in the following sense: the null space $N(S_\lambda)$ of S_λ is finite-dimensional and the range $R(S_\lambda)$ is of finite codimension in E.

An operator $S \in L(E, F)$ between arbitrary Banach spaces E and F satisfying these properties, that is

$$\dim(N(S)) < \infty \quad \text{and} \quad \text{codim}(R(S)) < \infty,$$

is called a *Fredholm operator*.

We now consider the quotient algebra $L(E, E)/K(E, E)$ of the operator algebra $L(E, E)$ of a Banach space E with respect to the two-sided and closed ideal $K(E, E)$ of compact operators, the so-called *Calkin algebra*. One can characterize Fredholm operators as those operators $S \in L(E, E)$ whose cosets $\bar{S}$ in $L(E, E)/K(E, E)$ are invertible. As for the operator $S_\lambda = \lambda I_E - T$ in $L(E, E)$, the invertibility of the coset $\bar{S}_\lambda = \lambda \bar{I}_E - \bar{T}$ in $L(E, E)/K(E, E)$ is granted by the convergence of the Neumann series

$\sum_{k=0}^{\infty} \bar{T}^k/\lambda^{k+1}$. Even in this modified situation the Neumann series converges for $|\lambda| > \lim_{k \to \infty} \| \bar{T}^k \|^{1/k}$ and diverges for $|\lambda| < \lim_{k \to \infty} \| \bar{T}^k \|^{1/k}$. The limit

$$r(\bar{T}) = \lim_{k \to \infty} \| \bar{T}^k \|^{1/k} \tag{4.1.7}$$

itself can be regarded as *the spectral radius of the element $\bar{T}$ in the Banach algebra $L(E,E)/K(E,E)$*. Indeed, introducing the *resolvent set*

$$\rho(\bar{T}) = \{\lambda \in \mathbb{C} : \lambda \bar{I}_E - \bar{T} \text{ is invertible in } L(E,E)/K(E,E)\}$$

of the coset $\bar{T}$ in $L(E,E)/K(E,E)$, and its complement

$$\sigma(\bar{T}) = \mathbb{C} \backslash \rho(\bar{T}),$$

the *spectrum* of $\bar{T}$, we can state that $\sigma(\bar{T})$ is completely contained in the circle of radius $r(\bar{T})$ with $\lambda = 0$ as centre. Moreover,

$$r(\bar{T}) = \sup_{\lambda \in \sigma(\bar{T})} |\lambda|$$

again turns out to be true. Since $\rho(T) \subseteq \rho(\bar{T})$ we have

$$\sigma(\bar{T}) \subseteq \sigma(T)$$

and, correspondingly,

$$r(\bar{T}) \leqslant r(T).$$

The set $\sigma(\bar{T})$ is called *the essential spectrum of T* and denoted by $\sigma_{\text{ess}}(T)$. In accordance with this the value

$$r_{\text{ess}}(T) = r(\bar{T})$$

is referred to as *the essential spectral radius of T*. Formula (4.1.7) expresses the essential spectral radius $r_{\text{ess}}(T)$ by means of the norm

$$\| \bar{T} \| = \inf \{ \| T - A \| : A \in K(E,E) \} \tag{4.1.8}$$

on the quotient algebra $L(E,E)/K(E,E)$. If we use the notation $\alpha(T) = \| \bar{T} \|$ and take into consideration that

$$\bar{T}^k = \overline{T^k}$$

we can rewrite the formula (4.1.7) as follows

$$r_{\text{ess}}(T) = \lim_{k \to \infty} \alpha^{1/k}(T^k). \tag{4.1.9}$$

An operator $T \in L(E,E)$ with $r_{\text{ess}}(T) = 0$ is called a *Riesz operator* (cf. Caradus, Pfaffenberger and Yood 1974; Pietsch 1978; König 1986; Pietsch 1987). Since $\alpha(T) = 0$ is a characteristic property of compact operators any compact operator $T: E \to E$ is a Riesz operator.

Obviously the expression

$$\alpha(T) = \inf \{ \| T - A \| : A \in K(E,F) \} \tag{4.1.10}$$

makes sense for operators $T \in L(E,F)$ acting between arbitrary Banach spaces E and F and hence, like the quantity $\beta(T)$ studied in section 3.5, is a measure of non-compactness. Let us compare $\alpha(T)$ with $\beta(T)$. The value $\alpha(T)$ represents the usual norm of the coset $\bar{T}$ of T in the quotient space $L(E,F)/K(E,F)$. The quantity $\beta(T)$ is uniquely defined on the cosets of $L(E,F)/K(E,F)$, too, for the additivity (E2) of the entropy numbers $\varepsilon_n(T)$ guarantees that

$$\beta(T) = \beta(T-A) \text{ for arbitrary } A \in K(E,F). \tag{4.1.11}$$

The measure of non-compactness $\beta(T)$ even represents a norm on the quotient space $L(E,F)/K(E,F)$. On the basis of (4.1.11) we may conclude that

$$\beta(T) \leqslant \varepsilon_n(T-A) \leqslant \|T-A\| \text{ for arbitrary } A \in K(E,F)$$

and thus

$$\beta(T) \leqslant \alpha(T). \tag{4.1.12}$$

However, it turns out that the two norms $\beta(T)$ and $\alpha(T)$ on $L(E,F)/K(E,F)$ are in general not equivalent (see Astala and Tylli 1987). Nevertheless for operators T mapping a Banach space E into itself $\beta(T)$ can be used instead of $\alpha(T)$ for calculating the essential spectral radius $r_{ess}(T)$ (see Zemánek 1983).

Proposition 4.1.1. *Let E be an arbitrary Banach space and $T \in L(E,E)$. Then*

$$r_{ess}(T) = \lim_{k \to \infty} \beta^{1/k}(T^k). \tag{4.1.13}$$

Proof. From (4.1.12) we have

$$\lim_{k \to \infty} \beta^{1/k}(T^k) \leqslant \lim_{k \to \infty} \alpha^{1/k}(T^k) = r_{ess}(T).$$

On the other hand, the measure of non-compactness $\alpha(T)$ is dominated by the measure of non-approximability $a(T)$, namely

$$\alpha(T) \leqslant \inf\{\|T-A\| : A \in L(E,E) \text{ with rank}(A) < \infty\} = a(T). \tag{4.1.14}$$

Therefore we have

$$\lim_{k \to \infty} \alpha^{1/k}(T^k) \leqslant \lim_{k \to \infty} a^{1/k}(T^k).$$

But Proposition 3.5.4 states that

$$\lim_{k \to \infty} a^{1/k}(T^k) = \lim_{k \to \infty} \beta^{1/k}(T^k). \tag{3.5.25}$$

This proves (4.1.13). ∎

Supplement. The multiplicativity of the entropy numbers implies

$$\beta(T^k) \leqslant \beta^k(T)$$

and hence

$$r_{ess}(T) \leqslant \beta(T) \leqslant \varepsilon_n(T) \text{ for } n = 1,2,3,\dots \tag{4.1.15}$$

Moreover, since

$$\beta(T) \leqslant \alpha(T) \qquad (4.1.12)$$

and

$$\alpha(T) \leqslant a(T), \qquad (4.1.14)$$

we also have

$$r_{\text{ess}}(T) \leqslant a(T) \leqslant a_n(T) \quad \text{for } n = 1, 2, 3, \ldots \qquad (4.1.16)$$

In view of (3.5.25) formula (4.1.13) involves various possibilities for representing the essential spectral radius $r_{\text{ess}}(T)$ of an operator $T \in L(E, E)$. We have

$$\lim_{k \to \infty} \beta^{1/k}(T^k) = \lim_{n \to \infty} \lim_{k \to \infty} g_n^{1/k}(T^k) \qquad (3.5.23)$$

by Proposition 3.5.3 and

$$\lim_{k \to \infty} a^{1/k}(T^k) = \lim_{n \to \infty} \lim_{k \to \infty} a_n^{1/k}(T^k)$$

from (3.5.26) and (3.5.27). Thus the common value of the iterated limits

$$\lim_{n \to \infty} \lim_{k \to \infty} g_n^{1/k}(T^k) = \lim_{n \to \infty} \lim_{k \to \infty} a_n^{1/k}(T^k), \qquad (3.5.10)$$

which first appeared in Proposition 3.5.2, is nothing other than the essential spectral radius $r_{\text{ess}}(T)$ of T. In section 4.3 we shall give analytical interpretations for the simple limits $\lim_{k \to \infty} a_n^{1/k}(T^k)$ and $\lim_{k \to \infty} g_n^{1/k}(T^k)$.

To illustrate how the formula

$$r_{\text{ess}}(T) = \lim_{k \to \infty} a^{1/k}(T^k) \qquad (4.1.17)$$

of $r_{\text{ess}}(T)$ in terms of the measure of non-approximability $a(T)$ works, we consider the diagonal operator

$$D(\xi_1, \xi_2, \ldots, \xi_n, \ldots) = (\sigma_1 \xi_1, \sigma_2 \xi_2, \ldots, \sigma_n \xi_n, \ldots)$$

with $\sigma_1 \geqslant \sigma_2 \geqslant \cdots \geqslant \sigma_n \geqslant \cdots \geqslant 0$ in l_p. Putting

$$\sigma_\infty = \lim_{n \to \infty} \sigma_n$$

we obtain

$$r_{\text{ess}}(D) = \sigma_\infty$$

by (3.5.29) and (4.1.17). In contrast with that we have

$$r(T) = \sigma_1$$

by (3.5.9) and (4.1.1). Hence $r_{\text{ess}}(D) < r(D)$ unless $\sigma_1 = \sigma_2 = \cdots = \sigma_n = \cdots = \sigma$.

The situation changes completely if D is combined with the shift operator S_+ in l_p defined by

$$S_+(\xi_1, \xi_2, \ldots, \xi_n, \ldots) = (0, \xi_1, \xi_2, \ldots, \xi_n, \ldots).$$

The product

$$D_+ = S_+ D \qquad (4.1.18)$$

results in

$$D_+(\xi_1, \xi_2, \ldots, \xi_n, \ldots) = (0, \sigma_1 \xi_1, \sigma_2 \xi_2, \ldots, \sigma_n \xi_n, \ldots)$$

so that

$$D_+^k(\xi_1, \xi_2, \ldots, \xi_n, \ldots)$$
$$= (\underbrace{0, 0, \ldots, 0}_{1 \; 2 \quad k}, \sigma_k \sigma_{k-1} \cdots \sigma_1 \xi_1, \sigma_{k+1} \cdots \sigma_2 \xi_2, \ldots, \sigma_{n+k-1} \cdots \sigma_n \xi_n, \ldots).$$

It follows that

$$\| D_+^k \| = \sigma_1 \sigma_2 \cdots \sigma_k$$

and thus

$$r(D_+) = \lim_{k \to \infty} \| D_+^k \|^{1/k} = \sigma_\infty.$$

Furthermore, one easily recognizes that

$$a_n(D_+^k) = \sigma_n \sigma_{n+1} \cdots \sigma_{n+k-1}$$

and, correspondingly,

$$a(D_+^k) = \sigma_\infty^k.$$

Therefore the essential spectral radius again is given by

$$r_{ess}(D_+) = \sigma_\infty.$$

As already mentioned above we shall restrict our investigations to that part of the spectrum $\sigma(T)$ of an operator $T : E \to E$ which lies outside the essential spectrum $\sigma_{ess}(T)$ of T, namely to the set

$$\Lambda(T) = \{ \lambda \in \sigma(T) : |\lambda| > r_{ess}(T) \}.$$

Since $\Lambda(T)$ exclusively consists of points λ belonging to the resolvent set $\rho(\bar{T})$ of the coset $\bar{T}$ of T in the Calkin algebra, the operator $S_\lambda = \lambda I_E - T$ is a Fredholm operator for all $\lambda \in \Lambda(T)$. Moreover, the values $\lambda \in \Lambda(T)$ with $\dim(N(S_\lambda)) > 0$ are isolated eigenvalues of T with finite algebraic multiplicities $m(T; \lambda)$ which do not have a point of accumulation except possibly for $|\lambda| = r_{ess}(T)$ (see Heuser 1975, Theorem 51.1.). The points λ with $|\lambda| > r_{ess}(T)$ of the spectrum $\sigma(T)$ of an arbitrary operator $T : E \to E$ are of the same nature as the non-zero points of the spectrum of a compact operator. In particular, the direct sum decomposition (4.1.5) of the Banach space E applies to any finite set of distinct eigenvalues $\mu_1, \mu_2, \ldots, \mu_r$ with $|\mu_i| > r_{ess}(T)$ for $1 \leq i \leq r$. Moreover, as in the case of an operator $T \in K(E, E)$ we can construct an *eigenvalue sequence* $\lambda_1(T), \lambda_2(T), \ldots, \lambda_n(T), \ldots$ for an arbitrary operator $T \in L(E, E)$ from the elements of the set $\Lambda(T) \cup \{ r_{ess}(T) \}$ (see Zemánek 1983):

(ES1) *The eigenvalues are arranged in order of non-increasing absolute values.*

(ES2) *Every eigenvalue $\lambda \in \Lambda(T)$ is counted according to its algebraic multiplicity $m(T, \lambda)$ so that it appears $m(T; \lambda)$ times.*

(ES3) *If T possesses less than n eigenvalues λ with $|\lambda| > r_{ess}(T)$, we put*

$$\lambda_n(T) = \lambda_{n+1}(T) = \cdots = r_{ess}(T).$$

Note that there are only finitely many eigenvalues λ of T with $|\lambda| \geqslant r$ for any $r > r_{ess}(T)$ because the eigenvalues do not have a point of accumulation λ_0 with $|\lambda_0| > r_{ess}(T)$. Accordingly, the construction of the eigenvalue sequence $\lambda_1(T)$, $\lambda_2(T), \ldots, \lambda_n(T), \ldots$ described above in fact makes sense. Finally let us emphasize that the convention (ES3) enables us to operate with an infinite eigenvalue sequence even if T possesses only finitely many eigenvalues or no eigenvalues λ at all with $|\lambda| > r_{ess}(T)$. The elements $\lambda_n(T)$ of the eigenvalue sequence with $\lambda_n(T) = r_{ess}(T)$ need not be eigenvalues. We remind the reader of the shifted diagonal operator D_+ with $r_{ess}(D_+) = r(D_+) = \sigma_\infty$ (see (4.1.18)). In this case the set $\Lambda(T)$ is empty, the eigenvalue sequence $(\lambda_n(D_+))$ by definition being given by the constant sequence $\sigma_\infty, \sigma_\infty, \ldots, \sigma_\infty, \ldots$ However, a value $\lambda \neq 0$ is never an eigenvalue of D_+ and $\lambda = 0$ is only if $\sigma_n = \sigma_{n+1} = \cdots = 0$ for some n. In what follows we shall nevertheless refer to the members of the eigenvalue sequence of an operator T as 'eigenvalues'. If $\lambda_n(T)$ is meant to be an eigenvalue in the proper sense, we shall indicate this by writing $\lambda_n(T) \in \Lambda(T)$. Under the supposition that at least $\lambda_1(T)$ is an eigenvalue in the proper sense the formulas (4.1.1) and (4.1.2) yield

$$|\lambda_1(T)| = \lim_{k \to \infty} \| T^k \|^{1/k}. \tag{4.1.19}$$

On the other hand, if $\Lambda(T) = \varnothing$ and therefore $\lambda_1(T) = r_{ess}(T)$, we have

$$r(T) = \sup_{\lambda \in \sigma(T)} |\lambda| \leqslant r_{ess}(T)$$

and, further, $r(T) = r_{ess}(T)$ since $r_{ess}(T) \leqslant r(T)$ so that (4.1.19) is always valid. This formula linking the absolute value of the first eigenvalue $\lambda_1(T)$ to the spectral radius $r(T) = \lim_{k \to \infty} \| T^k \|^{1/k}$ of T will be referred to as *the classical spectral radius formula.*

As a counterpart to (4.1.19) we have

$$\lim_{n \to \infty} |\lambda_n(T)| = r_{ess}(T) \tag{4.1.20}$$

from the definition of the eigenvalue sequence. Since $r_{ess}(T) = 0$ in the case of a compact operator, formula (4.1.20) implies the central result

$$\lim_{n \to \infty} \lambda_n(T) = 0 \tag{4.1.21}$$

of classical Riesz theory for compact operators.

4.2. Eigenvalues, entropy quantities, and Weyl type inequalities

The main aim of this section is to estimate the geometric mean $(\prod_{i=1}^{n}|\lambda_i(T)|)^{1/n}$ of the absolute values of the first n eigenvalues $\lambda_1(T)$, $\lambda_2(T),\ldots,\lambda_n(T)$ from above by the nth entropy modulus $g_n(T)$ (cf. Carl and Triebel 1980). Since entropy moduli $g_n(T)$ are in a sense geometrical quantities (see sections 1.2 and 1.4) the set of eigenvalues $\lambda_1(T)$, $\lambda_2(T),\ldots,\lambda_n(T)$ is put into a geometrical context as well. It is not surprising, therefore, that the proof of the corresponding inequality is also based on a geometrical background. This background consists in a relation between eigenvalues and invariant subspaces of an operator $T:E\to E$ and will be presented in the following lemma.

Lemma 4.2.1. *Let E be an arbitrary Banach space and $T\in L(E,E)$ an operator with $\lambda_n(T)\in\Lambda(T)$. Then there exists a sequence of ascending subspaces $E_1\subset E_2\subset\cdots E_n$ of E with $\dim(E_k)=k$ invariant under T such that the restriction of T to E_k has precisely $\lambda_1(T),\lambda_2(T),\ldots,\lambda_k(T)$ as its eigenvalues, $1\leqslant k\leqslant n$.*

Proof. If $\mu_1,\mu_2,\ldots,\mu_m$ are the distinct eigenvalues among $\lambda_1(T)$, $\lambda_2(T),\ldots,\lambda_n(T)$ in order of non-increasing absolute values, we have $\lambda_n(T)=\mu_m$ and

$$\sum_{j=1}^{m-1} m(T;\mu_j)<n\leqslant\sum_{j=1}^{m} m(T;\mu_j).$$

So let us consider the direct sum

$$E_l=N_\infty(\mu_1 I_E-T)\oplus\cdots\oplus N_\infty(\mu_m I_E-T) \tag{4.2.1}$$

appearing in the decomposition (4.1.5) of E. Obviously E_l is a subspace of E with $\dim(E_l)=l\geqslant n$ invariant under T, and the restriction of T to E_l has precisely $\mu_1,\mu_2,\ldots,\mu_m$ as its eigenvalues with the correct multiplicities $m(T;\mu_j)$, $1\leqslant j\leqslant m$. Hence it suffices to prove the assertion for operators acting in a finite-dimensional Banach space and, in the end, even for operators acting in a finite-dimensional Banach space with a single eigenvalue only. So let Z be finite-dimensional, say $\dim(Z)=d$, and let $A:Z\to Z$ be an operator with a single eigenvalue λ only, which may also be equal to zero. The existence of an ascending sequence $Z_1\subset Z_2\subset\cdots\subset Z_d=Z$ of invariant subspaces Z_k under A with $\dim(Z_k)=k$, $1\leqslant k\leqslant d$, is proved by induction with respect to d. If $d=1$ there is nothing to prove. We now assume the assertion is true for operators acting in Banach spaces of dimension $d-1$. Given $A:Z\to Z$ as above, we take the operator $S=\lambda I_Z-A$. The range $R(S)$ of S is at most $(d-1)$-dimensional. Let $Z_{d-1}\subset Z_d=Z$ be a subspace of Z of dimension $d-1$ with $R(S)\subseteq Z_{d-1}$. The restriction S_0 of S to Z_{d-1} satisfies the assumption since S_0 also has only one eigenvalue μ, namely $\mu = 0$. Therefore there exists an ascending sequence $Z_1 \subset Z_2 \subset \cdots \subset Z_{d-1}$ of subspaces invariant under S_0 with $\dim(Z_k) = k$, $1 \leqslant k \leqslant d - 1$. Supplementing this sequence by the whole space $Z = Z_d$ we obtain an ascending sequence of subspaces Z_k with $\dim(Z_k) = k$, $1 \leqslant k \leqslant d$, invariant under S. But since $A = \lambda I_Z - S$ the subspaces Z_k are also invariant under A. To complete the proof we return to the restriction of T to the subspace E_l of E given by (4.2.1). Since any of the spaces

$$Z^{(j)} = N_\infty(\mu_j I_E - T)$$

of dimension $d_j = m(T; \mu_j)$ is invariant under T, we can apply the statement just proved to the restrictions $A_j : Z^{(j)} \to Z^{(j)}$ of T to $Z^{(j)}$. Indeed, μ_j is the only eigenvalue of A_j and hence the space $Z^{(j)}$ is the termination of a chain $Z_1^{(j)} \subset Z_2^{(j)} \subset \cdots \subset Z_{d_j}^{(j)} = Z^{(j)}$ of subspaces with $\dim(Z_i^{(j)}) = i$ invariant under A_j. Joining these chains for $j = 1, 2, \ldots, m$ we obtain a sequence of ascending subspaces E_k invariant under T. In detail, given k with $k \leqslant n$, we determine r with $1 \leqslant r \leqslant m$ such that

$$\sum_{j=1}^{r-1} m(T; \mu_j) < k \leqslant \sum_{j=1}^{r} m(T; \mu_j)$$

and then choose s with $0 < s \leqslant m(T; \mu_r)$ such that

$$k = \sum_{j=1}^{r-1} m(T; \mu_j) + s.$$

Then

$$E_k = Z^{(1)} \oplus Z^{(2)} \oplus \cdots \oplus Z^{(r-1)} \oplus Z_s^{(r)}$$

is obviously a subspace of E of dimension k invariant under T and $E_1 \subset E_2 \subset \cdots \subset E_n$. According to the construction of E_k, $1 \leqslant k \leqslant n$, the restriction of T to E_k has precisely $\lambda_1(T), \lambda_2(T), \ldots, \lambda_k(T)$ as its eigenvalues, $\mu_1, \mu_2, \ldots, \mu_r$ being the distinct eigenvalues among them. ∎

Supplement 1. The ascending sequence $E_1 \subset E_2 \subset \cdots \subset E_k \subset \cdots \subset E_n$ of invariant subspaces of T with $\dim(E_k) = k$ can be used to construct a convenient matrix representation for the restriction T_n of T to E_n. We start with a vector $z_1 \neq 0$ in E_1 and complement it by a vector $z_2 \in E_2 \backslash E_1$ to give a basis of E_2. Proceeding by induction we in this way obtain a basis $z_1, z_2, \ldots, z_n$ of E_n with the particular property that $z_1, z_2, \ldots, z_k$ is a basis of E_k for any k, $1 \leqslant k \leqslant n$. Since E_k is invariant under T the image Tz_k of the kth basis vector z_k is a linear combination,

$$Tz_k = \sum_{j=1}^{k} \tau_{jk} z_j, \quad 1 \leqslant k \leqslant n, \tag{4.2.2}$$

of the first k basis vectors $z_1, z_2, \ldots, z_k$. This means that the (n, n)-matrix

(τ_{jk}) representing the restriction T_n of T to E_n has upper triangular shape, namely

$$\tau_{jk} = 0 \quad \text{for } j > k. \tag{4.2.3}$$

Accordingly, the diagonal elements τ_{kk} coincide with the eigenvalues $\lambda_k(T)$ of T for $1 \leqslant k \leqslant n$.

Supplement 2. For operators T_0 acting in a finite-dimensional Banach space E_0, the eigenvalue $\lambda = 0$, which was excluded by the assumption $\lambda_n(T) \in \Lambda(T)$ in Lemma 4.2.1, does not play a separate role as already observed in the proof of Lemma 4.2.1. Therefore an operator T_0 acting in a Banach space E_0 with $\dim(E_0) = n$ can always be represented by a triangular (n, n)-matrix (τ_{jk}) where the diagonal elements τ_{kk} are the eigenvalues $\lambda_k(T_0)$ of T_0. If $\mu_1, \mu_2, \ldots, \mu_r$ are the distinct eigenvalues among $\lambda_1(T_0), \lambda_2(T_0), \ldots, \lambda_n(T_0)$ then the space E_0 can be expressed as the direct sum

$$E_0 = N_\infty(\mu_1 I_{E_0} - T_0) \oplus \cdots \oplus N_\infty(\mu_r I_{E_0} - T_0) \tag{4.2.4}$$

of the spaces $Z^{(j)} = N_\infty(\mu_j I_{E_0} - T_0)$ of dimension $d_j = m(T_0; \mu_j)$ invariant under T_0. As an aside to this, the Jordan canonical form of an (n, n)-matrix representing an operator T_0 in an n-dimensional vector space is obtained by choosing a specific basis for each individual space $Z^{(j)}$ so that the result is a matrix, which is not only triangular like the matrix (τ_{jk}) constructed above, but of an even more special form (see Halmos 1955; Gantmacher 1958).

The very fact that the restriction T_n of an operator $T: E \to E$ to an appropriate n-dimensional subspace $E_n \subseteq E$ can be represented by a triangular matrix with the first n eigenvalues $\lambda_1(T), \lambda_2(T), \ldots, \lambda_n(T)$ on the diagonal will be the decisive point in the estimation of the geometric mean $(\prod_{i=1}^{n} |\lambda_i(T)|)^{1/n}$ from above by the nth entropy modulus $g_n(T)$. A more technical detail of the corresponding proof consists in the transition from T to the powers T^m. The relation between the eigenvalues $\lambda_i(T)$ of T and the eigenvalues $\lambda_i(T^m)$ of T^m will be revealed in the next lemma.

Lemma 4.2.2. *Let* $T \in L(E, E)$ *be an operator in an arbitrary Banach space E, and T^m a power of T, $m = 1, 2, 3, \ldots$ Then*

$$r_{\text{ess}}(T^m) = r_{\text{ess}}(T)^m \tag{4.2.5}$$

and

$$\lambda_i(T^m) = \lambda_i(T)^m \tag{4.2.6}$$

so long as the enumeration of the members of the eigenvalue sequence of T^m is adapted to the enumeration of the members of the eigenvalue sequence of T.

Proof. By using (4.1.13) we obtain

$$r_{ess}(T^m) = \lim_{k \to \infty} \beta^{1/k}(T^{mk})$$

which can also be interpreted as

$$r_{ess}(T^m) = \lim_{n \to \infty} \beta^{m/n}(T^n),$$

and hence amounts to (4.2.5). Now let $\mu \in \Lambda(T^m)$ be an eigenvalue of T^m. The subspace $E_0 = N_\infty(\mu I_E - T^m)$ of E spanned by the principal vectors belonging to μ is obviously invariant under T. Therefore we can consider the restriction T_0 of T to E_0 which possesses the properties of an operator acting in a finite-dimensional Banach space. This means that T_0 has at least one eigenvalue $\lambda \in \sigma(T_0)$, and the space E_0 can be given by the direct sum

$$E_0 = N_\infty(\lambda_1 I_{E_0} - T_0) \oplus \cdots \oplus N_\infty(\lambda_r I_{E_0} - T_0) \qquad (4.2.4)'$$

of the spaces $N_\infty(\lambda_i I_{E_0} - T_0)$ determined by the set $\{\lambda_1, \lambda_2, \ldots, \lambda_r\}$ of distinct eigenvalues of T_0. Let λ be any of the eigenvalues $\lambda_1, \lambda_2, \ldots, \lambda_r$ and $x_0 \in E_0$ an eigenvector associated with λ so that

$$T_0 x_0 = \lambda x_0. \qquad (4.2.7)$$

Obviously equation (4.2.7) can also be written as

$$T x_0 = \lambda x_0 \qquad (4.2.8)$$

which says that λ is also an eigenvalue of T. Since the eigenvector x_0 lies in the space $E_0 = N_\infty(\mu I_E - T^m)$ there exists a natural number k such that

$$(\mu I_E - T^m)^k x_0 = 0.$$

Applying the binomial formula

$$(\mu I_E - T^m)^k = \mu^k I_E - \binom{k}{1} \mu^{k-1} T^m + \cdots + (-1)^k T^{mk}$$

and using (4.2.8) we see that

$$(\mu - \lambda^m)^k x_0 = 0$$

which implies

$$\mu = \lambda^m.$$

Thus any eigenvalue $\mu \in \Lambda(T^m)$ of the mth power T^m of T arises from the mth power λ^m of an eigenvalue λ of T, the condition $|\mu| > r_{ess}(T^m)$ involving $|\lambda| > r_{ess}(T)$ by (4.2.5). Conversely, if λ is an eigenvalue of T with $|\lambda| > r_{ess}(T)$, then $\mu = \lambda^m$ is an eigenvalue of T^m with $|\mu| > r_{ess}(T)^m = r_{ess}(T^m)$. But what about the multiplicities? To answer this question we have to consider the spaces $N_\infty(\lambda I_E - T)$ and $N_\infty(\lambda^m I_E - T^m)$. The factorization

$$\lambda^m I_E - T^m = (\lambda^{m-1} I_E + \lambda^{m-2} T + \cdots + T^{m-1})(\lambda I_E - T)$$

makes it clear that

$$N_\infty(\lambda I_E - T) \subseteq N_\infty(\lambda^m I_E - T^m).$$

As a result of this, the space $N_\infty(\lambda I_E - T)$ spanned by the principal vectors of the eigenvalue λ of T can also be regarded as the space $N_\infty(\lambda I_{E_0} - T_0)$ spanned by the principal vectors of the eigenvalue λ of T_0, the restriction of T to $E_0 = N_\infty(\lambda^m I_E - T^m)$, that is

$$N_\infty(\lambda I_E - T) = N_\infty(\lambda I_{E_0} - T_0)$$

and, correspondingly,

$$m(T; \lambda) = m(T_0; \lambda).$$

But then the decomposition (4.2.4)' of the space E_0 tells us that

$$m(T^m; \mu) = \sum_{\substack{\lambda \in \Lambda(T), \\ \lambda^m = \mu}} m(T; \lambda).$$

Hence an eigenvalue $\mu \in \Lambda(T^m)$ appears in the eigenvalue sequence of the operator T^m as often as indicated by the different eigenvalues $\lambda \in \Lambda(T)$ of the operator T with $\lambda^m = \mu$, and their multiplicities $m(T; \lambda)$. An appropriate enumeration of the eigenvalues of T^m and T secures (4.2.6). From (4.2.5) the relation (4.2.6) remains true if the operator T^m possesses only finitely many eigenvalues, say less than n, so that

$$\lambda_i(T^m) = r_{\text{ess}}(T^m)$$

for $i \geqslant n$. This completes the proof of (4.2.6). ∎

Theorem 4.2.1. (cf. Carl and Triebel 1980). *Let $T \in L(E, E)$ be an operator in an arbitrary Banach space E and $\lambda_1(T), \lambda_2(T), \ldots, \lambda_n(T), \ldots$ its eigenvalue sequence. Then*

$$\left(\prod_{i=1}^{n} |\lambda_i(T)| \right)^{1/n} \leqslant g_n(T) \quad for \; n = 1, 2, 3, \ldots \tag{4.2.9}$$

Proof. First let us assume $\lambda_n(T) \in \Lambda(T)$. Then Lemma 4.2.1 and its Supplement 1 guarantee the existence of an n-dimensional subspace $E_n \subseteq E$ invariant under T such that the restriction T_n of T to E_n can be represented by an (upper) triangular (n, n)-matrix (τ_{jk}) with

$$\tau_{kk} = \lambda_k(T) \quad \text{for } 1 \leqslant k \leqslant n. \tag{4.2.10}$$

The matrix (τ_{jk}) is complex, say

$$\tau_{jk} = \alpha_{jk} + i\beta_{jk}. \tag{4.2.11}$$

It changes into a real $(2n, 2n)$-matrix if the underlying basis $z_1, z_2, \ldots, z_n$ of the complex n-dimensional space E_n is replaced by the system of $2n$ vectors

$$z_1, iz_1, z_2, iz_2, \ldots, z_n, iz_n, \tag{4.2.12}$$

which are linearly independent with respect to the field $\mathbb{R}$ of real numbers. The transformation formulas (4.2.2), for the restriction T_n of T to E_n, then yield

$$T_n z_k = \sum_{j=1}^{k} (\alpha_{jk} z_j + \beta_{jk}(iz_j)) \quad \text{for } 1 \leqslant k \leqslant n \qquad (4.2.2)(\text{a})$$

and

$$T_n(iz_k) = \sum_{j=1}^{k} (-\beta_{jk} z_j + \alpha_{jk}(iz_j)) \quad \text{for } 1 \leqslant k \leqslant n. \qquad (4.2.2)(\text{b})$$

Hence the change from the n-dimensional complex space E_n spanned by $z_1, z_2, \ldots, z_n$ to the $2n$-dimensional real space $E_{2n}^{\mathbb{R}}$ spanned by $z_1, iz_1, z_2, iz_2, \ldots, z_n, iz_n$ implies the change from the linear operator $T_n: E_n' \to E_n$ and its representing matrix (τ_{jk}) to the linear operator $T_{2n}^{\mathbb{R}}: E_{2n}^{\mathbb{R}} \to R_{2n}^{\mathbb{R}}$ described by (4.2.2)(a) and (4.2.2)(b). These formulas tell us that each element τ_{jk} of the complex matrix representing T_n has to be replaced by the real $(2,2)$-block

$$\begin{pmatrix} \alpha_{jk} & -\beta_{jk} \\ \beta_{jk} & \alpha_{jk} \end{pmatrix}.$$

From (4.2.3) we see that the real $(2n, 2n)$-matrix arising in this way has 'super-triangular form' so that the value of its determinant $\det T_{2n}^{\mathbb{R}}$ is given by the product of the determinants of the $(2,2)$-blocks appearing on the diagonal, namely

$$\det T_{2n}^{\mathbb{R}} = \prod_{k=1}^{n} (\alpha_{kk}^2 + \beta_{kk}^2) = \prod_{k=1}^{n} |\tau_{kk}|^2 = |\det T_{2n}^{\mathbb{R}}|.$$

By (4.2.10) this means

$$|\det T_{2n}^{\mathbb{R}}| = \prod_{k=1}^{n} |\lambda_k(T)|^2 = |\det T_n|^2. \qquad (4.2.13)$$

On the other hand, $|\det T_{2n}^{\mathbb{R}}|$ represents the ratio

$$|\det T_{2n}^{\mathbb{R}}| = \frac{\text{vol}_{2n}(T_{2n}^{\mathbb{R}}(U_{E_{2n}^{\mathbb{R}}}))}{\text{vol}_{2n}(U_{E_{2n}^{\mathbb{R}}})}$$

of the volumes of the image $T_{2n}^{\mathbb{R}}(U_{E_{2n}^{\mathbb{R}}})$ of the unit ball and of the unit ball $U_{E_{2n}^{\mathbb{R}}}$ itself. As we pointed out in section 1.2, the $2n$th root of this ratio can be estimated from above by the entropy modulus $g_{2n}(T_{2n}^{\mathbb{R}}(U_{E_{2n}^{\mathbb{R}}})) = g_{2n}(T_{2n}^{\mathbb{R}})$ (see (1.2.3) and (1.2.4)). By (4.2.13) we in this way arrive at

$$\left(\prod_{k=1}^{n} |\lambda_k(T)| \right)^{1/n} \leqslant g_{2n}(T_{2n}^{\mathbb{R}}).$$

But obviously

$$g_{2n}(T_{2n}^{\mathbb{R}}) = g_n(T_n).$$

In fact, the definition of the nth entropy modulus $g_n(M)$ of a bounded subset M of a complex n-dimensional Banach space is based on a comparison of volumes in a real (euclidean) space of dimension $2n$, the values of the entropy numbers $\varepsilon_k(M)$ being independent of the particular interpretation (see section 1.2).

It remains to estimate $g_n(T_n)$ from above by $g_n(T)$. Realizing that

$$J_n T_n = T J_n$$

where $J_n : E_n \to E$ is the natural embedding of E_n into E we see that

$$g_n(T_n) \leqslant 2g_n(J_n T_n) \leqslant 2g_n(T)$$

by the weak injectivity (1.4.3) of the entropy moduli. However, the inequality

$$\left(\prod_{k=1}^{n} |\lambda_k(T)| \right)^{1/n} \leqslant 2g_n(T) \qquad (4.2.14)$$

is not yet the result we are looking for. In order to get rid of the factor 2 we rewrite the left-hand side of (4.2.14) by taking the mth power $|\lambda_k(T)|^m$ of $|\lambda_k(T)|$ and using $|\lambda_k(T)|^m = |\lambda_k(T^m)|$, giving

$$\left(\prod_{k=1}^{n} |\lambda_k(T)| \right)^{1/n} = \left(\prod_{k=1}^{n} |\lambda_k(T^m)| \right)^{1/nm}.$$

If we now apply (4.2.14) with T^m instead of T we obtain

$$\left(\prod_{k=1}^{n} |\lambda_k(T)| \right)^{1/n} \leqslant 2^{1/m} g_n^{1/m}(T^m).$$

Finally, the multiplicativity (M2) of the entropy moduli yields

$$g_n^{1/m}(T^m) \leqslant g_n(T).$$

Hence the estimate (4.2.14) has been improved to

$$\left(\prod_{k=1}^{n} |\lambda_k(T)| \right)^{1/n} \leqslant 2^{1/m} g_n(T).$$

Letting $m \to \infty$ we arrive at the desired result (4.2.9).

So far we have assumed that $\lambda_n(T) \in \Lambda(T)$. Now let us suppose that

$$\lambda_m(T) \in \Lambda(T) \quad \text{but} \quad \lambda_i(T) = r_{\text{ess}}(T) \quad \text{for } m < i \leqslant n.$$

The inequality (4.2.9) just proved then tells us that

$$\prod_{i=1}^{m} |\lambda_i(T)| \leqslant g_m^{\,m}(T) \leqslant k^{1/2} \varepsilon_k^{\,m}(T)$$

with m instead of n and arbitrary $k = 1, 2, 3, \ldots$ Changing the product $\prod_{i=1}^{m} |\lambda_i(T)|$ to $\prod_{i=1}^{n} |\lambda_i(T)|$ we may conclude that

$$\prod_{i=1}^{n} |\lambda_i(T)| \leqslant k^{1/2} \varepsilon_k^{\,m}(T) \cdot r_{\text{ess}}^{\,n-m}(T) \leqslant k^{1/2} \varepsilon_k(T)^n$$

since $r_{\text{ess}}(T) \leqslant \varepsilon_k(T)$ (see (4.1.15)). Hence we have

$$\left(\prod_{i=1}^{n} |\lambda_i(T)| \right)^{1/n} \leqslant k^{1/2n} \varepsilon_k(T) \quad \text{for } k = 1, 2, 3, \dots,$$

which amounts to the validity of (4.2.9) even in the modified situation where $\lambda_i(T) = r_{\text{ess}}(T)$ for $i > m$. Moreover, if $\lambda_1(T) = \lambda_2(T) = \dots = \lambda_n(T) = r_{\text{ess}}$ it suffices to remark that

$$r_{\text{ess}}(T) \leqslant \beta(T) \leqslant g_n(T)$$

by (4.1.15) and (3.5.18). ∎

Supplement. Taking into account that

$$|\lambda_n(T)| \leqslant \left(\prod_{i=1}^{n} |\lambda_i(T)| \right)^{1/n}$$

and

$$g_n(T) \leqslant 2^{(k-1)/2n} \varepsilon_{2^{k-1}}(T) = \sqrt{2}^{(k-1)/n} e_k(T)$$

we may extract from (4.2.9) the original result

$$|\lambda_n(T)| \leqslant \sqrt{2}^{(k-1)/n} e_k(T) \quad \text{for } k, n = 1, 2, 3, \dots \quad (4.2.15)$$

obtained by Carl (1981) and its simplified but extremely useful version

$$|\lambda_n(T)| \leqslant \sqrt{2} e_n(T) \quad \text{for } n = 1, 2, 3, \dots \quad (4.2.16)$$

Combining (4.2.16) with the inequalities (3.1.13) and (3.1.14) of Bernstein type proved in section 3.1 we get the so-called *Weyl type inequalities*

$$\sup_{1 \leqslant k \leqslant m} k^{1/p} |\lambda_k(T)| \leqslant 2\sqrt{2} c_p \sup_{1 \leqslant k \leqslant m} k^{1/p} t_k(T), \quad (4.2.17)$$

$m = 1, 2, 3, \dots,$ and

$$\left(\sum_{k=1}^{m} k^{q/p-1} |\lambda_k(T)|^q \right)^{1/q} \leqslant 2\sqrt{2} C_{p,q} \left(\sum_{k=1}^{m} k^{q/p-1} t_k(T)^q \right)^{1/q}, \quad (4.2.18)$$

$m = 1, 2, 3, \dots,$ where $0 < p \leqslant \infty$ and $0 < q < \infty$. Weyl's original result says

$$\left(\sum_{k=1}^{m} |\lambda_k(T)|^p \right)^{1/p} \leqslant \left(\sum_{k=1}^{m} a_k(T)^p \right)^{1/p}, \quad m = 1, 2, 3, \dots,$$

for compact operators T acting in a Hilbert space (Weyl 1949; see section 4.4). The first Weyl type inequality in a Banach space setting was established by König (1977). This paper generated much further activity in this line of research. In particular, Johnson, König, Maurey and Retherford (1979) and König (1980) proved inequalities of the type

$$\sup_{1 \leqslant k \leqslant m} k^{1/p} |\lambda_k(T)| \leqslant \tilde{c}_p \sup_{1 \leqslant k \leqslant m} k^{1/p} s_k(T), \quad m = 1, 2, 3, \dots, \quad (4.2.19)$$

and

$$\left(\sum_{k=1}^{m} |\lambda_k(T)|^p \right)^{1/p} \leqslant \tilde{C}_p \left(\sum_{k=1}^{m} s_k(T)^p \right)^{1/p}, \quad m = 1, 2, 3, \ldots, \quad (4.2.20)$$

with the approximation numbers $a_k(T)$, the Kolmogorov numbers $d_k(T)$, or the Gelfand numbers $c_k(T)$ in place of $s_k(T)$. In this connection König (1980) for the first time used the Lorentz classification scheme $l_{p,q}$ instead of the l_p scale. Motivated by the arguments of König (1980) and König, Retherford and Tomczak-Jaegermann (1980), Pietsch (1980) worked with the so-called Weyl numbers $x_k(T)$ in place of $s_k(T)$ on the right-hand side of (4.2.19) and (4.2.20). Because $x_k(T) \leqslant c_k(T)$ this was a further improvement. Since $t_k(T) \leqslant d_k(T)$ and $t_k(T) \leqslant c_k(T)$ the Weyl type inequalities stated under (4.2.17) and (4.2.18) also represent strengthened versions of (4.2.19) and (4.2.20). In addition the approach to (4.2.17) and (4.2.18) through the Bernstein type inequalities and the relation (4.2.16) between entropy numbers and eigenvalues seems to be quite natural.

To end this section we point out that the estimate (4.2.9) of the geometric mean $(\prod_{i=1}^{n} |\lambda_i(T)|)^{1/n}$ of the absolute values of the first n eigenvalues from above by the entropy modulus $g_n(T)$ is a best possible in the following sense. Let I_E be the identity map of an arbitrary infinite-dimensional Banach space E. Then we have

$$g_n(I_E) = 1 \quad \text{for } n = 1, 2, 3, \ldots$$

from the norm determining property (1.4.4) of the entropy moduli and, simultaneously,

$$\lambda_i(I_E) = 1 \quad \text{for } i = 1, 2, 3, \ldots,$$

so that the equality

$$\left(\prod_{i=1}^{n} |\lambda_i(I_E)| \right)^{1/n} = g_n(I_E)$$

turns out to be true for all $n = 1, 2, 3, \ldots$ A less trivial example is given by the well-known diagonal operator $D: l_p \to l_p$ generated by a sequence $\sigma_1 \geqslant \sigma_2 \geqslant \cdots \geqslant \sigma_n \geqslant \cdots \geqslant 0$. Then we have $r_{\text{ess}}(D) = \lim_{n \to \infty} \sigma_n$ (see section 4.1) and $\lambda_i(D) = \sigma_i$ for $i = 1, 2, 3, \ldots$ Hence, the estimate (1.4.7) proved in Proposition 1.4.1 can be interpreted as

$$\left(\prod_{i=1}^{n} |\lambda_i(D)| \right)^{1/n} \leqslant g_n(D) \leqslant 6 \left(\prod_{i=1}^{n} |\lambda_i(D)| \right)^{1/n}.$$

This means that the asymptotic behaviour of the entropy moduli $g_n(D)$ is in fact determined by the asymptotic behaviour of the geometric means $(\prod_{i=1}^{n} |\lambda_i(D)|)^{1/n}$.

4.3. Generalizations of the classical spectral radius formula

The classical spectral radius formula

$$|\lambda_1(T)| = \lim_{k \to \infty} \| T^k \|^{1/k} \qquad (4.1.19)$$

will be generalized in two ways. One way consists in representations for the absolute values of the higher eigenvalues $\lambda_n(T)$ in terms of the approximation numbers $a_n(T^k)$ of the powers T^k of $T: E \to E$. The other way concerns representations for the geometric means $(\prod_{i=1}^{n} |\lambda_i(T)|)^{1/n}$ of the absolute values of the first n eigenvalues $\lambda_1(T), \lambda_2(T), \ldots, \lambda_n(T)$ in terms of the entropy moduli $g_n(T^k)$ of the powers T^k. The discovery of the representations for the $|\lambda_n(T)|$ can be attributed to König (1979; 1986), if we disregard a matrix version due to Yamamoto (1967). The representations for the geometric means $(\prod_{i=1}^{n} |\lambda_i(T)|)^{1/n}$ are due to Makai and Zemánek (1982). We shall give a unified proof which rests on the estimate (4.2.9) of $(\prod_{i=1}^{n} |\lambda_i(T)|)^{1/n}$ from above by $g_n(T)$ and on the interplay between entropy moduli and approximation numbers studied in Proposition 3.5.1.

Theorem 4.3.1. *Let $T \in L(E, E)$ be an operator in an arbitrary Banach space E and $\lambda_1(T), \lambda_2(T), \ldots, \lambda_n(T), \ldots$ its eigenvalue sequence. Then*

$$|\lambda_n(T)| = \lim_{k \to \infty} a_n^{1/k}(T^k) \qquad (4.3.1)$$

and

$$\left(\prod_{i=1}^{n} |\lambda_i(T)| \right)^{1/n} = \lim_{k \to \infty} g_n^{1/k}(T^k). \qquad (4.3.2)$$

Proof. For $n = 1$ formula (4.3.1) reduces to the classical spectral radius formula (4.1.19). To prove (4.3.1) for $n \geq 2$ we first show that

$$\lim_{k \to \infty} a_n^{1/k}(T^k) \leq |\lambda_n(T)|. \qquad (4.3.3)$$

If $\lambda_n(T) = \lambda_{n-1}(T) = \cdots = \lambda_1(T)$ the validity of (4.3.3) is obvious since in this case

$$\lim_{k \to \infty} a_n^{1/k}(T^k) \leq \lim_{k \to \infty} \| T^k \|^{1/k} = |\lambda_n(T)|.$$

So let us assume that $\lambda_n(T) \neq \lambda_1(T)$, say

$$\lambda_n(T) = \lambda_{n-1}(T) = \cdots = \lambda_{n-m+1}(T), \text{ but } \lambda_n(T) \neq \lambda_{n-m}(T). \qquad (4.3.4)$$

Denoting by $\mu_1, \mu_2, \ldots, \mu_r$ the different eigenvalues among $\lambda_1(T)$, $\lambda_2(T), \ldots, \lambda_{n-m}(T)$ we consider the projection

$$P = \sum_{i=1}^{r} P_{\mu_i}(T) \qquad (4.1.6)$$

of E onto the subspace $N_\infty(\mu_1 I_E - T) \oplus \cdots \oplus N_\infty(\mu_r I_E - T)$ with $N(P) = \bigcap_{i=1}^r R_\infty(\mu_i I_E - T)$. Then $\dim(R(P)) = n - m$ because every eigenvalue μ_i appears as often among $\lambda_1(T), \lambda_2(T), \ldots, \lambda_{n-m}(T)$ as indicated by its algebraic multiplicity. Consequently, we have

$$\text{rank}(PT^k) \leqslant n - m \quad \text{for } k = 1, 2, 3, \ldots$$

and thus may conclude that

$$a_{n-m+1}(T^k) \leqslant \|T^k - PT^k\| = \|(I_E - P)T^k\|.$$

But, by what was said in section 4.1, the projection P commutes with T. Therefore $I_E - P$ is also a projection that commutes with T so that

$$(I_E - P)T^k = (I_E - P)^k T^k = ((I_E - P)T)^k$$

and hence

$$a_{n-m+1}(T^k) \leqslant \|((I_E - P)T)^k\|. \tag{4.3.5}$$

Let us consider the eigenvalues λ of the operator

$$S = (I_E - P)T,$$

with $|\lambda| > r_{\text{ess}}(S)$. Note that

$$r_{\text{ess}}(S) = r_{\text{ess}}(T)$$

because S and T have the same coset in the Calkin algebra $L(E, E)/K(E, E)$. An eigenvector x associated with an eigenvalue λ of S with $|\lambda| > r_{\text{ess}}(S)$ is characterized by

$$(I_E - P)Tx = \lambda x. \tag{4.3.6}$$

This equation tells us that $x \in R(I_E - P)$ which means $x \in N(P)$. Hence it follows that

$$Tx = T(I_E - P)x = (I_E - P)Tx = \lambda x.$$

by (4.3.6). Conversely, an eigenvalue λ of T with $|\lambda| > r_{\text{ess}}(T)$, which has an eigenvector $x \in N(P)$, satisfies (4.3.6). Hence a complex number λ with $|\lambda| > r_{\text{ess}}(T)$ is an eigenvalue of S if and only if λ is an eigenvalue of T with an eigenvector $x \in N(P)$. This means $\lambda = \lambda_{n-m+j}(T)$ with $j = 1, 2, 3, \ldots$ Enumerating the eigenvalues of S in the usual way we can write

$$\lambda_j(S) = \lambda_{n-m+j}(T), \quad j = 1, 2, 3, \ldots$$

Of course, if $\lambda_{n-m+j}(T) = r_{\text{ess}}(T)$ for some $j \geqslant 1$, then also $\lambda_j(S) = r_{\text{ess}}(T)$. In any case, the classical spectral radius formula applies to $\lambda_1(S)$, namely

$$\lim_{k \to \infty} \|S^k\|^{1/k} = |\lambda_1(S)|,$$

that is to say

$$\lim_{k \to \infty} \|((I_E - P)T)^k\|^{1/k} = |\lambda_{n-m+1}(T)|.$$

From (4.3.4), (4.3.5) and

$$a_n(T^k) \leqslant a_{n-m+1}(T^k)$$

we finally reach the desired result

$$\lim_{k \to \infty} a_n^{1/k}(T^k) \leqslant |\lambda_1(S)| = |\lambda_n(T)|. \tag{4.3.3}$$

Next we turn to the proof of (4.3.2). For the limit on the right-hand side of (4.3.2) we have the representation

$$\lim_{k \to \infty} g_n^{1/k}(T^k) = \left(\prod_{i=1}^{n} \lim_{k \to \infty} a_i^{1/k}(T^k) \right)^{1/n} \tag{3.5.2}$$

(see Proposition 3.5.1). Using result (4.3.3) we obtain

$$\lim_{k \to \infty} g_n^{1/k}(T^k) \leqslant \left(\prod_{i=1}^{n} |\lambda_i(T)| \right)^{1/n}.$$

For the proof of the converse inequality we make use of

$$|\lambda_i(T)|^k = |\lambda_i(T^k)| \tag{4.2.6}$$

and the result (4.2.9) with T replaced by T^k, that is

$$\left(\prod_{i=1}^{n} |\lambda_i(T^k)| \right)^{1/n} \leqslant g_n(T^k).$$

In this way we get

$$\left(\prod_{i=1}^{n} |\lambda_i(T)| \right)^{1/n} = \left(\prod_{i=1}^{n} |\lambda_i(T^k)| \right)^{1/nk} \leqslant g_n^{1/k}(T^k) \quad \text{for } k = 1, 2, 3, \dots$$

If we take the limit with respect to k on the right-hand side of the last inequality we arrive at

$$\left(\prod_{i=1}^{n} |\lambda_i(T)| \right)^{1/n} \leqslant \lim_{k \to \infty} g_n^{1/k}(T^k)$$

which completes the proof of (4.3.2).

By using the relation

$$\lim_{k \to \infty} a_n^{1/k}(T^k) = \frac{\lim_{k \to \infty} g_n^{n/k}(T^k)}{\lim_{k \to \infty} g_{n-1}^{(n-1)/k}(T^k)} \tag{3.5.1}$$

(see Proposition 3.5.1) between the limits $\lim_{k \to \infty} a_n^{1/k}(T^k)$ and

$$\lim_{k \to \infty} g_n^{n/k}(T^k) = \prod_{i=1}^{n} |\lambda_i(T)|, \tag{4.3.2}'$$

we now immediately see that

$$\lim_{k \to \infty} a_n^{1/k}(T^k) = \frac{\prod_{i=1}^{n} |\lambda_i(T)|}{\prod_{i=1}^{n-1} |\lambda_i(T)|} = |\lambda_n(T)|.$$

The last conclusion is based on the assumption $\prod_{i=1}^{n-1} |\lambda_i(T)| \neq 0$ which amounts to $\lambda_{n-1}(T) \neq 0$. The final result (4.3.1), however, is true in any case. For, if $\lambda_{n-1}(T) = 0$, then $\lim_{k \to \infty} g_{n-1}^{1/k}(T^k) = 0$ by (4.3.2) and thus

$\lim_{k \to \infty} a_n^{1/k}(T^k) = 0$ by Proposition 3.5.1. But since $\lambda_n(T) = 0$ in this situation, we are done. ∎

In contrast to the simple limits $\lim_{k \to \infty} a_n^{1/k}(T^k)$ and $\lim_{k \to \infty} g_n^{1/k}(T^k)$, the iterated limits $\lim_{n \to \infty} \lim_{k \to \infty} a_n^{1/k}(T^k)$ and $\lim_{n \to \infty} \lim_{k \to \infty} g_n^{1/k}(T^k)$ have already been related to the spectrum of the operator T in section 4.1, namely by the statement that their common value coincides with the essential spectral radius $r_{\text{ess}}(T)$ of T (see Proposition 4.1.1 and (3.5.23), (3.5.10)). Taking the limits $n \to \infty$ in the formulas (4.3.1) and (4.3.2) we can confirm this fact because

$$\lim_{n \to \infty} |\lambda_n(T)| = r_{\text{ess}}(T) \tag{4.1.20}$$

as well as

$$\lim_{n \to \infty} \left(\prod_{i=1}^{n} |\lambda_i(T)| \right)^{1/n} = r_{\text{ess}}(T)$$

from the definition of the eigenvalue sequence.

Corollary. *Let E be an arbitrary Banach space. If $R \in L(E, E)$ and $S \in L(E, E)$ are commuting operators, then*

$$|\lambda_{m+n-1}(RS)| \leqslant |\lambda_m(R)| \cdot |\lambda_n(S)| \tag{4.3.7}$$

and

$$\prod_{i=1}^{n} |\lambda_i(RS)| \leqslant \prod_{i=1}^{n} |\lambda_i(R)| \cdot |\lambda_i(S)|. \tag{4.3.8}$$

Proof. Since $RS = SR$ we have

$$a_{m+n-1}((RS)^k) = a_{m+n-1}(R^k S^k) \leqslant a_m(R^k) a_n(S^k)$$

and

$$g_n((RS)^k) = g_n(R^k S^k) \leqslant g_n(R^k) g_n(S^k)$$

by the multiplicativity of the approximation numbers and entropy moduli, respectively. Hence it follows that

$$\lim_{k \to \infty} a_{m+n-1}^{1/k}((RS)^k) \leqslant \lim_{k \to \infty} a_m^{1/k}(R^k) \cdot \lim_{k \to \infty} a_n^{1/k}(S^k)$$

and

$$\lim_{k \to \infty} g_n^{1/k}((RS)^k) \leqslant \lim_{k \to \infty} g_n^{1/k}(R^k) \cdot \lim_{k \to \infty} g_n^{1/k}(S^k)$$

which, by (4.3.1) and (4.3.2), proves the inequalities (4.3.7) and (4.3.8). ∎

4.4. The Hilbert space setting

Like the inequalities of Bernstein–Jackson type (see section 3.4), the relations between approximation and spectral properties also become simpler for operators acting in a Hilbert space. Particularly satisfactory results are achieved for self-adjoint operators.

Proposition 4.4.1. *Let* $T \in L(H, H)$ *be a self-adjoint operator acting in a Hilbert space H. Then*

$$|\lambda_n(T)| = a_n(T) \quad for \ n = 1, 2, 3, \ldots \tag{4.4.1}$$

Proof. We start from the inequality

$$a_n^2(T) \leqslant a_n(T^*T) \tag{3.4.12}$$

already used for the proof of Theorem 3.4.1. For a self-adjoint operator, (3.4.12) reduces to

$$a_n^2(T) \leqslant a_n(T^2).$$

We check by induction that

$$a_n^{2^m}(T) \leqslant a_n(T^{2^m}) \quad \text{for } m = 1, 2, 3, \ldots$$

This inequality can be rewritten as

$$a_n(T) \leqslant a_n^{1/2^m}(T^{2^m}) \quad \text{for } m = 1, 2, 3, \ldots$$

and thus implies

$$a_n(T) \leqslant \lim_{m \to \infty} a_n^{1/2^m}(T^{2^m}) = \lim_{k \to \infty} a_n^{1/k}(T^k).$$

On the basis of the formula (4.3.1) for $|\lambda_n(T)|$ we now recognize that

$$a_n(T) \leqslant |\lambda_n(T)|. \tag{4.4.2}$$

For the proof of the converse inequality we first suppose $\lambda_n(T) \in \Lambda(T)$. Then by Lemma 4.2.1 there exists an n-dimensional subspace H_n of H invariant under T such that the restriction T_n of T to H_n has precisely $\lambda_1(T), \lambda_2(T), \ldots, \lambda_n(T)$ as its eigenvalues. Since $|\lambda_n(T)| > r_{\text{ess}}(T) \geqslant 0$ the operator T_n possesses an inverse T_n^{-1} on H_n and hence we have

$$\lambda_1(T_n^{-1}) = \lambda_n(T_n)^{-1} = \lambda_n(T)^{-1}. \tag{4.4.3}$$

Because the self-adjointness of the operator $T : H \to H$ involves the self-adjointness of both the operators T_n and T_n^{-1} we can estimate $|\lambda_1(T_n^{-1})|$ from below according to (4.4.2). If we still take into consideration that

$$a_1(T_n^{-1}) = \| T_n^{-1} \|$$

and make use of (4.4.3), we in this way arrive at

$$\| T_n^{-1} \| \leqslant |\lambda_n(T)|^{-1}. \tag{4.4.4}$$

Furthermore, the norm $\| T_n^{-1} \|$ of the inverse operator T_n^{-1} can be related to the nth approximation number $a_n(T_n)$ of T_n by using the representation

$$I_n = T_n T_n^{-1}$$

of the identity map I_n of H_n. The norm determining property (A5) of the approximation numbers then yields

$$1 \leqslant a_n(T_n) \cdot \| T_n^{-1} \|. \qquad (4.4.5)$$

Combining (4.4.4) with (4.4.5) we obtain

$$|\lambda_n(T)| \leqslant a_n(T_n). \qquad (4.4.6)$$

Finally the term $a_n(T_n)$ on the right-hand side of (4.4.6) can be replaced by $a_n(T)$. This is due to the Hilbert space situation which guarantees the existence of a norm 1 (orthogonal) projection onto every subspace. Let $P:H \to H$ be the orthogonal projection of H onto H_n so that

$$P = J_n P_n$$

with $P_n:H \to H_n$, $\| P_n \| = 1$, and the natural embedding $J_n:H_n \to H$ of H_n into H. Then T_n can be represented as

$$T_n = P_n T J_n$$

which enables us to conclude that

$$a_i(T_n) \leqslant a_i(T), \quad i = 1, 2, 3, \ldots \qquad (4.4.7)$$

Hence (4.4.6) can be extended to

$$|\lambda_n(T)| \leqslant a_n(T).$$

To see the case $\lambda_n(T) = r_{\text{ess}}(T)$ we refer to

$$r_{\text{ess}}(T) \leqslant a_n(T)$$

(see (4.1.16)) and (4.4.2) with $r_{\text{ess}}(T)$ in place of $|\lambda_n(T)|$. This completes the proof. ∎

Remark. We remind the reader of the approximation numbers

$$a_n(D) = \sigma_n \qquad (2.1.8)$$

of a diagonal operator $D:l_p \to l_p$ generated by a non-increasing sequence $\sigma_1 \geqslant \sigma_2 \geqslant \cdots \geqslant \sigma_n \geqslant \cdots \geqslant 0$ of non-negative numbers. The basic idea for the proof of (2.1.8) was the same as that for the proof of $|\lambda_n(T)| \leqslant a_n(T)$ under the suppositions of Proposition 4.4.1. Indeed, since the restriction T_n of the self-adjoint operator $T:H \to H$ to the invariant subspace H_n of H is also self-adjoint, it can be represented by a matrix, which is not only triangular, but also diagonal.

Corollary. *In the case of a self-adjoint operator $T \in L(H,H)$ in a Hilbert space H the representations (4.1.13) and (4.1.17) for the essential spectral radius $r_{\text{ess}}(T)$ reduce to*

$$r_{\text{ess}}(T) = \beta(T) \qquad (4.4.8)$$

and

$$r_{\text{ess}}(T) = a(T), \qquad (4.4.9)$$

respectively.

Proof. The representation (4.4.1) for $|\lambda_n(T)|$ implies

$$\lim_{n \to \infty} |\lambda_n(T)| = a(T)$$

which, by (4.1.20), amounts to (4.4.9). Furthermore, the measure of non-approximability $a(T)$ is related to the measure of non-compactness $\beta(T) = e(T)$ by $\beta(T) \leqslant a(T)$ (see (3.5.14)). On the other hand, we have $r_{\text{ess}}(T) \leqslant \beta(T)$ by (4.1.15). This proves (4.4.8).   ∎

We once more recall that

$$r_{\text{ess}}(T) \leqslant a(T) \leqslant a_n(T) \tag{4.1.16}$$

is valid for operators T acting in an arbitrary Banach space E. The claim

$$|\lambda_n(T)| \leqslant a_n(T), \tag{4.4.10}$$

however, which results from (4.1.16) if $r_{\text{ess}}(T)$ is replaced by $|\lambda_n(T)| \geqslant r_{\text{ess}}(T)$, in general is not even true for operators T acting in a Hilbert space H. Following König (1986) we consider the operator $T: l_2^n \to l_2^n$ defined by the (n, n)-matrix

$$\begin{pmatrix} 0 & 1 & 0 & \cdots & & 0 \\ . & 0 & 1 & \cdots & & 0 \\ . & . & 0 & \cdots & & 0 \\ . & . & . & & . & \vdots \\ 0 & . & & & . & 1 \\ \sigma & 0 & & \cdots & & 0 \end{pmatrix}$$

with $\sigma > 1$. The adjoint operator T^* is then represented by the (n, n)-matrix

$$\begin{pmatrix} 0 & & \cdots & & 0 & \sigma \\ 1 & 0 & \cdots & & & 0 \\ 0 & 1 & . & & & \\ . & 0 & & . & & \vdots & \vdots \\ \vdots & \vdots & & & 0 & \\ 0 & 0 & \cdots & 0 & 1 & 0 \end{pmatrix}$$

The product T^*T is obviously the diagonal operator with $\sigma^2, 1, 1, \ldots, 1$ as diagonal elements and, correspondingly, the absolute value $|T| = (T^*T)^{1/2}$ is the diagonal operator with the generating sequence $\sigma, 1, \ldots, 1$. Since $a_i(T) = a_i(|T|)$ (see 3.4.10) we therefore have

$$a_1(T) = \sigma \quad \text{and} \quad a_i(T) = 1 \quad \text{for } 2 \leqslant i \leqslant n.$$

Now let us calculate the absolute values $|\lambda_i(T)|$ of the eigenvalues $\lambda_i(T)$ of T. Checking that

$$T^n = \sigma I_n,$$

where I_n denotes the identity map of l_2^n, we recognize that

$$\lambda_i(T^n) = \sigma \qquad \text{for } 1 \leqslant i \leqslant n$$

and hence

$$|\lambda_i(T)| = \sigma^{1/n} \quad \text{for } 1 \leqslant i \leqslant n$$

by (4.2.6). Comparing $|\lambda_n(T)|$ with $a_n(T)$ we see that

$$|\lambda_n(T)| = \sigma^{1/n} > 1 = a_n(T),$$

which contradicts (4.4.10) and simultaneously makes it clear that there is no constant C making an estimate

$$|\lambda_n(T)| \leqslant C \cdot a_n(T)$$

possible for all operators $T: l_2^n \to l_2^n$. However, if we take the product $\prod_{i=1}^n |\lambda_i(T)| = \sigma$ the equality

$$\prod_{i=1}^n |\lambda_i(T)| = \prod_{i=1}^n a_i(T)$$

turns out to be true. This is because T is an operator acting in an n-dimensional Hilbert space, as we shall see later (see Proposition 4.4.2). But at any rate we can prove

$$\prod_{i=1}^n |\lambda_i(T)| \leqslant \prod_{i=1}^n a_i(T), \quad n = 1, 2, 3, \ldots, \qquad (4.4.11)$$

for operators T acting in an arbitrary Hilbert space H.

The inequality (4.4.11) was established first by Weyl (1949) for compact operators $T \in L(H, H)$. Certainly Weyl did not make use of the approximation numbers, but operated with the eigenvalues $\lambda_i(|T|)$ of the absolute value $|T| = (T^*T)^{1/2}$ of the operator T (see (3.4.6)), which were referred to as the *singular numbers* $s_i(T)$ *of* T. It was Allakhverdiev (1957) who discovered the coincidence

$$\lambda_i(|T|) = a_i(T) \qquad (4.4.12)$$

of the singular numbers of T with its approximation numbers. Equation (4.4.12) became the origin of the theory of approximation numbers in the Banach space setting and, more generally, of the theory of so-called s-numbers (see Pietsch 1974, 1978, 1987). The approach to (4.4.12) within this new framework is based on the claim

$$a_i(T) = a_i(|T|) \qquad (3.4.10)$$

already used for the proof of Theorem 3.4.1, and on the equality

$$a_i(|T|) = \lambda_i(|T|)$$

guaranteed by Proposition 4.4.1 because of the positivity of the operator $|T|$.

In his 1949 paper Weyl also proved the inequality

$$\sum_{i=1}^{n} |\lambda_i(T)|^p \leqslant \sum_{i=1}^{n} a_i(T)^p, \quad n = 1, 2, 3, \ldots, \tag{4.4.13}$$

for compact operators $T \in L(H, H)$ and arbitrary $p > 0$, as already mentioned in section 4.2. In fact, (4.4.13) results from (4.4.11) by purely analytical considerations.

Lemma 4.4.1 *Let* $\alpha_1 \geqslant \alpha_2 \geqslant \cdots \geqslant \alpha_n \geqslant 0$ *and* $\beta_1, \beta_2, \ldots, \beta_n$ *be arbitrary non-negative numbers such that*

$$\prod_{i=1}^{k} \alpha_i \leqslant \prod_{i=1}^{k} \beta_i \quad \text{for } k = 1, 2, \ldots, n. \tag{4.4.14}$$

Then

$$\sum_{i=1}^{n} \alpha_i^p \leqslant \sum_{i=1}^{n} \beta_i^p \quad \text{for any exponent } p > 0. \tag{4.4.15}$$

Proof (communicated to the authors by *A.* Hess). Without loss of generality we can assume $\alpha_i > 0$ and $\beta_i > 0$ for $1 \leqslant i \leqslant n$. The supposition (4.4.14) then implies

$$\prod_{i=1}^{k} \alpha_i^p \leqslant \prod_{i=1}^{k} \beta_i^p \quad \text{for } k = 1, 2, \ldots, n$$

and any $p > 0$. Hence it suffices to prove (4.4.15) for $p = 1$. The main tool is a generalized version of the inequality between the geometric and the arithmetic mean of positive numbers $a_1, a_2, \ldots, a_n$, that is the inequality

$$\prod_{i=1}^{n} a_i^{p_i} \leqslant \sum_{i=1}^{n} p_i a_i \quad \text{with } \sum_{i=1}^{n} p_i = 1.$$

Putting $p_i = \alpha_i / \sum_{j=1}^{n} \alpha_j$ and $a_i = \beta_i / \alpha_i$ we thus obtain

$$\prod_{i=1}^{n} \left(\frac{\beta_i}{\alpha_i} \right)^{\alpha_i / \sum_{j=1}^{n} \alpha_j} \leqslant \sum_{i=1}^{n} \frac{\alpha_i}{\sum_{j=1}^{n} \alpha_j} \cdot \frac{\beta_i}{\alpha_i} = \frac{\sum_{i=1}^{n} \beta_i}{\sum_{i=1}^{n} \alpha_i}.$$

It remains to show that

$$1 \leqslant \prod_{i=1}^{n} \left(\frac{\beta_i}{\alpha_i} \right)^{\alpha_i / \sum_{j=1}^{n} \alpha_j}$$

which amounts to

$$1 \leqslant \prod_{i=1}^{n} \left(\frac{\beta_i}{\alpha_i} \right)^{\alpha_i}. \tag{4.4.16}$$

But (4.4.16) immediately becomes clear by rewriting the product $\prod_{i=1}^{n} (\beta_i / \alpha_i)^{\alpha_i}$ in the form

$$\prod_{i=1}^{n} \left(\frac{\beta_i}{\alpha_i} \right)^{\alpha_i} = \prod_{k=1}^{n} \left(\frac{\beta_1 \beta_2 \cdots \beta_k}{\alpha_1 \alpha_2 \cdots \alpha_k} \right)^{\alpha_k - \alpha_{k+1}}$$

with $\alpha_{n+1} = 0$ and making use of the supposition (4.4.14) and $\alpha_k - \alpha_{k+1} \geqslant 0$. Hence the proof of

$$1 \leqslant \sum_{i=1}^{n} \beta_i \bigg/ \sum_{i=1}^{n} \alpha_i$$

is finished. ∎

On the basis of the result (4.4.1) derived for self-adjoint operators we now prove both the multiplicative and the additive Weyl inequalities (4.4.11) and (4.4.13), respectively, which serve as a substitute for (4.4.1) in case of an arbitrary operator $T \in L(H, H)$ (cf. Gohberg and Krein 1969).

Proposition 4.4.2. (Weyl's inequalities). *Let* $T \in L(H, H)$ *be an arbitrary operator acting in a Hilbert space H. Then*

$$\prod_{i=1}^{n} |\lambda_i(T)| \leqslant \prod_{i=1}^{n} a_i(T) \quad \text{for } n = 1, 2, 3, \ldots \tag{4.4.11}$$

and

$$\sum_{i=1}^{n} |\lambda_i(T)|^p \leqslant \sum_{i=1}^{n} a_i(T)^p \quad \text{for } n = 1, 2, 3, \ldots \tag{4.4.13}$$

and any exponent $p > 0$. If, in particular, $T: H_n \to H_n$ is an operator acting in an n-dimensional Hilbert space H_n, then we even have equality in (4.4.11), that is

$$\prod_{i=1}^{n} |\lambda_i(T)| = \prod_{i=1}^{n} a_i(T). \tag{4.4.17}$$

Proof. We start with the proof of (4.4.17) for an operator $T: H_n \to H_n$. If rank $(T) < n$ we have $a_n(T) = 0$ as well as $\lambda_n(T) = 0$ and there is nothing left to prove. So let us suppose that rank $(T) = n$. Then T is invertible and the partial isometry U in the polar decomposition

$$T = U|T| \tag{3.4.7}$$

of T turns out to be an isometry. This implies

$$|\det(\omega_{ik})| = 1$$

for the determinant of any representing matrix (ω_{ik}) of U and hence

$$|\det(\tau_{ik})| = |\det(\sigma_{ik})|$$

for the determinants of the corresponding matrices (τ_{ik}) and (σ_{ik}) representing the operators T and $|T|$, respectively. Furthermore, according to Lemma 4.2.1, Supplement 2, the determinants $\det(\tau_{ik})$ and $\det(\sigma_{ik})$ are given by the products of the eigenvalues $\lambda_i(T)$ and $\lambda_i(|T|)$ of the underlying operators T and $|T|$, respectively, so that

$$\prod_{i=1}^{n} |\lambda_i(T)| = \prod_{i=1}^{n} \lambda_i(|T|).$$

Allakhverdiev's result (4.4.12) finally yields

$$\prod_{i=1}^{n} |\lambda_i(T)| = \prod_{i=1}^{n} a_i(T). \tag{4.4.17}$$

Now let us consider an operator $T:H \to H$ acting in an infinite-dimensional Hilbert space H. If $\lambda_n(T) \in \Lambda(T)$ then Lemma 4.2.1 guarantees the existence of an n-dimensional subspace $H_n \subset H$ invariant under T such that the restriction T_n of T to H_n has precisely $\lambda_1(T), \lambda_2(T), \ldots, \lambda_n(T)$ as its eigenvalues. Therefore (4.4.17) applies, namely

$$\prod_{i=1}^{n} |\lambda_i(T)| = \prod_{i=1}^{n} a_i(T_n).$$

However, the approximation numbers $a_i(T_n)$ of the restriction T_n of T to H_n are subject to the estimates

$$a_i(T_n) \leqslant a_i(T), \quad i = 1, 2, 3, \ldots \tag{4.4.7}$$

This completes the proof of (4.4.11) under the assumption $\lambda_n(T) \in \Lambda(T)$. On the other hand, if

$$\lambda_m(T) \in \Lambda(T), \text{ but } \lambda_i(T) = r_{\text{ess}}(T) \quad \text{for } m < i \leqslant n,$$

we have

$$\prod_{i=1}^{m} |\lambda_i(T)| \leqslant \prod_{i=1}^{m} a_i(T)$$

and finally even

$$\prod_{i=1}^{n} |\lambda_i(T)| \leqslant \prod_{i=1}^{n} a_i(T) \tag{4.4.11}$$

since $r_{\text{ess}}(T) \leqslant a_i(T)$ by (4.1.16).

The additive Weyl inequality (4.4.13) follows from (4.4.11) by Lemma 4.4.1.   ∎

5
Operators with values in C(X)

5.1. Why $C(X)$-valued operators?

From the very beginning of functional analysis the space $C[a, b]$ of real- or complex-valued continuous functions on a closed interval $[a, b]$ has been one of the most popular and important Banach spaces. The significance of $C[a, b]$ is, not least, based on its universality in the sense that any separable Banach space E is isometrically isomorphic to a subspace of $C[a, b]$ (see Kantorowitsch and Akilow 1978).

The *universality of the Banach space $C[a, b]$* implies a *universality of the class of $C[a, b]$-valued compact operators* in the following sense. Given a compact operator T between arbitrary Banach spaces E and F there exists a compact operator $S: E \to C[a, b]$ such that the injective compactness quantities of S and T coincide, namely

$$c_n(T) = c_n(S), \ t_n(T) = t_n(S), \ f_n(T) = f_n(S), \tag{5.1.1}$$

and

$$\tfrac{1}{2} e_n(T) \leqslant e_n(S) \leqslant 2 e_n(T). \tag{5.1.2}$$

This results from the separability of the range of a compact operator $T: E \to F$. Indeed, considering the compact operator $T_0: E \to F_0$ induced by T, where $F_0 = \overline{T(E)}$, we obtain

$$c_n(T) = c_n(T_0), \ t_n(T) = t_n(T_0), f_n(T) = f_n(T_0), \tag{5.1.3}$$

and

$$e_n(T) \leqslant e_n(T_0) \leqslant 2 e_n(T) \tag{5.1.4}$$

by the injectivity of the Gelfand numbers $c_n(T)$, the symmetrized approximation numbers $t_n(T)$, the inner entropy numbers $f_n(T)$, and by the injectivity of the outer entropy numbers $e_n(T)$ up to a factor 2 (see (1.3.6)). Owing to the separability of F_0 there is a metric injection $J: F_0 \to C[a, b]$ of F_0 into $C[a, b]$. By the injectivity arguments already used we thus get the relations (5.1.1) and (5.1.2) from (5.1.3) and (5.1.4), respectively, with

$$S = J T_0$$

as the composed compact operator from E into $C[a, b]$.

Hence the properties of an arbitrary compact operator T reflected by injective compactness quantities are always shared by a compact operator

S with values in $C[a,b]$. In this sense compact $C[a,b]$-valued operators represent a model for compact operators.

In the following we shall consider an arbitrary compact metric space (X,d) and study compact operators $T:E \to C(X)$ with values in the space $C(X)$ of continuous functions on X. Then it is obvious to ask how the behaviour of the sequence $\varepsilon_n(X)$ of the entropy numbers of X influences the sequences $a_n(T)$ and $e_n(T)$ of the approximation numbers and the dyadic entropy numbers of the operator $T:E \to C(X)$, respectively. For the approximation numbers $a_n(T)$, a universal estimate from above can be derived in analogy to the classical *Jackson's inequality*

$$E_n(f) \leqslant C \cdot \omega(f;(b-a)/2n) \qquad (3.0.5)$$

for the approximation numbers $E_n(f)$ of functions $f \in C[a,b]$ by polynomials p of degree $\deg(p) \leqslant n$, as already mentioned in the introduction to chapter 3. Note that

$$(b-a)/2n = \varepsilon_n([a,b])$$

is the nth entropy number of the closed interval $[a,b]$. The crucial point in the operator situation $T:E \to C[a,b]$ or, more generally $T:E \to C(X)$, is the *definition of a modulus of continuity $\omega(T;\delta)$*. This definition presents itself in quite a natural way (see (3.0.6) and section 5.5) and then allows the investigation of the approximability of compact operators $T:E \to C(X)$ in an amazing analogy to the approximability of continuous functions on a closed interval $[a,b]$. In particular, the well-known *Hölder classes $C^\alpha[a,b]$ of continuous functions* give rise to corresponding classes $L^\alpha(E,C(X))$ of $C(X)$-valued operators (see section 5.9). On the basis of the universal estimate of Jackson type

$$a_{n+1}(T) \leqslant \omega(T;\varepsilon_n(X)) \qquad (3.0.8)$$

to be proved in section 5.6, we shall be able to describe the decrease of the approximation numbers $a_n(T)$ of a Hölder continuous operator $T \in L^\alpha(E,C,(X))$ in terms of the entropy numbers $\varepsilon_n(X)$ of X. The dyadic entropy numbers $e_n(T)$ of an operator $T \in L^\alpha(E,C(X))$ can be shown to have the same rate of decrease if we assume that the compact metric space (X,d) is connected and consists of more than one point (see section 5.9). This is due to the inequality of Bernstein type

$$e_m(T) \leqslant c_p \cdot m^{-1/p} \sup_{1 \leqslant k \leqslant m} k^{1/p} a_k(T) \qquad (3.1.8)$$

considered in section 3.1. Hence the scope of the general methods developed in the previous chapters is demonstrated in an impressive way.

Specific methods for a more detailed investigation of approximation and entropy quantities of Hölder continuous operators $T:E \to C(X)$ are developed in section 5.10. Since they are based on special suppositions for the finite-dimensional operators $S:E \to l_\infty^n$ defined on E they are referred to as *local techniques*.

For arbitrary $C(X)$-valued operators $T:E \to C(X)$ the combination of the Bernstein type inequality (3.1.8) with the estimate (3.0.8) of Jackson type is more involved. However, under appropriate simultaneous constraints for both the compact metric space X and the operator $T:E \to C(X)$ we can derive an estimate for the entropy numbers $e_n(T)$ from above similar to the estimate (3.0.8) for the approximation numbers (see Theorem 5.7.1). If we confine ourselves to compact and convex subsets X of a normed space with more than one point, this estimate for the $e_n(T)$ applies to all compact operators $T:E \to C(X)$. When rank$(T) > 1$ it can be simplified in such a way that the analogy to (3.0.8) becomes even more obvious (see Proposition 5.8.4).

Sections 5.6, 5.7, 5.8, and 5.9 do not impose any particular conditions on the domain E. In contrast with that, sections 5.11 and 5.12 are devoted to *integral operators* and thus automatically require E to be a function space. We shall deal with the space $E = C(X)$ and with spaces $E = L_p(X;\mu)$ of functions f whose pth power $|f|^p$ is integrable with respect to a Borel measure μ on X. The integral operators $T_{K,\mu}:C(X) \to C(X)$ and $T_{K,\mu}:L_p(X;\mu) \to C(X)$ to be considered will be generated by a continuous kernel $K(s,t)$ on $X \times X$. These operators turn out to be compact and, moreover, their modulus of continuity $\omega(T_{K,\mu};\delta)$ can be related to a *modulus of continuity* $\Omega_1(K;\delta)$ or $\Omega_p(K;\delta)$, respectively, *associated with the kernel* $K \in C(X \times X)$ (see Propositions 5.11.1 and 5.11.2). Hence in particular the Hölder continuity of an integral operator $T_{K,\mu}$ can be expressed by a corresponding Hölder condition for the generating kernel $K(s,t)$. Certain approximation and compactness properties of $T_{K,\mu}$ appear as immediate consequences of corresponding Hölder continuity properties of $K(s,t)$ (see section 5.12).

Having learnt to work with integral operators and their kernels, we develop a method for generating operators from an arbitrary Banach space E into $C(X)$ by a kind of '*abstract kernels*' (see section 5.13). In the end, however, we see that every compact operator from E into $C(X)$ is obtained in this way. This result makes it clear that an arbitrary compact operator with values in $C(X)$ may be regarded as a kind of *generalized integral operator* generated by a continuous generalized kernel. The special results of sections 5.11 and 5.12 can be classified into this general framework, and the proofs given in these sections subsequently appear as concrete versions of the abstract proofs in section 5.13. But since we preferred a direct approach to integral operators we have derived separately entropy and compactness properties of these operators, which are generated by kernels in the classical sense.

5.2. Local properties of $C(X)$

We recall that a metric space (X, d) is compact if it is precompact and complete. Given a compact metric space (X, d) we denote by $C(X)$ the linear space of all real- or complex-valued continuous functions f on X. If equipped with the norm

$$\| f \| = \sup \{ |f(s)| : s \in X \}, \tag{5.2.1}$$

the space $C(X)$ becomes a Banach space.

Local properties of a Banach space quite generally are understood to be *geometrical properties of its finite-dimensional subspaces*. For the space $C(X)$, every finite-dimensional subspace M can be embedded into a finite-dimensional subspace N which is arbitrarily close to an l_∞^m-space. This results from a fundamental relation between coverings of X by finitely many open subsets $G_1, G_2, \ldots, G_n$ and finite systems of continuous functions $\varphi_1, \varphi_2, \ldots, \varphi_n$ on X forming a so-called partition of unity.

A *partition of unity* $\{ \varphi_1, \varphi_2, \ldots, \varphi_n \}$ *on* X is characterized by the properties

$$\varphi_i \in C(X) \text{ and } 0 \leqslant \varphi_i(s) \leqslant 1 \quad \text{for } s \in X, \ 1 \leqslant i \leqslant n, \tag{5.2.2}$$

and

$$\sum_{i=1}^{n} \varphi_i(s) = 1 \quad \text{for } s \in X. \tag{5.2.3}$$

The relation between partitions of unity and open coverings of X that we are going to prove is based on the concept of the *support*

$$\operatorname{supp}(\varphi) = \{ s \in X : \varphi(s) \neq 0 \}$$

of a continuous function φ on X and on the concept of the *distance*

$$\operatorname{dist}(s; M) = \inf \{ d(s, t) : t \in M \}$$

of a point s from a non-empty subset $M \subseteq X$. For a fixed subset $M \subseteq X$ the function $\operatorname{dist}(s; M)$ is uniformly continuous on X since

$$|\operatorname{dist}(s; M) - \operatorname{dist}(t; M)| \leqslant d(s, t)$$

for arbitrary points $s, t \in X$. This inequality, which is an immediate consequence of the triangle inequality, also implies that

$$\operatorname{dist}(s; M) = 0 \text{ if and only if } s \in \bar{M}.$$

Hence, if $A_1, A_2, \ldots, A_n$ are closed subsets of X with $\bigcap_{i=1}^{n} A_i = \emptyset$ we have

$$\sum_{i=1}^{n} \operatorname{dist}(s; A_i) > 0 \quad \text{for all } s \in X. \tag{5.2.4}$$

Lemma 5.2.1. *Any partition of unity* $\{\varphi_1, \varphi_2, \ldots, \varphi_n\}$ *on X gives rise to a covering*

$$X = \bigcup_{i=1}^{n} G_i \qquad (5.2.5)$$

of X by open subsets

$$G_i = \text{supp}(\varphi_i). \qquad (5.2.6)$$

Conversely, any covering (5.2.5) of X by open subsets $G_1, G_2, \ldots, G_n$ *can be satisfied by the supports (5.2.6) of a partition of unity* $\{\varphi_i, \varphi_2, \ldots, \varphi_n\}$ *on X.*

Proof. For any partition of unity $\{\varphi_i, \varphi_2, \ldots, \varphi_n\}$ on X the sets (5.2.6) are open, because of the continuity of the functions φ_i, and, by (5.2.3), cover X. On the other hand, given a covering (5.2.5) of X by open sets $G_1, G_2, \ldots, G_n$ we take the closed subsets

$$A_i = X \backslash G_i, \quad 1 \leqslant i \leqslant n,$$

and remark that

$$\bigcap_{i=1}^{n} A_i = \emptyset.$$

Hence we are sure of (5.2.4) and can define functions $\varphi_i(s)$ by

$$\varphi_i(s) = \frac{\text{dist}(s; A_i)}{\sum_{j=1}^{n} \text{dist}(s; A_j)} \quad \text{for } 1 \leqslant i \leqslant n.$$

Obviously the system $\{\varphi_1, \varphi_2, \ldots, \varphi_n\}$ possesses the properties (5.2.2) and (5.2.3) and thus represents a partition of unity on X. Moreover,

$$\varphi_i(s) \neq 0 \text{ if and only if } s \notin A_i.$$

But this amounts to (5.2.6). ∎

A partition of unity $\{\varphi_1, \varphi_2, \ldots, \varphi_n\}$ on X, related to the sets $G_1, G_2, \ldots, G_n$ of an open covering (5.2.5) of X by

$$\text{supp}(\varphi_i) \subseteq G_i \quad \text{for } 1 \leqslant i \leqslant n, \qquad (5.2.7)$$

is quite generally said to be *subordinate to the open covering* (5.2.5). In this terminology the part of Lemma 5.2.1 that is essential for what follows says that every covering of X by finitely many open sets gives rise to a partition of unity subordinate to it. A strengthened version of this result, namely the existence of a partition of unity $\{\varphi_1, \varphi_2, \ldots, \varphi_n\}$ with

$$\overline{\text{supp}(\varphi_i)} \subseteq G_i \quad \text{for } 1 \leqslant i \leqslant n$$

even turns out to be true, but will not be needed for what follows. In this connection we mention that the definition of the support $\text{supp}(\varphi)$ of a function $\varphi \in C(X)$ has been adapted to suit the interests of this book.

Usually the support of a continuous function φ on X is understood to be the closure of the set of points $s \in X$ with $\varphi(s) \neq 0$.

Proposition 5.2.1. (cf. Lindenstrauss and Tzafriri 1977, 1979). *Let* (X, d) *be a compact metric space and* $M \subseteq C(X)$ *a finite-dimensional subspace of the space* $C(X)$. *Then for every* $\varepsilon > 0$ *there exists a finite-dimensional subspace* $N \subseteq C(X)$ *containing* M *such that the Banach–Mazur distance* $d(N, l_\infty^n)$ *between* N *and* l_∞^n *with* $n = \dim(N)$ *satisfies the inequality*

$$d(N, l_\infty^n) < 1 + \varepsilon. \tag{5.2.8}$$

Proof. On a finite-dimensional Banach space all norms are equivalent. Hence, given a basis $f_1, f_2, \ldots, f_m$ of M there exists a constant $C > 0$ such that

$$\sup_{1 \leqslant i \leqslant m} |\xi_i| \leqslant C \left\| \sum_{i=1}^m \xi_i f_i \right\| \tag{5.2.9}$$

for all scalars $\xi_1, \xi_2, \ldots, \xi_m$. We put

$$\eta = \frac{\varepsilon}{2mC(2 + \varepsilon)} \tag{5.2.10}$$

and, taking account of the uniform continuity of the functions $f_1, f_2, \ldots, f_m$ on the compact metric space (X, d), determine $\delta > 0$ such that

$$|f_i(s) - f_i(t)| < \eta \quad \text{for } d(s, t) < \delta \text{ and } 1 \leqslant i \leqslant m.$$

Then we choose a minimal system of open balls $\mathring{U}(t_k; \delta)$ of radius δ which cover X, $1 \leqslant k \leqslant n$. A partition of unity $\{\varphi_1, \varphi_2, \ldots, \varphi_n\}$ subordinate to the covering

$$X = \bigcup_{k=1}^n \mathring{U}(t_k; \delta) \tag{5.2.11}$$

in the sense of

$$\operatorname{supp}(\varphi_k) \subseteq \mathring{U}(t_k; \delta) \quad \text{for } 1 \leqslant k \leqslant n \tag{5.2.7'}$$

spans a space

$$L = \operatorname{span}\{\varphi_1, \varphi_2, \ldots, \varphi_n\}$$

isometrically isomorphic to l_∞^n. Indeed, since none of the balls $\mathring{U}(t_k; \delta)$ of the covering (5.2.11) can be omitted there are elements $s_k \in \mathring{U}(t_k; \delta)$ with $s_k \notin \mathring{U}(t_j; \delta)$ for $j \neq k$. From (5.2.7') and (5.2.3) we therefore have

$$\varphi_j(s_k) = \delta_{jk} \quad \text{for } 1 \leqslant j, k \leqslant n.$$

This implies

$$\sup_{1 \leqslant j \leqslant n} |\lambda_j| = \sup_{1 \leqslant j \leqslant n} |\lambda_j \varphi_j(s_j)| \leqslant \sup_{s \in X} \left| \sum_{k=1}^n \lambda_k \varphi_k(s) \right| = \|\varphi\|$$

for $\varphi = \sum_{k=1}^{n} \lambda_k \varphi_k$ with arbitrary scalars $\lambda_1, \lambda_2, \ldots, \lambda_n$. On the other hand,

$$\|\varphi\| = \sup_{s \in X} \left| \sum_{k=1}^{n} \lambda_k \varphi_k(s) \right| \leqslant \sup_{1 \leqslant j \leqslant n} |\lambda_j|$$

is true for any partition of unity $\{\varphi_1, \varphi_2, \ldots, \varphi_n\}$ on X.

The isometry of L to l_{∞}^n having already been proved, it suffices to construct an n-dimensional subspace $N \subseteq C(X)$ which contains M and satisfies the condition

$$d(N, L) < 1 + \varepsilon.$$

We first consider the subspace H of L spanned by the functions

$$h_i(s) = \sum_{k=1}^{n} f_i(t_k) \varphi_k(s), \quad 1 \leqslant i \leqslant m.$$

Since these elements $h_1, h_2, \ldots, h_m$ are relatively 'close' to the basis $f_1, f_2, \ldots, f_m$ of M, they also turn out to be linearly independent. In fact, we have

$$\|f_i - h_i\| = \sup_{s \in X} \left| \sum_{k=1}^{n} (f_i(s) - f_i(t_k)) \varphi_k(s) \right| < \eta$$

because

$$|f_i(s) - f_i(t_k)| < \eta \quad \text{for } s \in \mathring{U}(t_k; \delta)$$

and

$$\varphi_k(s) = 0 \quad \text{for } s \notin \mathring{U}(t_k; \delta),$$

and thus obtain

$$\sup_{1 \leqslant i \leqslant m} |\xi_i| \leqslant C \left(\left\| \sum_{i=1}^{m} \xi_i (f_i - h_i) \right\| + \left\| \sum_{i=1}^{m} \xi_i h_i \right\| \right)$$

$$\leqslant Cm\eta \sup_{1 \leqslant i \leqslant m} |\xi_i| + C \left\| \sum_{i=1}^{m} \xi_i h_i \right\|$$

because of (5.2.9). The choice (5.2.10) of η guarantees that

$$\eta < \frac{1}{2mC} \tag{5.2.12}$$

and hence

$$\sup_{1 \leqslant i \leqslant m} |\xi_i| \leqslant 2C \left\| \sum_{i=1}^{m} \xi_i h_i \right\|. \tag{5.2.13}$$

This proves the linear independence of $h_1, h_2, \ldots, h_m$. In the next step we employ the coordinate functionals

$$\zeta_i = \langle h, a_i^{(0)} \rangle, \quad \text{where } h = \sum_{i=1}^{m} \zeta_i h_i,$$

with respect to the basis $h_1, h_2, \ldots, h_m$ of H in order to define an isomorphism A_0 of H onto M by

$$A_0 h = \sum_{i=1}^{m} \langle h, a_i^{(0)} \rangle f_i.$$

If we then extend the functionals $a_i^{(0)}$ over H to linear functionals a_i over L with the same norm and put

$$Ah = h + \sum_{i=1}^{m} \langle h, a_i \rangle (f_i - h_i) \quad \text{for } h \in L,$$

we obtain an isomorphism $A: L \to A(L) = N$ which is an extension of $A_0: H \to M$ and has the desired property

$$\|A\| \, \|A^{-1}\| < 1 + \varepsilon. \tag{5.2.14}$$

This follows from

$$h = \sum_{i=1}^{m} \langle h, a_i \rangle h_i \quad \text{for } h \in H$$

and a norm estimate for Ah from above and below. The crucial point is the norm estimate

$$\|a_i\| = \|a_i^{(0)}\| \leqslant 2C$$

for the functionals a_i which results from (5.2.13). Then a straight-forward estimate for $\|Ah\|$ yields

$$\|Ah\| \leqslant \|h\| + \sum_{i=1}^{m} |\langle h, a_i \rangle| \, \|f_i - h_i\|$$

$$< \|h\| (1 + 2mC\eta) \quad \text{for } h \neq 0$$

and thus

$$\|A\| < 1 + 2mC\eta.$$

Similarly we get

$$\|h\| < \|Ah\| + 2mC\eta \|h\| \quad \text{for } h \neq 0$$

which means

$$\|h\| (1 - 2mC\eta) < \|Ah\| \quad \text{for } h \neq 0.$$

Note, that $1 - 2mC\eta > 0$ by (5.2.12), so that the above inequality can also be written as

$$\|A^{-1}\| < \frac{1}{1 - 2mC\eta}.$$

If we now use (5.2.10) to substitute for η on the right-hand side of the inequality

$$\|A\| \, \|A^{-1}\| < \frac{1 + 2mC\eta}{1 - 2mC\eta}$$

we actually arrive at (5.2.14). The proof of the assertion (5.2.8) is completed by the remark that

$$d(N, l^n_\infty) = d(A(L), L) \leqslant \|A\| \, \|A^{-1}\|. \qquad \blacksquare$$

The *space* $l_1(\Gamma)$ *of summable number families over an arbitrary index set* T possesses a similar local property. This property has already been used in section 2.7 when studying the 'local Kolmogorov numbers' $\hat{d}_n(T)$. The proof will be sketched now according to the pattern of the proof of Proposition 5.2.1.

Proposition 5.2.2. *Let* Γ *be an arbitrary index set and* $M \subseteq l_1(\Gamma)$ *a finite-dimensional subspace of the space* $l_1(\Gamma)$. *Then for every* $\varepsilon > 0$ *there exists a finite-dimensional subspace* $N \subseteq l_1(\Gamma)$ *containing* M *such that*

$$d(N, l^n_1) < 1 + \varepsilon \quad \text{with } n = \dim(N). \qquad (5.2.15)$$

Proof. Given a basis $f_1, f_2, \dots, f_m$ of M we again have an estimate (5.2.9) with a constant $C > 0$ and can then introduce $\eta > 0$ related to $\varepsilon > 0$ by (5.2.10). Since any element $\{\lambda_\gamma\}_{\gamma \in \Gamma} \in l_1(\Gamma)$ has at most countably many components $\lambda_\gamma \neq 0$ there is a countable subset $\Gamma_0 \subseteq \Gamma$ such that the components $\lambda_\gamma^{(i)}$ of the basis vectors f_i, $1 \leqslant i \leqslant m$, vanish for $\gamma \notin \Gamma_0$. Hence M can be considered as a subspace of the space $l_1(\Gamma_0)$ which is isometric to l_1. We simply write

$$f_i = (\lambda_1^{(i)}, \lambda_2^{(i)}, \dots, \lambda_k^{(i)}, \dots)$$

for the m basis vectors $f_1, f_2, \dots, f_m$ of M. Choosing a natural number n such that

$$\sum_{j=n+1}^{\infty} |\lambda_j^{(i)}| < \eta \quad \text{for } 1 \leqslant i \leqslant m$$

we can use the elements

$$h_i = (\lambda_1^{(i)}, \lambda_2^{(i)}, \dots, \lambda_n^{(i)}, 0, 0, \dots), \quad 1 \leqslant i \leqslant m,$$

to approximate the basis vectors $f_1, f_2, \dots, f_m$ in the sense that

$$\|f_i - h_i\| < \eta \quad \text{for } 1 \leqslant i \leqslant m.$$

From here on all arguments in the proof of Proposition 5.2.1 can be carried over to this case. The n-dimensional subspace $N \subseteq l_1(\Gamma)$ that we are looking for is the image $A(L) = N$ of a subspace L isometric to l_1^n, where A is an isomorphism which extends an isomorphism A_0 of $H = \text{span}\{h_1, h_2, \dots, h_m\}$ onto M. $\qquad \blacksquare$

Remark. In the terminology of Lindenstrauss and Pełczyński (1968) the spaces $C(X)$ and $l_1(\Gamma)$ are $\mathscr{L}_{\infty,\lambda}$-*spaces* and $\mathscr{L}_{1,\lambda}$-*spaces*, respectively, for any $\lambda > 1$.

5.3. Approximation quantities of $C(X)$-valued operators

Knowledge of the local structure of the space $C(X)$ is the basis for calculating the *local approximation numbers* $\hat{a}_n(T)$ (see section 2.7) of $C(X)$-valued operators. The actual approximation numbers $a_n(T)$ that we were interested in originally can be related to the local ones by using the general results of section 2.7. In this way we obtain universal inequalities between approximation numbers and Gelfand numbers on one side and Kolmogorov numbers and symmetrized approximation numbers on the other side.

Proposition 5.3.1. *Let (X, d) be a compact metric space, E an arbitrary Banach space, and $T \in L(E, C(X))$. Then*

$$\hat{a}_n(T) = c_n(T) \quad for \ n = 1, 2, 3, \ldots \tag{5.3.1}$$

Proof. According to the definition

$$\hat{a}_n(T) = \sup \{a_n(TI_M^E): M \subseteq E, \dim(M) < \infty \}$$

of the local approximation numbers $\hat{a}_n(T)$ we have to deal with the restrictions TI_M^E of T to finite-dimensional subspaces $M \subseteq E$. Let $M \subseteq E$ with $\dim(M) < \infty$ be fixed and let T_0 denote the operator from M onto $L = (TI_M^E)(M)$ induced by TI_M^E so that

$$TI_M^E = I_L^{C(X)} T_0$$

with $I_L^{C(X)}$ as the natural embedding of L into $C(X)$. Given $\varepsilon > 0$, Proposition 5.2.1 guarantees the existence of a finite-dimensional subspace $N \subseteq C(X)$ containing L and of an isomorphism $A: N \to l_\infty^{\dim(N)}$ with

$$\|A\| \|A^{-1}\| < 1 + \varepsilon. \tag{5.2.14}$$

Splitting up the embedding $I_L^{C(X)}$ into the two embeddings $I_L^N: L \to N$ and $I_N^{C(X)}: N \to C(X)$ and inserting the isomorphisms $A: N \to l_\infty^{\dim(N)}$ and $A^{-1}: l_\infty^{\dim(N)} \to N$ we thus obtain the factorization

$$TI_M^E = I_N^{C(X)} A^{-1} A I_L^N T_0$$

of the restriction map TI_M^E. Now we estimate the nth approximation number $a_n(TI_M^E)$ by

$$a_n(TI_M^E) \leqslant \|I_N^{C(X)} A^{-1}\| a_n(AI_L^N T_0). \tag{5.3.2}$$

In view of the fact that the range $l_\infty^{\dim(N)}$ of A has the metric extension property we may replace $a_n(AI_L^N T_0)$ by $c_n(AI_L^N T_0)$ (see Proposition 2.3.3) and extend the estimate (5.3.2) to

$$a_n(TI_M^E) \leqslant \|A^{-1}\| \|A\| c_n(T_0) \leqslant (1 + \varepsilon)c_n(T_0).$$

Because of the injectivity of the Gelfand numbers we have

$$c_n(T_0) = c_n(I_L^{C(X)} T_0) = c_n(TI_M^E)$$

and, as a consequence of this,

$$a_n(TI_M^E) \leqslant (1 + \varepsilon)c_n(TI_M^E) \leqslant (1 + \varepsilon)c_n(T).$$

Therefore the local approximation number $\hat{a}_n(T)$ can also be estimated from above by

$$\hat{a}_n(T) \leqslant (1 + \varepsilon)c_n(T)$$

and finally even by

$$\hat{a}_n(T) \leqslant c_n(T).$$

On the other hand, Proposition 2.7.5 tells us that

$$c_n(T) = \hat{c}_n(T).$$

But the local Gelfand numbers

$$\hat{c}_n(T) = \sup \{c_n(TI_M^E) : M \subseteq E, \dim(M)\} \qquad (2.7.17)$$

are obviously related to the local approximation numbers by

$$\hat{c}_n(T) \leqslant \hat{a}_n(T).$$

This completes the proof of (5.3.1). ∎

Theorem 5.3.1. *Let* (X, d) *be a compact metric space, E an arbitrary Banach space, and $T \in L(E, C(X))$. Then*

$$a_n(T) \leqslant 5c_n(T) \qquad (5.3.3)$$

and

$$d_n(T) \leqslant 5t_n(T) \quad for \ n = 1, 2, 3, \ldots \qquad (5.3.4)$$

Proof. By Proposition 2.7.4 we have

$$a_n(T) \leqslant 5\hat{a}_n(T).$$

Hence the assertion (5.3.3) immediately results from (5.3.1). For the proof of (5.3.4) we now only need to refer to $d_n(T) = a_n(TQ_E)$ and $t_n(T) = c_n(TQ_E)$. ∎

In the case of a compact operator $T : E \to C(X)$ the situation is even more satisfactory.

Theorem 5.3.2. *Let* (X, d) *be a compact metric space, E an arbitrary Banach space, and $T \in K(E, C(X))$ a compact operator. Then*

$$a_n(T) = c_n(T) \qquad (5.3.5)$$

and

$$d_n(T) = t_n(T) \quad for \ n = 1, 2, 3, \ldots \qquad (5.3.6)$$

Proof. By Proposition 2.7.3 we have $\hat{a}_n(T) = a_n(T)$ for a compact operator T. Hence the assertion (5.3.5) follows from (5.3.1). For the proof of (5.3.6) we make use of Proposition 2.7.3 and Proposition 5.3.1 with TQ_E instead of T, concluding that

$$d_n(T) = a_n(TQ_E) = \hat{a}_n(TQ_E) = c_n(TQ_E) = t_n(T).$$ ∎

5.4. The modulus of continuity of functions

As we have already pointed out in section 5.1, compact $C(X)$-valued operators can be studied in analogy to continuous functions on X, a central notion being that of the modulus of continuity.

The *modulus of continuity* $\omega(f;\delta)$ *of a bounded function* f on a compact metric space (X,d) is defined by

$$\omega(f;\delta) = \sup\{|f(s) - f(t)|: s, t \in X, d(s,t) \leq \delta\} \quad (5.4.1)$$

for $0 \leq \delta < \infty$ (see (3.0.7)). The following properties of $\omega(f;\delta)$ can easily be checked:

$$\omega(f;\delta) \leq \omega(f;\delta') \quad \text{for } 0 \leq \delta \leq \delta', \quad (5.4.2)$$

$$\omega(f + g;\delta) \leq \omega(f;\delta) + \omega(g;\delta), \quad (5.4.3)$$

$$\omega(\lambda f;\delta) = |\lambda|\omega(f;\delta), \quad (5.4.4)$$

$$\omega(f;\delta) \leq 2\|f\|. \quad (5.4.5)$$

We recall that the norm of a function $f \in l_\infty(X)$ is given by the same expression

$$\|f\| = \sup\{|f(s)|: s \in X\}$$

as the norm of a continuous function f on X (see section 2.3). Since

$$|\omega(f;\delta) - \omega(g;\delta)| \leq \omega(f - g;\delta)$$

as a result of (5.4.3), we obtain

$$|\omega(f;\delta) - \omega(g;\delta)| \leq 2\|f - g\| \quad (5.4.6)$$

from (5.4.5). This inequality expresses the uniform continuity of the map $\omega(\cdot;\delta): l_\infty(X) \to \mathbb{R}$ of $l_\infty(X)$ into the real line $\mathbb{R}$ for fixed $\delta \geq 0$.

What about the continuity of $\omega(f;\delta)$ as a function of $\delta \geq 0$ for fixed $f \in l_\infty(X)$? Since

$$\omega(f;0) = 0$$

the continuity from the right at $\delta = 0$ amounts to

$$\lim_{\delta \to +0} \omega(f;\delta) = 0 \quad (5.4.7)$$

or, equivalently,

$$\inf\{\omega(f;\delta): \delta > 0\} = 0 \quad (5.4.8)$$

because $\omega(f;\delta)$ is non-increasing (see (5.4.2)). Obviously (5.4.7) or (5.4.8) is a characteristic property of continuous functions f on X, for continuity of a function f on a compact metric space implies uniform continuity and thus

$$\omega(f;\delta) < \varepsilon$$

for sufficiently small $\delta < \delta(\varepsilon)$. But, what is more, the continuity of a function f on a compact metric space (X,d) implies the continuity of $\omega(f;\delta)$ from the right for arbitrary $\delta \geqslant 0$.

Proposition 5.4.1. *Let (X,d) be a compact metric space, f a continuous function on X, and $\omega(f;\delta)$ its modulus of continuity. Then $\omega(f;\delta)$ is continuous from the right for all $\delta \geqslant 0$, that is*

$$\inf\{\omega(f;\delta'):\delta < \delta'\} = \omega(f;\delta) \tag{5.4.9}$$

for $\delta \geqslant 0$.

Proof. We assume that $\omega(f;\delta)$ is not continuous from the right at $\delta = \delta_0$. Then there exist $\varepsilon_0 > 0$ and a sequence δ_n with $\delta_0 < \delta_n < \delta_0 + 1/n$ such that

$$\omega(f;\delta_n) > \omega(f;\delta_0) + \varepsilon_0 \quad \text{for } n = 1,2,3,\ldots$$

From the definition of $\omega(f;\delta_n)$ we can now determine elements $s_n, t_n \in X$ with $d(s_n,t_n) \leqslant \delta_n$ and

$$\omega(f;\delta_n) \geqslant |f(s_n) - f(t_n)| > \omega(f;\delta_0) + \varepsilon_0 \quad \text{for } n = 1,2,3,\ldots$$

Furthermore, since the metric space (X,d) is compact the sequences $s_n, t_n \in X$ possess convergent subsequences s_{n_k} and t_{n_k}, such that, say,

$$\lim_{k \to \infty} s_{n_k} = s_0 \quad \text{and} \quad \lim_{k \to \infty} t_{n_k} = t_0.$$

Owing to the continuity of $d(s,t)$ on $X \times X$ we have

$$d(s_0,t_0) \leqslant \delta_0. \tag{5.4.10}$$

On the other hand, the continuity of the function f on X implies

$$\lim_{k \to \infty} |f(s_{n_k}) - f(t_{n_k})| = |f(s_0) - f(t_0)| \tag{5.4.11}$$

$$\geqslant \omega(f;\delta_0) + \varepsilon_0.$$

The two claims (5.4.10) and (5.4.11), however, contradict the definition of $\omega(f;\delta_0)$. This proves the continuity of $\omega(f;\delta)$ from the right at $\delta = \delta_0$ and hence (5.4.9) for all $\delta \geqslant 0$. ∎

The modulus of continuity $\omega(f;\delta)$ of a continuous function f on a compact metric space (X,d) need not be continuous from the left. The simplest examples are given by continuous functions f on disconnected compact metric spaces (X,d). For instance, let X be the subset $X = \{0,1\}$ of the real line $\mathbb{R}$ and $f \in C(X)$ the function with

$$f(0) = 0 \quad \text{and} \quad f(1) = 1.$$

The corresponding modulus of continuity

$$\omega(f;\delta) = \begin{cases} 1 & \text{for } \delta \geqslant 1 \\ 0 & \text{for } \delta < 1 \end{cases}$$

is obviously discontinuous at $\delta = 1$. But there are even examples of continuous functions f on connected compact metric spaces with a discontinuous modulus of continuity $\omega(f;\delta)$. Let us consider the subset

$$X = \{(x,0):0 \leqslant x \leqslant 1\} \cup \{(0,y):0 \leqslant y \leqslant 1\} \cup \{(x,1):0 \leqslant x \leqslant 1\} \quad (5.4.12)$$

of the euclidean plane $\mathbb{R}^2$ equipped with the euclidean metric and the continuous function

$$f(x,y) = x(y - \tfrac{1}{2})$$

on X. The values of $f(x,y)$ on X are displayed in the following diagram:

$$
\begin{array}{ll}
(0,1) \bullet\!\!\rule[0.5ex]{3em}{0.4pt}\!\!\bullet (1,1) & \\
\quad\left| \quad f(x,1) = \tfrac{1}{2}x \right. & \\
f(0,y) = 0 & \\
\quad\left| \quad f(x,0) = -\tfrac{1}{2}x \right. & \\
(0,0) \bullet\!\!\rule[0.5ex]{3em}{0.4pt}\!\!\bullet (1,0) &
\end{array}
$$

One can easily check that

$$\omega(f;\delta) = \begin{cases} \delta/2 & \text{for } 0 \leqslant \delta < 1, \\ 1 & \text{for } 1 \leqslant \delta < \infty. \end{cases}$$

Hence even in this case the modulus of continuity $\omega(f;\delta)$ is discontinuous at $\delta = 1$.

The continuity of $\omega(f;\delta)$ for arbitrary functions $f \in C(X)$ can be enforced by imposing additional conditions upon the compact metric space (X,d). These conditions mean that the closures $\overline{\mathring{U}(s;\delta)}$ of the open balls $\mathring{U}(s;\delta)$ should coincide with the closed balls $U(s;\delta)$ for all $s \in X$. In the case of the compact subset (5.4.12) of the euclidean plane they are violated: for example

$$U((1,0);1) = \overline{\mathring{U}((1,0);1)} \cup \{(1,1)\}.$$

Proposition 5.4.2. *Let (X,d) be a compact metric space such that*

$$\overline{\mathring{U}(s;\delta)} = U(s;\delta) \qquad (5.4.13)$$

for all $s \in X$ and all δ with $0 < \delta \leqslant \delta^*$. Then the modulus of continuity $\omega(f;\delta)$ of any function $f \in C(X)$ is continuous for $0 \leqslant \delta \leqslant \delta^*$.

Proof. It suffices to prove the continuity of $\omega(f;\delta)$ from the left for any δ_0 with $0 < \delta_0 \leqslant \delta^*$ which is equivalent to

$$\sup \{\omega(f;\delta'):0 < \delta' < \delta_0\} = \omega(f;\delta_0). \qquad (5.4.14)$$

We start from an arbitrary pair of elements $s_0, t_0 \in X$ with $d(s_0, t_0) \leqslant \delta_0$. Interpreting this condition as $t_0 \in U(s_0; \delta_0)$ we can make use of the supposition (5.4.13) and determine a sequence $t_n \in \mathring{U}(s_0; \delta_0)$ such that $\lim_{n \to \infty} t_n = t_0$. In this way we obtain

$$|f(s_0) - f(t_0)| = \lim_{n \to \infty} |f(s_0) - f(t_n)| \qquad (5.4.15)$$

$$\leqslant \sup_{1 \leqslant n < \infty} \omega(f; \delta_n),$$

where

$$\delta_n = d(s_0, t_n) < \delta_0.$$

The supremum on the right-hand side of (5.4.15) can obviously be replaced by the supremum of $\omega(f; \delta')$ with respect to all δ' with $0 < \delta' < \delta_0$. Letting s_0 and t_0 vary in X such that $d(s_0, t_0) \leqslant \delta_0$ and taking account of the definition (5.4.1) of the modulus of continuity $\omega(f; \delta_0)$ we finally arrive at

$$\omega(f; \delta_0) \leqslant \sup \{\omega(f; \delta') : 0 < \delta' < \delta_0\}.$$

This proves (5.4.14), the converse inequality being trivial.   ■

Convex and compact subsets of normed spaces represent examples of compact metric spaces (X, d) satisfying (5.4.13) for all $s \in X$ and all δ with $0 < \delta < \infty$. Indeed, in this situation $t \in U(s; \delta)$ means $\|s - t\| \leqslant \delta$. The convexity of X guarantees that for $s, t \in X$ the points

$$t_n = \frac{1}{n} s + \left(1 - \frac{1}{n}\right) t, \quad n = 1, 2, 3, \ldots,$$

also belong to X. If $t \in U(s; \delta)$, then $t_n \in \mathring{U}(s; \delta)$. Since $\lim_{n \to \infty} t_n = t$ we can confirm (5.4.13).

Hence from Proposition 5.4.2 the modulus of continuity $\omega(f; \delta)$ of a continuous function f defined on a convex and compact subset X of a normed space turns out to be continuous for all $\delta \geqslant 0$. This result can also be derived from the *subadditivity*

$$\omega(f; \delta_1 + \delta_2) \leqslant \omega(f; \delta_1) + \omega(f; \delta_2) \qquad (5.4.16)$$

of the modulus of continuity $\omega(f; \delta)$ *of a bounded function* f defined on a convex and compact subset X of a normed space, as we shall see at once.

To check (5.4.16) we fix $\varepsilon > 0$ and choose elements $s, t \in X$ with $\|s - t\| \leqslant \delta_1 + \delta_2$ such that

$$\omega(f; \delta_1 + \delta_2) \leqslant |f(s) - f(t)| + \varepsilon.$$

Since X is convex, the element

$$u = \frac{\delta_1}{\delta_1 + \delta_2} t + \frac{\delta_2}{\delta_1 + \delta_2} s$$

is also contained in X and, moreover, satisfies the conditions

$$\|s - u\| \leqslant \delta_1 \quad \text{and} \quad \|t - u\| \leqslant \delta_2.$$

Therefore we may conclude that

$$|f(s) - f(t)| \leqslant |f(s) - f(u)| + |f(u) - f(t)|$$
$$\leqslant \omega(f; \delta_1) + \omega(f; \delta_2)$$

and finally

$$\omega(f; \delta_1 + \delta_2) \leqslant \omega(f; \delta_1) + \omega(f; \delta_2) + \varepsilon$$

which proves (5.4.16).

The inequality (5.4.16) implies

$$\omega(f; \delta_1) \leqslant \omega(f; \delta_1 - \delta_2) + \omega(f; \delta_2) \quad \text{for } \delta_1 \geqslant \delta_2$$

and

$$\omega(f; \delta_2) \leqslant \omega(f; \delta_2 - \delta_1) + \omega(f; \delta_1) \quad \text{for } \delta_2 \geqslant \delta_1$$

and hence, together,

$$|\omega(f; \delta_1) - \omega(f; \delta_2)| \leqslant \omega(f; |\delta_1 - \delta_2|). \tag{5.4.17}$$

Since $\lim_{\delta \to +0} \omega(f; \delta) = 0$ for every continuous function f on X, the inequality (5.4.17) tells us that the modulus of continuity $\omega(f; \delta)$ of a continuous function f on a convex and compact subset X of a normed space is uniformly continuous for $0 \leqslant \delta < \infty$.

In section 5.8 we shall see that the entropy behaviour of a compact $C(X)$-valued operator $T: E \to C(X)$ becomes quite clear if X is assumed to be a convex and compact subset of a normed space (see Proposition 5.8.4). This is because the modulus of continuity $\omega(T; \delta)$ of a $C(X)$-valued operator, like the modulus of continuity $\omega(f; \delta)$ of a function $f \in l_\infty(X)$, possesses the property of subadditivity in the case of a convex and compact subset X of a normed space (see section 5.5, (5.5.14)).

5.5. The modulus of continuity of $C(X)$-valued operators

Given a compact metric space (X, d) and an operator $T: E \to C(X)$ from an arbitrary Banach space E into $C(X)$ we put

$$\omega(T; \delta) = \sup_{\|x\| \leqslant 1} \omega(Tx; \delta), \tag{3.0.6}$$

calling $\omega(T; \delta)$ *the modulus of continuity of the operator T*. This definition in fact makes sense for all $\delta \geqslant 0$ since

$$\omega(Tx; \delta) \leqslant 2\|Tx\| \quad \text{for arbitrary } x \in E$$

(see (5.4.5)) and, consequently,

$$\omega(T; \delta) \leqslant 2\|T\|. \tag{5.5.1}$$

Moreover, we have

$$\omega(Tx;\delta) \leqslant \omega(T;\delta)\|x\| \qquad (5.5.2)$$

by inserting $f = T(x/\|x\|)$ and $\lambda = \|x\|$ into (5.4.4).

Taking the composition of the operator $T:E \to C(X)$ with an operator $S:Z \to E$ we obtain

$$\omega(TSz;\delta) \leqslant \omega(T;\delta)\|S\|\|z\| \quad \text{for arbitrary } z \in Z$$

from (5.5.2) and hence

$$\omega(TS;\delta) \leqslant \omega(T;\delta)\|S\|. \qquad (5.5.3)$$

In a similar way the properties (5.4.2), (5.4.3) and (5.4.4), like (5.4.5), also immediately transfer from the modulus of continuity $\omega(f;\delta)$ of bounded functions f on X to the modulus of continuity $\omega(T;\delta)$ of $C(X)$-valued operators $T:E \to C(X)$, that is to say

$$\omega(T;\delta) \leqslant \omega(T;\delta') \quad \text{for } 0 \leqslant \delta \leqslant \delta', \qquad (5.5.4)$$

$$\omega(T_1 + T_2;\delta) \leqslant \omega(T_1;\delta) + \omega(T_2;\delta), \qquad (5.5.5)$$

$$\omega(\lambda T;\delta) = |\lambda|\omega(T;\delta). \qquad (5.5.6)$$

From (5.5.5) we have

$$|\omega(T_1;\delta) - \omega(T_2;\delta)| \leqslant \omega(T_1 - T_2;\delta)$$

and hence also

$$|\omega(T_1;\delta) - \omega(T_2;\delta)| \leqslant 2\|T_1 - T_2\|. \qquad (5.5.7)$$

Furthermore, using (5.5.6) we can ascertain that the definition (3.0.6) of $\omega(T;\delta)$ is equivalent to

$$\omega(T;\delta) = \sup_{\|x\| < 1} \omega(Tx;\delta). \qquad (3.0.6)'$$

Therefore the inequality (5.5.3) in the case of a metric surjection $S = Q$ from $Z = \tilde{E}$ onto E (see section 1.3, (1.3.2)) reduces to an equality

$$\omega(TQ;\delta) = \omega(T;\delta). \qquad (5.5.8)$$

This property (5.5.8) of *surjectivity of the modulus of continuity* $\omega(T;\delta)$ of a $C(X)$-valued operator $T:E \to C(X)$ makes it clear that $\omega(T;\delta)$, as a function of δ, reflects *intrinsic properties of the set* $T(U_E)$ and hence *structural properties of* T in the sense of our interpretation. Nice structural properties of the operator T imply nice structural properties of the real-valued function $\omega(T;\delta)$ of the real variable $\delta \geqslant 0$ and vice versa.

Let $T:E \to C(X)$ be a compact operator defined on an arbitrary Banach space E. This means that the set $T(U_E)$ is a precompact subset of $C(X)$. Thus we are reminded of the *Arzelà–Ascoli theorem* (cf. Dunford and Schwartz 1958). It characterizes *precompact subsets M of the Banach space*

$C(X)$ of continuous functions f on a compact metric space (X, d) by the properties of *boundedness and equicontinuity*, namely

$$\|f\| \leqslant K \quad \text{for all } f \in M \text{ with an appropriate constant } K$$

and

$$|f(s) - f(t)| \leqslant \varepsilon \quad \text{for } s, t \in X \text{ with } d(s, t) \leqslant \delta_{\varepsilon}$$

and all $f \in M$, respectively. The emphasis on equicontinuity lies in the fact that for any $\varepsilon > 0$ a positive number δ_{ε} can be chosen uniformly for all $f \in M$. By using the modulus of continuity $\omega(f; \delta)$ this condition can be given the more concise form

$$\sup \{\omega(f; \delta) : f \in M\} \leqslant \varepsilon \quad \text{for } \delta \leqslant \delta_{\varepsilon}.$$

In the limit version it reads as

$$\lim_{\delta \to +0} \sup \{\omega(f; \delta) : f \in M\} = 0.$$

With $M = T(U_E)$ we now can characterize the compactness of the operator $T: E \to C(X)$ in terms of its modulus of continuity $\omega(T; \delta)$.

Proposition 5.5.1. *An operator $T: E \to C(X)$ mapping an arbitrary Banach space E into the space $C(X)$ of continuous functions on a compact metric space (X, d) is compact if and only if*

$$\lim_{\delta \to +0} \omega(T; \delta) = 0. \tag{5.5.9}$$

Since $\omega(T; 0) = 0$, the limit relation (5.5.9) amounts to the continuity of $\omega(T; \delta)$ from the right at $\delta = 0$. In section 5.4 we saw that the modulus of continuity $\omega(f; \delta)$ of a continuous function f on a compact metric space (X, d) is continuous from the right for arbitrary $\delta \geqslant 0$. On the basis of this result we shall now prove that the modulus of continuity $\omega(T; \delta)$ of a compact operator $T: E \to C(X)$ is also continuous from the right for all $\delta \geqslant 0$.

Proposition 5.5.2. *Let (X, d) be a compact metric space, E an arbitrary Banach space, and $T \in K(E, C(X))$ a compact operator. Then the modulus of continuity $\omega(T; \delta)$ of T is continuous from the right for all $\delta \geqslant 0$, that is*

$$\inf \{\omega(T; \delta') : \delta < \delta'\} = \omega(T; \delta). \tag{5.5.10}$$

Proof. We assume $\omega(T; \delta)$ is not continuous from the right at $\delta = \delta_0$ so that

$$\omega(T; \delta_n) > \omega(T; \delta_0) + \varepsilon_0, \quad n = 1, 2, 3, \ldots,$$

where ε_0 is an appropriate positive number and $\delta_n > \delta_0$ is a sequence tending to δ_0 for $n \to \infty$. According to the definition of the modulus of continuity $\omega(T;\delta)$ we can find elements $x_n \in U_E$ such that

$$\omega(Tx_n;\delta_n) > \omega(Tx_n;\delta_0) + \varepsilon_0 \quad \text{for } n = 1,2,3,\ldots \tag{5.5.11}$$

Furthermore, the compactness of the operator T enables us to select a subsequence x_{n_k} from x_n with a convergent sequence of images Tx_{n_k} in $C(X)$, say

$$\lim_{k \to \infty} Tx_{n_k} = f,$$

so that

$$\| Tx_{n_k} - f \| < \frac{\varepsilon_0}{6}$$

for $k \geqslant k_0$. By using the inequality (5.4.6), which expresses the uniform continuity of the map $\omega(\cdot,\delta)\colon l_\infty(X) \to \mathbb{R}$ for fixed $\delta \geqslant 0$, we may in this way conclude that

$$\omega(Tx_{n_k};\delta_{n_k}) - \omega(f;\delta_{n_k}) < \frac{\varepsilon_0}{3} \tag{5.5.12}$$

and

$$\omega(f;\delta_0) - \omega(Tx_{n_k};\delta_0) < \frac{\varepsilon_0}{3}. \tag{5.5.13}$$

Combining (5.5.12) and (5.5.13) with (5.5.11) for $n = n_k$ we obtain

$$\omega(f;\delta_{n_k}) > \omega(Tx_{n_k};\delta_{n_k}) - \frac{\varepsilon_0}{3} > \omega(Tx_{n_k};\delta_0) + \tfrac{2}{3}\varepsilon_0 > \omega(f;\delta_0) + \frac{\varepsilon_0}{3}.$$

But this contradicts the continuity from the right of the modulus of continuity $\omega(f;\delta)$ of the function $f \in C(X)$ at $\delta = \delta_0$ (see Proposition 5.4.1) and thus proves (5.5.10). ∎

If the closures of the open balls of the compact metric space (X,d) coincide with the closed balls, the modulus of continuity $\omega(T;\delta)$ of a compact $C(X)$-valued operator T is continuous from the left, too.

Proposition 5.5.3. *Let* (X,d) *be a compact metric space such that*

$$\overset{\circ}{U}(s;\delta) = U(s;\delta) \tag{5.4.13}$$

for all $s \in X$ *and all* δ *with* $0 < \delta \leqslant \delta^*$. *Then the modulus of continuity* $\omega(T;\delta)$ *of a compact operator* $T\colon E \to C(X)$ *from an arbitrary Banach space* E *into* $C(X)$ *is continuous for* $0 \leqslant \delta \leqslant \delta^*$.

Proof. By Proposition 5.5.2 the modulus of continuity $\omega(T;\delta)$ is continuous from the right for all $\delta \geqslant 0$. To prove the continuity of $\omega(T;\delta)$

from the left at $\delta = \delta_0$ with $0 < \delta_0 \leqslant \delta^*$ we consider an arbitrary element $x \in U_E$ and use the continuity of $\omega(Tx; \delta)$ from the left at $\delta = \delta_0$, guaranteed by Proposition 5.4.2, that is to say

$$\sup\{\omega(Tx; \delta'): 0 < \delta' < \delta_0\} = \omega(Tx; \delta_0). \qquad (5.4.14)'$$

Taking the supremum with respect to $x \in U_E$ on both sides of (5.4.14)' we obtain

$$\sup\{\omega(T; \delta'): 0 < \delta' < \delta_0\} = \omega(T; \delta_0),$$

which proves the assertion.   ∎

As we have already observed in section 5.4 a convex and compact subset X of a normed space satisfies condition (5.4.13) for all $s \in X$ and all $\delta > 0$. Hence, the modulus of continuity $\omega(T; \delta)$ of a compact $C(X)$-valued operator T in this case is continuous for all $\delta \geqslant 0$. Furthermore it is *subadditive*, that is to say

$$\omega(T; \delta_1 + \delta_2) \leqslant \omega(T; \delta_1) + \omega(T; \delta_2), \qquad (5.5.14)$$

which immediately follows from the subadditivity (5.4.16) of the modulus of continuity $\omega(f; \delta)$ of a continuous function f on a convex and compact subset X of a normed space. Also in the operator case the subadditivity (5.5.14) of the modulus of continuity $\omega(T; \delta)$ implies

$$|\omega(T; \delta_1) - \omega(T; \delta_2)| \leqslant \omega(T; |\delta_1 - \delta_2|) \qquad (5.5.15)$$

and hence the uniform continuity of $\omega(T; \delta)$ for $0 \leqslant \delta < \infty$, the operator T being compact.

5.6. Approximation numbers and the modulus of continuity

In this section we relate the *degree of approximability*, in terms of approximation numbers, to the *degree of compactness*, in terms of the modulus of continuity of operators with values in $C(X)$. We shall see how the entropy numbers $\varepsilon_n(X)$ of the underlying compact metric space X enter the inequalities (3.0.8) of Jackson type already given at the beginning of chapter 3. The methods to be employed in this connection are based on Jörgens (1970) and Heinrich and Kühn (1985).

Theorem 5.6.1. *Let (X, d) be a compact metric space, E an arbitrary Banach space, and $T \in K(E, C(X))$ a compact operator. Then*

$$a_{n+1}(T) \leqslant \omega(T; \varepsilon_n(X)) \quad \textit{for } n = 1, 2, 3, \ldots \qquad (3.0.8)$$

Proof. Fix a natural number $n \geqslant 1$ and choose $\delta > \varepsilon_n(X)$. Then there exist $t_1, t_2, \ldots, t_n \in X$ such that the open balls $\mathring{U}(t_k; \delta)$, $k = 1, 2, \ldots, n$, cover X.

Furthermore, let $\{\varphi_1, \varphi_2, \ldots, \varphi_n\} \subset C(X)$ be a partition of unity subordinate to this open covering (see section 5.2, (5.2.7)). Given $T \in K(E, C(X))$ we now define an operator $A \in L(E, C(X))$ by

$$Ax = \sum_{k=1}^{n} (Tx)(t_k)\varphi_k \quad \text{for } x \in E. \tag{5.6.1}$$

Property (5.2.3) of the partition of unity guarantees that

$$(Tx)(t) - (Ax)(t) = \sum_{k=1}^{n} ((Tx)(t) - (Tx)(t_k))\varphi_k(t)$$

and hence

$$|(Tx)(t) - (Ax)(t)| \leqslant \sum_{k=1}^{n} |(Tx)(t) - (Tx)(t_k)| \, \varphi_k(t)$$

for all $t \in X$. Since

$$|(Tx)(t) - (Tx)(t_k)| \leqslant \omega(Tx; \delta) \quad \text{for } t \in \overset{\circ}{U}(t_k; \delta)$$

and

$$\varphi_k(t) = 0 \quad \text{for } t \notin \overset{\circ}{U}(t_k; \delta)$$

we may conclude that

$$|(Tx)(t) - (Ax)(t)| \leqslant \omega(Tx; \delta) \cdot \sum_{k=1}^{n} \varphi_k(t) = \omega(Tx; \delta)$$

and finally that

$$\|(T - A)x\| \leqslant \omega(Tx; \delta) \tag{5.6.2}$$

for arbitrary $\delta > \varepsilon_n(X)$. If we take the supremum with respect to $x \in U_E$ on both sides of the inequality (5.6.2) we obtain

$$\|(T - A)\| \leqslant \omega(T; \delta).$$

This implies

$$a_{n+1}(T) \leqslant \omega(T; \delta),$$

since $\operatorname{rank}(A) < n + 1$. By the continuity of $\omega(T; \delta)$ from the right in the case of a compact operator T (see Proposition 5.5.2) it follows that

$$a_{n+1}(T) \leqslant \omega(T; \varepsilon_n(X)). \qquad \blacksquare$$

Theorem 5.6.1 in particular shows that a compact operator T with values in a Banach space $C(X)$ of continuous functions on a compact metric space (X, d) is approximable.

We now want to give an application of Theorem 5.6.1. For this purpose we introduce the *Banach space $C^\alpha(X)$ of Hölder continuous functions of type α* with $0 < \alpha \leqslant 1$ on the compact metric space (X, d).

A function $f \in C(X)$ is called Hölder continuous of type α, $0 < \alpha \leqslant 1$, if

$$\sup_{\substack{s,t \in X \\ s \neq t}} \frac{|f(s) - f(t)|}{d^\alpha(s,t)} < \infty.$$

Putting

$$|f|_\alpha = \sup_{\substack{s,t \in X \\ s \neq t}} \frac{|f(s) - f(t)|}{d^\alpha(s,t)} \tag{5.6.3}$$

we may state

$$|f(s) - f(t)| \leqslant |f|_\alpha \cdot d^\alpha(s,t) \quad \text{for all } s, t \in X$$

and, moreover, characterize the value $|f|_\alpha$ as the smallest possible constant M such that

$$|f(s) - f(t)| \leqslant M \cdot d^\alpha(s,t) \quad \text{for all } s, t \in X.$$

Another characterization of $|f|_\alpha$ can be given in terms of the modulus of continuity.

Lemma 5.6.1. *Let $f \in C(X)$ be Hölder continuous of type α. Then*

$$|f|_\alpha = \sup_{\delta > 0} \frac{\omega(f;\delta)}{\delta^\alpha}. \tag{5.6.4}$$

Proof. The definition (5.6.3) of $|f|_\alpha$ implies

$$|f|_\alpha \leqslant \sup_{\substack{s,t \in X \\ s \neq t}} \frac{\omega(f;d(s,t))}{d^\alpha(s,t)} \leqslant \sup_{\delta > 0} \frac{\omega(f;\delta)}{\delta^\alpha}.$$

Conversely, if we insert the definition (5.4.1) of the modulus of continuity $\omega(f;\delta)$ we obtain

$$\sup_{\delta > 0} \frac{\omega(f;\delta)}{\delta^\alpha} = \sup_{\delta > 0} \sup_{\substack{s,t \in X \\ d(s,t) \leqslant \delta}} \frac{|f(s) - f(t)|}{\delta^\alpha} \leqslant \sup_{\substack{s,t \in X \\ s \neq t}} \frac{|f(s) - f(t)|}{d^\alpha(s,t)} = |f|_\alpha. \quad \blacksquare$$

Hence a function $f \in C(X)$ is Hölder continuous of type α if and only if

$$\sup_{\delta > 0} \frac{\omega(f;\delta)}{\delta^\alpha} < \infty. \tag{5.6.5}$$

The class $C^\alpha(X)$ of all Hölder continuous functions of type α on X forms a linear space which turns out to be a Banach space if it is equipped with the norm

$$\| f \|_\alpha = \max(\| f \|, |f|_\alpha). \tag{5.6.6}$$

Let us consider the *identity map* $I_\alpha: C^\alpha(X) \to C(X)$ of $C^\alpha(X)$ into $C(X)$. From

$$\omega(I_\alpha f;\delta) \leqslant |f|_\alpha \delta^\alpha \leqslant \| f \|_\alpha \delta^\alpha$$

and from the definition of the modulus of continuity $\omega(I_\alpha; \delta)$ (see (3.0.6)) we get

$$\omega(I_\alpha; \delta) \leqslant \delta^\alpha. \tag{5.6.7}$$

By Proposition 5.5.1 we may therefore conclude that the operator I_α is compact. Thus Theorem 5.6.1 applies and tells us that

$$a_{n+1}(I_\alpha) \leqslant \varepsilon_n^\alpha(X) \quad \text{for } n = 1, 2, 3, \ldots \tag{5.6.8}$$

For $a_1(I_\alpha) = \| I_\alpha \|$, we can easily check that

$$\| I_\alpha \| = 1. \tag{5.6.9}$$

The estimate (5.6.8) that we are principally interested in even reflects the exact asymptotic decrease of the approximation numbers $a_n(I_\alpha)$ and also of the symmetrized approximation numbers $t_n(I_\alpha) \leqslant a_n(I_\alpha)$. Before proving this we remind the reader of the proof of the inequality

$$a_{n+1}(T) \leqslant \omega(T; \varepsilon_n(X)) \tag{3.0.8}$$

in Theorem 5.6.1. Partitions of unity $\{\varphi_1, \varphi_2, \ldots, \varphi_n\} \subset C(X)$ subordinate to a covering of X by open balls $\mathring{U}(t_k; \delta)$, $1 \leqslant k \leqslant n$, turned out to be the essential tool. In contrast with that, for an estimate of $a_{n+1}(I_\alpha)$ or $t_{n+1}(I_\alpha)$ from below, we shall now employ *finite systems of functions* $\{\psi_1, \psi_2, \ldots, \psi_m\} \subset C^\alpha(X)$ with the property

$$0 \leqslant \psi_i(t) \leqslant 1 \quad \text{for } t \in X, 1 \leqslant i \leqslant m,$$

subordinate to a packing of X with disjoint open balls $\mathring{U}(s_i; \varepsilon)$, $1 \leqslant i \leqslant m$, in the sense that

$$\operatorname{supp}(\psi_i) \subseteq \mathring{U}(s_i; \varepsilon) \text{ and } \psi_i(s_i) = 1 \quad \text{for } 1 \leqslant i \leqslant m.$$

For the sake of convenience we introduce the universal auxiliary function

$$\psi_{s,\varepsilon}^{(\alpha)}(t) = \left(\max\left(1 - \frac{d(s, t)}{\varepsilon}, 0 \right) \right)^\alpha \quad \text{for } t \in X,$$

related to the ball $\mathring{U}(s; \varepsilon)$ of radius ε with centre $s \in X$. Given elements $s_1, s_2, \ldots, s_m$ in X with $d(s_i, s_k) \geqslant 2\varepsilon$ for $i \neq k$ the functions

$$\psi_i(t) = \psi_{s_i, \varepsilon}^{(\alpha)}(t), \quad 1 \leqslant i \leqslant m,$$

will then provide a finite system in $C^\alpha(X)$ with the required properties. These properties are listed in detail in the following lemma.

Lemma 5.6.2. *The functions $\psi_{s,\varepsilon}^{(\alpha)}$ for arbitrary $s \in X$, $\varepsilon > 0$, and $0 < \alpha \leqslant 1$ satisfy the following properties:*

$$0 \leqslant \psi_{s,\varepsilon}^{(\alpha)}(t) \leqslant 1 \quad \textit{for } t \in X \textit{ and } \psi_{s,\varepsilon}^{(\alpha)}(s) = 1 \tag{5.6.10}$$

$$\operatorname{supp}(\psi_{s,\varepsilon}^{(\alpha)}(t)) \subseteq \mathring{U}(s; \varepsilon) \tag{5.6.11}$$

$$\psi_{s,\varepsilon}^{(\alpha)} \in C^\alpha(X) \textit{ and } \| \psi_{s,\varepsilon}^{(\alpha)} \|_\alpha \leqslant \max(1, \varepsilon^{-\alpha}) \tag{5.6.12}$$

> *If* $s_1, s_2, \ldots, s_m$ *is a finite system of elements
> in* X *such that* $d(s_i, s_k) \geqslant 2\varepsilon$ *for* $i \neq k$, *then*
>
> $$\left\| \sum_{i=1}^{m} \xi_i \psi_{s_i, \varepsilon}^{(\alpha)} \right\|_\alpha \leqslant \max(1, 2\varepsilon^{-\alpha}) \cdot \sup_{1 \leqslant i \leqslant m} |\xi_i|$$
>
> *for all m-tuples of scalars* $(\xi_1, \xi_2, \ldots, \xi_m)$.   (5.6.13)

Proof. Properties (5.6.10) and (5.6.11) follow immediately from the definition of $\psi_{s,\varepsilon}^{(\alpha)}(t)$. To prove (5.6.12) we refer to the inequalities

$$|a^\alpha - b^\alpha| \leqslant |a - b|^\alpha \quad \text{for } 0 \leqslant \alpha \leqslant 1 \text{ and } a, b \geqslant 0 \qquad (5.6.14)$$

and

$$|\max(a, 0) - \max(b, 0)| \leqslant |a - b| \qquad (5.6.15)$$

for arbitrary real numbers a and b. Combining (5.6.14) and (5.6.15) we obtain

$$|(\max(a, 0))^\alpha - (\max(b, 0))^\alpha| \leqslant |a - b|^\alpha \qquad (5.6.16)$$

for $0 < \alpha \leqslant 1$ and arbitrary real numbers a and b. Inserting $a = 1 - d(s, t_1)/\varepsilon$ and $b = 1 - d(s, t_2)/\varepsilon$ into (5.6.16) and using the definition of $\psi_{s,\varepsilon}^{(\alpha)}(t)$ we see that

$$|\psi_{s,\varepsilon}^{(\alpha)}(t_1) - \psi_{s,\varepsilon}^{(\alpha)}(t_2)| \leqslant \varepsilon^{-\alpha} \cdot |d(s, t_1) - d(s, t_2)|^\alpha.$$

Observe that

$$|d(s, t_1) - d(s, t_2)| \leqslant d(t_1, t_2)$$

and hence

$$|\psi_{s,\varepsilon}^{(\alpha)}(t_1) - \psi_{s,\varepsilon}^{(\alpha)}(t_2)| \leqslant \varepsilon^{-\alpha} \cdot d^\alpha(t_1, t_2) \qquad (5.6.17)$$

for arbitrary $t_1, t_2 \in X$, which means $\psi_{s,\varepsilon}^{(\alpha)} \in C^\alpha(X)$ and, moreover,

$$|\psi_{s,\varepsilon}^{(\alpha)}|_\alpha \leqslant \varepsilon^{-\alpha}.$$

Since $|\psi_{s,\varepsilon}^{(\alpha)}(t)| \leqslant 1$ by (5.6.10) definition (5.6.6) of the norm in $C^\alpha(X)$ yields

$$\| \psi_{s,\varepsilon}^{(\alpha)} \|_\alpha \leqslant \max(1, \varepsilon^{-\alpha}).$$

This completes the proof of (5.6.12). For the proof of the remaining property (5.6.13) we consider the open balls $\overset{\circ}{U}(s_i; \varepsilon)$ determined by a system of elements $s_1, s_2, \ldots, s_m$ with $d(s_i, s_k) \geqslant 2\varepsilon$ for $i \neq k$. Obviously

$$\overset{\circ}{U}(s_i; \varepsilon) \cap \overset{\circ}{U}(s_k; \varepsilon) = \varnothing \quad \text{for } i \neq k.$$

From (5.6.11) the sum $\sum_{i=1}^{m} \xi_i \psi_{s_i, \varepsilon}^{(\alpha)}(t)$ therefore reduces to one item or even to zero and, because of (5.6.10), can be estimated as

$$\left| \sum_{i=1}^{m} \xi_i \psi_{s_i, \varepsilon}^{(\alpha)}(t) \right| \leqslant \sup_{1 \leqslant i \leqslant m} |\xi_i|.$$

Similar arguments make it clear that for arbitrary $t_1, t_2 \in X$ the sum $\sum_{i=1}^{m} \xi_i(\psi_{s_i, \varepsilon}^{(\alpha)}(t_1) - \psi_{s_i, \varepsilon}^{(\alpha)}(t_2))$ comprises at most two items so that its absolute value, by (5.6.17), can be estimated by

$$\left| \sum_{i=1}^{m} \xi_i(\psi_{s_i,\varepsilon}^{(\alpha)}(t_1) - \psi_{s_i,\varepsilon}^{(\alpha)}(t_2)) \right| \leqslant 2 \cdot \varepsilon^{-\alpha} \cdot d^{\alpha}(t_1, t_2) \cdot \sup_{1 \leqslant i \leqslant m} |\xi_i|.$$

Thus, from the definition (5.6.6) of the norm in $C^{\alpha}(X)$, we finally gain

$$\left\| \sum_{i=1}^{m} \xi_i \psi_{s_i,\varepsilon}^{(\alpha)} \right\|_{\alpha} \leqslant \max(1, 2\varepsilon^{-\alpha}) \cdot \sup_{1 \leqslant i \leqslant m} |\xi_i|,$$

which finishes the proof of Lemma 5.6.2. ∎

We are now in a position to derive the desired estimate for the symmetrized approximation numbers $t_n(I_\alpha)$ from below.

Proposition 5.6.1. *Let (X, d) be a compact metric space. Then the symmetrized approximation numbers $t_n(I_\alpha)$ of the identity map $I_\alpha : C^{\alpha}(X) \rightarrow C(X)$ are subject to the inequalities*

$$\frac{1}{2}\left(\min\left(1, \frac{\varepsilon_n(X)}{2}\right) \right)^{\alpha} \leqslant t_{n+1}(I_\alpha), \quad n = 1, 2, 3, \ldots \qquad (5.6.18)$$

Proof. We choose ε with $0 < 2\varepsilon < \varepsilon_n(X)$ and a maximal system of elements $s_1, s_2, \ldots, s_m$ in X with $d(s_i, s_k) \geqslant 2\varepsilon$ for $i \neq k$. Then

$$X \subseteq \bigcup_{i=1}^{m} \overset{\circ}{U}(s_i; 2\varepsilon)$$

and therefore $m \geqslant n + 1$. We consider the m-dimensional Banach space l_∞^m and its identity map $I_\infty^{(m)}$. The norm determining property (T5) of the symmetrized approximation numbers guarantees that

$$t_{n+1}(I_\infty^{(m)}) = 1 \qquad (5.6.19)$$

since $m \geqslant n + 1$ (see section 2.6). To obtain the desired estimate for $t_{n+1}(I_\alpha)$ from below we factorize $I_\infty^{(m)}$ by I_α. For this purpose we introduce the operators

$$R^{(m)}(\xi_1, \xi_2, \ldots, \xi_m) = \sum_{i=1}^{m} \xi_i \psi_{s_i,\varepsilon}^{(\alpha)}$$

from l_∞^m into $C^{\alpha}(X)$ and

$$S^{(m)}(f) = (f(s_1), f(s_2), \ldots, f(s_m))$$

from $C(X)$ into l_∞^m. Because of (5.6.10), (5.6.11), and

$$\overset{\circ}{U}(s_i; \varepsilon) \cap \overset{\circ}{U}(s_k; \varepsilon) = \varnothing \quad \text{for } i \neq k$$

the function $f(t) = \sum_{i=1}^{m} \xi_i \psi_{s_i,\varepsilon}^{(\alpha)}(t)$ satisfies the conditions

$$f(s_1) = \xi_1, f(s_2) = \xi_2, \ldots, f(s_m) = \xi_m.$$

Hence, the product $S^{(m)} I_\alpha R^{(m)}$ represents the identity map

$$I_\infty^{(m)} = S^{(m)} I_\alpha R^{(m)} \qquad (5.6.20)$$

of l_∞^m. This factorization of $I_\infty^{(m)}$ and the equality (5.6.19) give rise to the estimate

$$1 \leqslant \| S^{(m)} \| t_{n+1}(I_\alpha) \| R^{(m)} \|.$$

Obviously

$$\| S^{(m)} \| \leqslant 1. \qquad (5.6.21)$$

Furthermore, (5.6.13) tells us that

$$\| R^{(m)} \| \leqslant \max{(1, 2\varepsilon^{-\alpha})}. \qquad (5.6.22)$$

But since

$$\max{(1, 2\varepsilon^{-\alpha})} \leqslant 2 \cdot \max{(1, \varepsilon^{-\alpha})} = \frac{2}{(\min{(1, \varepsilon)})^\alpha} \qquad (5.6.23)$$

we obtain

$$\tfrac{1}{2}(\min{(1, \varepsilon)})^\alpha \leqslant t_{n+1}(I_\alpha) \quad \text{for } 0 < 2\varepsilon < \varepsilon_n(X).$$

Changing ε to $\varepsilon_n(X)/2$ we finally arrive at (5.6.18). ∎

Proposition 5.6.2. *Let (X, d) be a compact metric space and $I_\alpha : C^\alpha(X) \to C(X)$ the identity map of the space $C^\alpha(X)$ of Hölder continuous functions of type α into the space $C(X)$. Then the behaviour of the approximation quantities $t_n(I_\alpha)$ and $a_n(I_\alpha)$, for sufficiently large n, is determined by*

$$2^{-1-\alpha}\varepsilon_n^\alpha(X) \leqslant t_{n+1}(I_\alpha) \leqslant a_{n+1}(I_\alpha) \leqslant \varepsilon_n^\alpha(X). \qquad (5.6.24)$$

Moreover,

$$c_n(I_\alpha) = a_n(I_\alpha) \text{ and } d_n(I_\alpha) = t_n(I_\alpha) \qquad (5.6.25)$$

for $n = 1, 2, 3, \ldots$

Proof. The compactness of X implies $\varepsilon_n(X) < 2$ for sufficiently large n so that the left-hand side of (5.6.18) takes the form $2^{-1-\alpha}\varepsilon_n^\alpha(X)$. The estimate for $a_{n+1}(I_\alpha)$ from above has already been stated in (5.6.8). Since the operator I_α is compact, Theorem 5.3.2 applies, which proves (5.6.25). ∎

Apart from being an instructive example, the identity map $I_\alpha : C^\alpha(X) \to C(X)$ of the space $C^\alpha(X)$ of Hölder continuous functions of type α into the space $C(X)$ is of considerable significance. In section 5.9 we shall introduce a class $L^\alpha(E, C(X))$ of $C(X)$-valued operators to be called Hölder continuous of type α in analogy to the class $C^\alpha(X)$ of Hölder continuous functions of type α on X. It will turn out that this class in a sense is generated by the identity map $I_\alpha : C^\alpha(X) \to C(X)$.

5.7. Entropy numbers and the modulus of continuity

Having estimated the approximation numbers $a_{n+1}(T)$ of a compact operator $T : E \to C(X)$ from above by the modulus of continuity $\omega(T; \delta)$

with $\delta = \varepsilon_n(X)$ we are also in a position to derive estimates for the entropy numbers $e_n(T)$ from above by expressions involving the modulus of continuity $\omega(T;\delta)$ for various values of δ. In fact, the inequality

$$e_n(T) \leqslant c_p \cdot n^{-1/p} \sup_{1 \leqslant k \leqslant n} k^{1/p} a_k(T) \tag{3.1.8}'$$

obtained from the general Bernstein type inequality (3.1.1), combined with (3.0.8) (see Theorem 5.6.1) enables us to conclude that

$$e_n(T) \leqslant c_p \cdot n^{-1/p} \left(\|T\| + \sup_{1 \leqslant k \leqslant n-1} (k+1)^{1/p} \omega(T; \varepsilon_k(X)) \right)$$

for a compact operator $T: E \to C(X)$. This implies

$$e_n(T) \leqslant c_p \left(\|T\| n^{-1/p} + 2^{1/p} \sup_{1 \leqslant k \leqslant n} \left(\frac{k}{n} \right)^{1/p} \omega(T; \varepsilon_k(X)) \right) \tag{5.7.1}$$

for arbitrary $p > 0$. However, the estimate (5.7.1) for $e_n(T)$ is not as clear as the estimate (3.0.8) for $a_{n+1}(T)$ that we have been starting from. To simplify it we shall impose an additional condition on the modulus of continuity $\omega(T; \varepsilon_k(X))$. Motivated by the expression $(k/n)^{1/p} \omega(T; \varepsilon_k(X))$ in the supremum on the right-hand side of (5.7.1) we claim the existence of constants $\sigma > 0$ and $\rho \geqslant 1$ such that

$$\left(\frac{k}{n} \right)^{\sigma} \omega(T; \varepsilon_k(X)) \leqslant \rho \cdot \omega(T; \varepsilon_n(X)) \tag{5.7.2}$$

for all $n = 1, 2, 3, \ldots$ and $k = 1, 2, \ldots, n$. This claim involves *simultaneous constraints for the compact metric space* (X, d) *and the operator* $T: E \to C(X)$. Before deriving conditions which guarantee (5.7.2) we demonstrate that this claim does in fact simplify the estimate (5.7.1) for $e_n(T)$.

Theorem 5.7.1. *Let* (X, d) *be a compact metric space, E a Banach space, and* $T \in K(E, C(X))$ *a compact operator such that the condition* (5.7.2) *is satisfied with appropriate constants $\sigma > 0$ and $\rho \geqslant 1$. Then*

$$e_n(T) \leqslant C(p, \rho, \sigma) \cdot (\|T\| n^{-1/p} + \omega(T; \varepsilon_n(X))) \tag{5.7.3}$$

for $n = 1, 2, 3, \ldots$ and arbitrary $p > 0$ with a constant $C(p, \rho, \sigma)$ depending on the parameter p and on ρ and σ.

Proof. If $\sigma \leqslant 1/p$ we can estimate the supremum on the right-hand side of (5.7.1) by

$$\sup_{1 \leqslant k \leqslant n} \left(\frac{k}{n} \right)^{1/p} \omega(T; \varepsilon_k(X)) \leqslant \sup_{1 \leqslant k \leqslant n} \left(\frac{k}{n} \right)^{\sigma} \omega(T; \varepsilon_k(X))$$

$$\leqslant \rho \cdot \omega(T; \varepsilon_n(X))$$

since $k \leqslant n$. Thus we arrive at

$$e_n(T) \leqslant \rho c_p 2^{1/p} (\|T\| \cdot n^{-1/p} + \omega(T; \varepsilon_n(X))).$$

If $\sigma > 1/p$ we employ (5.7.1) with $p = 1/\sigma$ and obtain

$$e_n(T) \leqslant c_{1/\sigma}\left(\|T\| n^{-\sigma} + 2^\sigma \sup_{1 \leqslant k \leqslant n} \left(\frac{k}{n}\right)^\sigma \omega(T; \varepsilon_k(X)) \right)$$

$$\leqslant \rho c_{1/\sigma} 2^\sigma (\|T\| n^{-1/p} + \omega(T; \varepsilon_n(X))).$$

Altogether we see that the estimate (5.7.3) holds with

$$C(p, \rho, \sigma) = \begin{cases} \rho c_p 2^{1/p} & \text{for } p \leqslant 1/\sigma, \\ \rho c_{1/\sigma} 2^\sigma & \text{for } p > 1/\sigma. \end{cases} \qquad \blacksquare$$

A condition

$$\left(\frac{k}{n}\right)^\sigma \varepsilon_k(X) \leqslant \rho \varepsilon_n(X) \qquad (5.7.4)$$

for $n = 1, 2, 3, \ldots$ and $k = 1, 2, \ldots, n$ analogous to (5.7.2) with constants $\sigma > 0$ and $\rho \geqslant 1$, but concerning the sequence of entropy numbers $\varepsilon_k(X)$ of X itself, for special operators $T: E \to C(X)$ implies the corresponding behaviour (5.7.2) of the modulus of continuity $\omega(T; \varepsilon_k(X))$. This is true in particular for the identity map $I_\alpha: C^\alpha(X) \to C(X)$ already dealt with in section 5.6.

Lemma 5.7.1. *Let (X, d) be a compact metric space such that the sequence of entropy numbers $\varepsilon_k(X)$ satisfies the condition (5.7.4) with constants $\sigma > 0$ and $\rho \geqslant 1$. Then the modulus of continuity $\omega(I_\alpha; \varepsilon_k(X))$ of the identity map $I_\alpha: C^\alpha(X) \to C(X)$ is subject to a corresponding condition, namely*

$$\left(\frac{k}{n}\right)^{\tilde\sigma} \omega(I_\alpha; \varepsilon_k(X)) \leqslant \tilde\rho \cdot \omega(I_\alpha; \varepsilon_n(X)) \qquad (5.7.2)'$$

for all $n = 1, 2, 3, \ldots$ and $k = 1, 2, \ldots, n$ with appropriate constants $\tilde\sigma > 0$ and $\tilde\rho \geqslant 1$.

Proof. We combine the estimate (5.6.7) for $\omega(I_\alpha; \delta)$ with the condition (5.7.4), and obtain

$$\omega(I_\alpha; \varepsilon_k(X)) \leqslant \varepsilon_k^\alpha(X) \leqslant \rho^\alpha \left(\frac{n}{k}\right)^{\sigma\alpha} \varepsilon_n^\alpha(X).$$

We then make use of the estimate (5.6.24) valid for sufficiently large n, say $n \geqslant n_0$, which gives

$$\omega(I_\alpha; \varepsilon_k(X)) \leqslant 2^{1+\alpha} \rho^\alpha \left(\frac{n}{k}\right)^{\sigma\alpha} a_{n+1}(I_\alpha).$$

Finally we refer to (3.0.8) (see Theorem 5.6.1). The result is

$$\left(\frac{k}{n}\right)^{\sigma\alpha} \omega(I_\alpha; \varepsilon_k(X)) \leqslant 2^{1+\alpha} \rho^\alpha \omega(I_\alpha; \varepsilon_n(X))$$

for $n \geqslant n_0$. But by an appropriate choice of $\tilde{\rho} \geqslant 2^{1+\alpha}\rho^{\alpha}$ the validity of (5.7.2)' can be brought about with $\tilde{\sigma} = \sigma\alpha$ for all $n = 1, 2, 3, \ldots$ and $k = 1, 2, \ldots, n$. ∎

We emphasize that the condition (5.7.4) implies

$$\frac{\varepsilon_1(X)}{\rho} n^{-\sigma} \leqslant \varepsilon_n(X) \quad \text{for } n = 1, 2, 3, \ldots, \qquad (5.7.5)$$

which means that the decay of the sequence $\varepsilon_n(X)$ is at most of order $n^{-\sigma}$ if $\varepsilon_1(X) \neq 0$. But the condition $\varepsilon_1(X) \neq 0$ is equivalent to stating that the metric space (X, d) consists of more than one point. In this case we have $\varepsilon_n(X) \neq 0$ for all $n = 1, 2, 3, \ldots$ according to (5.7.5), which means that X contains infinitely many points.

Proposition 5.7.1. *Let (X, d) be a compact metric space with more than one point and with the property (5.7.4). Then the dyadic entropy numbers $e_n(I_\alpha)$ of the identity map $I_\alpha : C^\alpha(X) \to C(X)$ can be estimated by*

$$c \cdot \varepsilon_n^\alpha(X) \leqslant e_n(I_\alpha) \leqslant C \cdot \varepsilon_n^\alpha(X) \quad \text{for } n = 1, 2, 3, \ldots \qquad (5.7.6)$$

with certain positive constants c and C.

Proof. By Lemma 5.7.1 the modulus of continuity $\omega(I_\alpha; \varepsilon_k(X))$ satisfies condition (5.7.2)' with certain constants $\tilde{\sigma} > 0$ and $\tilde{\rho} \geqslant 1$. Therefore, Theorem 5.7.1 yields

$$e_n(I_\alpha) \leqslant C(p, \tilde{\rho}, \tilde{\sigma}) \cdot (n^{-1/p} + \omega(I_\alpha; \varepsilon_n(X))) \qquad (5.7.7)$$

for $n = 1, 2, 3, \ldots$ and arbitrary $p > 0$. To obtain the required estimate for $e_n(I_\alpha)$ from above we link the parameter p to the constant σ in condition (5.7.4) and the Hölder index α by putting $p = 1/\sigma\alpha$. If we now take into consideration that $\varepsilon_1(X) \neq 0$ by assumption, we may employ (5.7.5) to estimate the term $n^{-1/p}$ on the right-hand side of (5.7.7) from above, namely

$$n^{-1/p} = n^{-\sigma\alpha} \leqslant \frac{\rho^\alpha}{\varepsilon_1^\alpha(X)} \varepsilon_n^\alpha(X) \quad \text{for } n = 1, 2, 3, \ldots$$

Since $\omega(I_\alpha; \varepsilon_n(X))$ is also dominated by $\varepsilon_n^\alpha(X)$ (see (5.6.7)) the estimate (5.7.7) of $e_n(I_\alpha)$ from above can be extended to

$$e_n(I_\alpha) \leqslant C \cdot \varepsilon_n^\alpha(X) \quad \text{for } n = 1, 2, 3, \ldots$$

with an appropriate constant C.

The estimate of $e_n(I_\alpha)$ from below is reached in the same way as the estimate (5.6.18) of $t_{n+1}(I_\alpha)$ from below, that is, by using the factorization (5.6.20) of the identity map $I_\infty^{(m)}$ of l_∞^m with $m \geqslant n+1$ by I_α. Obviously

$$2^{-(n-1)/m} \leqslant e_n(I_\infty^{(m)})$$

(see (1.1.10)) and hence

$$\tfrac{1}{2} \leqslant e_n(I_\infty^{(m)})$$

under the above condition for m and n. On the other hand, we have

$$e_n(I_\infty^{(m)}) \leqslant \| S^{(m)} \| e_n(I_\alpha) \| R^{(m)} \|$$

$$\leqslant \frac{2}{(\min(1,\varepsilon))^\alpha} e_n(I_\alpha)$$

by (5.6.20), (5.6.21), (5.6.22) and (5.6.23) for each ε with $0 < 2\varepsilon < \varepsilon_n(X)$. Thus we obtain

$$\tfrac{1}{4}\left(\min\left(1, \frac{\varepsilon_n(X)}{2} \right) \right)^\alpha \leqslant e_n(I_\alpha) \quad \text{for } n = 1, 2, 3, \ldots,$$

which reduces to

$$2^{-2-\alpha} \varepsilon_n^\alpha(X) \leqslant e_n(I_\alpha)$$

for sufficiently large n and hence amounts to

$$c \cdot \varepsilon_n^\alpha(X) \leqslant e_n(I_\alpha) \quad \text{for } n = 1, 2, 3, \ldots$$

with an appropriate constant $c > 0$. ∎

Remark. The lower estimate (5.7.6) for $e_n(I_\alpha)$ could be proved without any additional assumptions on the compact metric space (X, d). If (X, d) actually satisfies the conditions formulated above, Proposition 5.7.1 tells us that the dyadic entropy numbers $e_n(I_\alpha)$ behave asymptotically in the same way as the approximation quantities $a_n(I_\alpha) = c_n(I_\alpha)$ and $t_n(I_\alpha) = d_n(I_\alpha)$, namely like $\varepsilon_n^\alpha(X)$ (see Proposition 5.6.2).

5.8. Entropy properties of connected compact metric spaces

We recall that by Proposition 5.6.2 the asymptotic behaviour of the approximation quantities $a_n(I_\alpha)$, $c_n(I_\alpha)$, $t_n(I_\alpha)$, and $d_n(I_\alpha)$ is generally determined by $\varepsilon_n^\alpha(X)$. A complete description of the asymptotic behaviour of the dyadic entropy numbers $e_n(I_\alpha)$ in the case of an arbitrary compact metric space (X, d) has been given by Heinrich and Kühn (1985). However, the result is rather complicated. Instead of reproducing it here we shall show that the restriction (5.7.4) on (X, d) is not as strong as one might think at first glance. Indeed, every compact and connected metric space (X, d) satisfies (5.7.4) with $\sigma = 1$ and $\rho = 12$. Before proving this we remind the reader of the concept of connectedness.

A metric space (X, d) is said to be *connected* if the empty set $\varnothing$ and the whole space X are the only subsets of X that are both open and closed.

This definition of connectedness makes sense in arbitrary topological spaces (see Köthe 1960). But when operating with a compact metric space

(X, d) it is more convenient to use a more intuitive characterization of connectedness. Such a characterization can be obtained by using *a weakened version of the notion of a 'connected metric space'* (see Mangoldt and Knopp 1973; Heinrich and Kühn 1985).

A metric space (X, d) is said to be *ε-connected* if, for any two points s and t in X, there exists a finite chain of elements $s_0, s_1, \ldots, s_n$ with $s_0 = s$, $s_n = t$, and $d(s_{i-1}, s_i) \leqslant \varepsilon$ for $i = 1, 2, \ldots, n$, so-called *ε-chain*.

Lemma 5.8.1. *A compact metric space (X, d) is connected if and only if it is ε-connected for every $\varepsilon > 0$.*

Proof. First let us assume that (X, d) is a connected metric space in the sense of the original definition. For a given $\varepsilon > 0$ and an arbitrary $s \in X$ we then form the set

$$X_\varepsilon(s) = \{t \in X: \text{ there exists an ε-chain between } s \text{ and } t\}.$$

If t belongs to $X_\varepsilon(s)$, the open ball

$$\mathring{U}(t; \varepsilon) = \{u \in X : d(u, t) < \varepsilon\}$$

is obviously contained in $X_\varepsilon(s)$ as well. Therefore $X_\varepsilon(s)$ is an open subset of X. On the other hand, if t belongs to the complement $X \backslash X_\varepsilon(s)$ of $X_\varepsilon(s)$ in X there is no ε-chain between s and t and hence likewise none between s and any point u with $d(t; u) < \varepsilon$. This means that the complement $X \backslash X_\varepsilon(s)$ is open, too. Consequently, the subset $X_\varepsilon(s)$ of X is open and closed simultaneously. Since $s \in X_\varepsilon(s)$ we have $X_\varepsilon(s) \neq \varnothing$. By the assumption then $X_\varepsilon(s)$ necessarily coincides with X. In other words, X is ε-connected for any $\varepsilon > 0$.

Next we suppose (X, d) to be a compact metric space which is not connected. Then there exists a subset M of X, different from the empty set and from the whole space X, which is both open and closed. The same properties of course apply to the complement $X \backslash M$ of M in X. Particularly, the two closed subsets M and $X \backslash M$ of the compact metric space X are themselves compact. Since they are disjoint we can ensure the existence of an $\varepsilon_0 > 0$ such that

$$d(s, t) \geqslant \varepsilon_0 \quad \text{for all } s \in M \text{ and } t \in X \backslash M \tag{5.8.1}$$

by an ordinary compactness argument. As a result of (5.8.1), for $0 < \varepsilon < \varepsilon_0$, no point $s \in M$ is ε-connected with a point $t \in X \backslash M$. This proves that a compact metric space which is ε-connected for every $\varepsilon > 0$, is connected. ∎

Remark. A non-compact metric space may be ε-connected for every $\varepsilon > 0$ without being connected. As an example we mention the space of rational numbers in the interval $[0, 1]$ with the natural metric.

Proposition 5.8.1. *Let* (X, d) *be a compact and connected metric space. Then*

$$\frac{k}{n}\varepsilon_k(X) \leqslant 12\varepsilon_n(X) \tag{5.8.2}$$

for $n = 1, 2, 3, \ldots$ *and* $k = 1, 2, \ldots, n$.

Proof. Since X is compact it has a finite diameter

$$\text{diam}(X) = \sup\{d(s, t) : s, t \in X\}.$$

If X consists of one point only, we have $\text{diam}(X) = 0$. But then $\varepsilon_n(X) = 0$ for $n = 1, 2, 3, \ldots$ so that (5.8.2) is satisfied trivially. Hence let us assume that X contains more than one point so that $\text{diam}(X) > 0$. Given a natural number $m \geqslant 1$ we can then choose $\varepsilon_0 > 0$ such that

$$6m\varepsilon_0 < \text{diam}(X) \tag{5.8.3}$$

and determine a maximal system of elements $s_1, s_2, \ldots, s_k$ in X with

$$d(s_i, s_j) \geqslant 6m\varepsilon_0 \quad \text{for } i, j = 1, 2, \ldots, k, \ i \neq j. \tag{5.8.4}$$

This implies

$$X \subseteq \bigcup_{i=1}^{k} \mathring{U}(s_i; 6m\varepsilon_0)$$

and thus

$$\varepsilon_k(X) \leqslant 6m\varepsilon_0. \tag{5.8.5}$$

On the other hand, if we consider the balls $\mathring{U}(s_i; 3m\varepsilon_0)$ of half the radius with centres at s_i, we recognize that, from (5.8.3), the whole space X is contained in none of them. Hence none of the sets

$$X \backslash \mathring{U}(s_i; 3m\varepsilon_0) = \{s \in X : d(s_i, s) \geqslant 3m\varepsilon_0\}, \quad 1 \leqslant i \leqslant k,$$

is empty. Moreover, not even the shells

$$S_{ij} = \{s \in X : 3j\varepsilon_0 \leqslant d(s_i, s) < (3j + 1)\varepsilon_0\},$$

$j = 1, 2, \ldots, m - 1$, of width ε_0 situated within the balls $\mathring{U}(s_i; 3m\varepsilon_0)$, $i = 1, 2, \ldots, k$, are empty. This is because the space (X, d), by Lemma 5.8.1, is ε-connected for arbitrary $\varepsilon > 0$. In detail we can argue as follows: a point $s \in X \backslash \mathring{U}(s_i; 3m\varepsilon_0)$ is connected with the point s_i by an ε_0-chain. If the shell S_{ij} were empty there would be two subsequent members t_r and t_{r+1} in an ε_0-chain between s and s_i with

$$d(s_i, t_r) \geqslant (3j + 1)\varepsilon_0 \text{ and } d(s_i, t_{r+1}) < 3j\varepsilon_0.$$

The distance $d(t_r, t_{r+1})$ could then be estimated from below by

$$d(t_r, t_{r+1}) \geqslant d(t_r, s_i) - d(s_i, t_{r+1}) > \varepsilon_0.$$

But this contradicts the assumption that t_r and t_{r+1} are subsequent members of an ε_0-chain. Hence we can pick an element $s_{i,j}$ from each shell S_{ij}. Supplementing the system of elements s_{ij} by the centres s_i of the shells

S_{ij} and putting $s_{i0} = s_i$ we obtain a set $\{s_{ij} : 1 \leqslant i \leqslant k, 0 \leqslant j \leqslant m - 1\}$ of km elements whose mutual distance $d(s_{i_1 j_1}, s_{i_2 j_2})$ is limited from below by $2\varepsilon_0$. Indeed, under the assumption $i_1 \neq i_2$ the inequality

$$d(s_{i_1 j_1}, s_{i_2 j_2}) \geqslant d(s_{i_1 0}, s_{i_2 0}) - d(s_{i_1 j_1}, s_{i_1 0}) - d(s_{i_2 j_2}, s_{i_2 0})$$

yields

$$d(s_{i_1 j_1}, s_{i_2 j_2}) > 6m\varepsilon_0 - (3j_1 + 1)\varepsilon_0 - (3j_2 + 1)\varepsilon_0$$
$$= (6m - 3(j_1 + j_2) - 2)\varepsilon_0 \geqslant 4\varepsilon_0$$

from (5.8.4) and the definition of S_{ij}. On the other hand, if $i_1 = i_2 = i$ we can suppose $j_1 > j_2$ and make use of the estimate

$$d(s_{ij_1}, s_{ij_2}) \geqslant d(s_{ij_1}, s_{i0}) - d(s_{ij_2}, s_{i0})$$

which amounts to

$$d(s_{ij_1}, s_{ij_2}) > 3j_1 \varepsilon_0 - (3j_2 + 1)\varepsilon_0$$
$$= (3(j_1 - j_2) - 1)\varepsilon_0 \geqslant 2\varepsilon_0.$$

Thus we can state

$$d(s_{i_1 j_1}, s_{i_2 j_2}) > 2\varepsilon_0 \quad \text{for all pairs } (i_1, j_1) \neq (i_2, j_2)$$

as claimed above. Hence the $(km - 1)$th inner entropy number $\varphi_{km-1}(X)$ of X is subject to the estimate

$$\varphi_{km-1}(X) \geqslant \varepsilon_0.$$

But since $\varphi_{km-1}(X) \leqslant \varepsilon_{km-1}(X)$ (see (1.1.3)) we obtain

$$\varepsilon_0 \leqslant \varepsilon_{km-1}(X). \tag{5.8.6}$$

Combining (5.8.6) with (5.8.5) we arrive at

$$\varepsilon_k(X) \leqslant 6m\varepsilon_{km-1}(X).$$

This inequality is the key for the estimate (5.8.2) that we are looking for. Given $n \geqslant 1$ and k with $1 \leqslant k \leqslant n$ we choose the natural number m such that

$$(m - 1)k \leqslant n < mk.$$

Then

$$\varepsilon_{km-1}(X) \leqslant \varepsilon_n(X)$$

and simultaneously

$$m \leqslant 2\frac{n}{k}$$

which provides the desired result

$$\varepsilon_k(X) \leqslant 6 \cdot 2\frac{n}{k} \varepsilon_{km-1}(X) \leqslant 12\frac{n}{k}\varepsilon_n(X)$$

for $n = 1, 2, 3, \ldots$ and $k = 1, 2, \ldots, n$. ∎

Proposition 5.8.2. (cf. Timan 1964, 1977) *Let* (X, d) *be a compact connected metric space with more than one point. Then the dyadic entropy numbers* $e_n(I_\alpha)$ *of the identity map* $I_\alpha : C^\alpha(X) \to C(X)$ *behave asymptotically like* $\varepsilon_n^\alpha(X)$.

Proof. By Proposition 5.8.1 the entropy numbers $\varepsilon_k(X)$ of the compact metric space (X, d) satisfy condition (5.7.4) with $\sigma = 1$ and $\rho = 12$. Proposition 5.7.1 then guarantees the asymptotic behaviour of the dyadic entropy numbers $e_n(I_\alpha)$ claimed above. ∎

According to Proposition 5.8.2 the identity map $I_\alpha : C^\alpha(X) \to C(X)$ of the space $C^\alpha(X)$ of Hölder continuous functions of type α on X into $C(X)$ has clear entropy behaviour under quite general conditions. However, if we want to obtain satisfactory estimates for the dyadic entropy numbers $e_n(T)$ of an arbitrary compact operator $T : E \to C(X)$ we have to impose stronger restrictions on the compact metric space (X, d) than connectedness. The next proposition will state properties for (X, d) which guarantee that

$$\left(\frac{k}{n}\right)^\sigma \omega(T; \varepsilon_k(X)) \leqslant \rho \omega(T; \varepsilon_n(X)), \tag{5.7.2}$$

$n = 1, 2, 3, \dots$ and $k = 1, 2, \dots, n$, for all operators $T : E \to C(X)$ with appropriate constants $\sigma > 0$ and $\rho \geqslant 1$ and thus finally will enable us to estimate $e_n(T)$ from above on the basis of Theorem 5.7.1.

Proposition 5.8.3. *Let X be a compact and convex subset of a normed space, E an arbitrary Banach space, and $T \in L(E, C(X))$. Then*

$$\frac{k}{n} \omega(T; \varepsilon_k(X)) \leqslant 13 \omega(T; \varepsilon_n(X)) \tag{5.8.7}$$

for $n = 1, 2, 3, \dots$ and $k = 1, 2, \dots, n$.

Proof. Since a convex subset of a normed space is connected we have

$$\frac{k}{n} \varepsilon_k(X) \leqslant 12 \varepsilon_n(X) \tag{5.8.2}$$

for $n = 1, 2, 3, \dots$ and $k = 1, 2, \dots, n$ by Proposition 5.8.1. Because of the monotonicity (5.5.4) of the modulus of continuity $\omega(T; \delta)$ we may therefore conclude that

$$\omega(T; \varepsilon_k(X)) \leqslant \omega\left(T; 12\frac{n}{k}\varepsilon_n(X)\right).$$

Furthermore, under the above assumptions the modulus of continuity $\omega(T; \delta)$ satisfies property (5.5.14) of subadditivity, which implies

$$\omega(T; m\delta) \leqslant m\omega(T; \delta)$$

for any integer multiple $m\delta$ of δ. Monotonicity and subadditivity together yield

$$\omega\left(T; 12\frac{n}{k}\varepsilon_n(X)\right) \leqslant \omega\left(T; \left(\left[12\frac{n}{k}\right]+1\right)\varepsilon_n(X)\right)$$

$$\leqslant \left(12\frac{n}{k}+1\right)\omega(T; \varepsilon_n(X))$$

$$\leqslant 13\frac{n}{k}\omega(T; \varepsilon_n(X)),$$

which proves the assertion (5.8.7). ∎

Proposition 5.8.4. *Let X be a compact and convex subset of a normed space with more than one point, let E be an arbitrary Banach space, and $T \in K(E, C(X))$ a compact operator with rank $(T) > 1$. Then*

$$e_n(T) \leqslant C_0 \frac{\|T\|}{\omega(T; \varepsilon_1(X))}\omega(T; \varepsilon_n(X)) \tag{5.8.8}$$

for $n = 1, 2, 3, \ldots,$ where C_0 is a universal constant.

Proof. Because X consists of more than one point we have $\varepsilon_1(X) \neq 0$. On the other hand, since rank $(T) > 1$, there exists an element $x_0 \in U_E$ which is mapped onto a non-constant function $f_0 = Tx_0$ in $C(X)$. But for a non-constant continuous function f_0 on a compact connected metric space the modulus of continuity $\omega(f_0; \delta)$ is positive whenever $\delta > 0$. Hence it follows that $\omega(T; \varepsilon_1(X)) > 0$ so that the inequality (5.8.7) can be used for estimating $1/n$ from above by $\omega(T; \varepsilon_n(X))$, namely

$$\frac{1}{n} \leqslant \frac{13}{\omega(T; \varepsilon_1(X))}\omega(T; \varepsilon_n(X)), \quad n = 1, 2, 3, \ldots \tag{5.8.9}$$

If we now apply Theorem 5.7.1 with $p = 1$, realizing that condition (5.7.2) is satisfied with $\sigma = 1$ and $\rho = 13$, we obtain

$$e_n(T) \leqslant C(1, 13, 1) \cdot \left(13\frac{\|T\|}{\omega(T; \varepsilon_1(X))}+1\right)\omega(T; \varepsilon_n(X))$$

for $n = 1, 2, 3, \ldots$ Finally we refer to property (5.5.1) of the modulus of continuity $\omega(T; \delta)$ which states

$$1 \leqslant 2\frac{\|T\|}{\omega(T; \varepsilon_1(X))}$$

and thus leads us to the desired result (5.8.8) with $C_0 = 15C(1, 13, 1)$. ∎

At first glance the condition that X be a compact and convex subset of a normed space with more than one point seems to be rather restrictive.

However, if we consider *the space C[a,b]* of continuous functions on a closed interval $[a,b]$ and, more generally, *the space C(X) of continuous functions on a closed, bounded and convex subset of a finite-dimensional space* to be *the most important examples of Banach spaces of continuous functions*, we shall no longer be in any doubt about the significance of the estimates (5.8.7) and (5.8.8) obtained under these additional conditions. The estimate (5.8.2) for the entropy numbers $\varepsilon_k(X)$ of an arbitrary compact and connected metric space (X,d), which was fundamental to the proof of the analogous estimate (5.8.7) for $\omega(T;\varepsilon_k(X))$, can even be improved in the case where X is a closed, bounded and convex subset of a finite-dimensional Banach space. For instance, let X be the closed unit ball U_Z of a real Banach space Z with $\dim(Z) = m$. Then

$$k^{-1/m} \leqslant \varepsilon_k(U_Z) \leqslant 4 \cdot k^{-1/m} \quad \text{for } k = 1,2,3,\ldots \qquad (1.1.10)'$$

and hence

$$\varepsilon_k(U_Z) \leqslant 4\left(\frac{n}{k}\right)^{1/m} \varepsilon_n(U_Z) \qquad (5.8.10)$$

for $n = 1,2,3,\ldots$ and $k = 1,2,3,\ldots$ which implies

$$\omega(T;\varepsilon_k(U_Z)) \leqslant 5\left(\frac{n}{k}\right)^{1/m} \omega(T;\varepsilon_n(U_Z)) \qquad (5.8.11)$$

for $n = 1,2,3,\ldots$ and $k = 1,2,\ldots,n$, $T:E \to C(U_Z)$ being an arbitrary operator with values in $C(U_Z)$.

The property of connectedness of the underlying compact metric space (X,d), which turned out to be sufficient for an estimate of type (5.7.4) for the entropy numbers $\varepsilon_k(X)$, is not necessary for (5.7.4). An *example* is given by the following compact subset X of the unit interval $[0,1]$ *which generalizes Cantor's ternary set* from $p = 3$ to an arbitrary natural number $p \geqslant 3$: The set X is meant to consist of all real numbers $s \in [0,1]$ admitting a p-adic representation

$$s = \sum_{k=1}^{\infty} \delta_k p^{-k} \quad \text{with } \delta_k = 0 \text{ or } \delta_k = p - 1. \qquad (5.8.12)$$

In a similar way to Cantor's ternary set the set X is obtained by removing consecutively certain open intervals from the closed interval $[0,1]$. Therefore the set X is closed and hence also compact. To determine the entropy properties of X we consider the partial sums

$$s_n = \sum_{k=1}^{n} \delta_k p^{-k} \quad \text{with } \delta_k = 0 \text{ or } \delta_k = p - 1 \qquad (5.8.13)$$

of the infinite series (5.8.12). Obviously

$$|s - s_n| = \sum_{k=n+1}^{\infty} \delta_k p^{-k} \leqslant \sum_{k=n+1}^{\infty} (p-1)p^{-k} = p^{-n}.$$

Since there are 2^n elements of the form (5.8.13) in the set X it follows that

$$\varepsilon_{2^n}(X) \leqslant p^{-n}.$$

Furthermore, if k is an arbitrary natural number we choose a natural number n such that

$$2^{n-1} \leqslant k < 2^n \tag{5.8.14}$$

and conclude that

$$\varepsilon_k(X) \leqslant \varepsilon_{2^{n-1}}(X) \leqslant p^{-n+1} \leqslant p \cdot p^{-\log_2 k}. \tag{5.8.15}$$

If we now note that $p^{\log_2 k} = k^{\log_2 p}$ we can generate a power of k on the right-hand side of (5.8.15), namely

$$\varepsilon_k(X) \leqslant p \cdot k^{-\log_2 p}.$$

With the natural logarithm instead of the dyadic one this inequality reads as

$$\varepsilon_k(X) \leqslant p \cdot k^{-\log p / \log 2}. \tag{5.8.16}$$

To obtain an estimate for $\varepsilon_k(X)$ from below we consider two different elements s and t of type (5.8.13), say

$$s = \sum_{k=1}^{n} \delta_k(s) p^{-k} \quad \text{and} \quad t = \sum_{k=1}^{n} \delta_k(t) p^{-k},$$

and denote by m the smallest natural number k with $\delta_k(s) \neq \delta_k(t)$, so that

$$|\delta_m(s) - \delta_m(t)| = p - 1$$

and, consequently,

$$|s - t| = \left| \sum_{k=m}^{n} (\delta_k(s) - \delta_k(t)) p^{-k} \right|$$

$$\geqslant (p-1)p^{-m} - \sum_{k=m+1}^{n} |\delta_k(s) - \delta_k(t)| p^{-k}$$

$$> (p-1)p^{-m} - \sum_{k=m+1}^{\infty} (p-1)p^{-k} = (p-2)p^{-m} \geqslant p^{-n}.$$

Since there are 2^n elements of the form (5.8.13) with a mutual distance larger than p^{-n} we may conclude that

$$\varphi_{2^n-1}(X) \geqslant \tfrac{1}{2} p^{-n}$$

and hence also

$$\varepsilon_{2^n-1}(X) \geqslant \tfrac{1}{2} p^{-n}.$$

Given an arbitrary natural number k we again choose n according to (5.8.14) and then proceed in an analogous way to that of the proof of (5.8.16), namely

$$\varepsilon_k(X) \geqslant \varepsilon_{2^n-1}(X) \geqslant \frac{1}{2p} p^{-(n-1)} \geqslant \frac{1}{2p} p^{-\log_2 k}.$$

This result can be expressed as

$$\varepsilon_k(X) \geqslant \frac{1}{2p} k^{-\log p/\log 2}, \quad k = 1, 2, 3, \ldots,$$

and finally can be used to estimate a power $n^{-\log p/\log 2}$ from above, namely

$$n^{-\log p/\log 2} \leqslant 2p\varepsilon_n(X). \tag{5.8.17}$$

Combining (5.8.16) and (5.8.17) we obtain an estimate

$$\left(\frac{k}{n}\right)^{\log p/\log 2} \varepsilon_k(X) \leqslant 2p^2 \varepsilon_n(X) \tag{5.8.18}$$

of type (5.7.4), valid for arbitrary pairs of natural numbers k, n. But the set X is not connected since it is not even ε-connected for $\varepsilon < (p-2)/p$. In other words, for any ε with $0 < \varepsilon < 1$ there exists a compact metric space (X, d) with the property (5.7.4), which is not ε-connected. To see this, one only has to refer to the generalized Cantor set related to a natural number p with $p > 2/(1 - \varepsilon)$ in the above sense. Note that constants $\rho \geqslant 1$, $\sigma > 0$ giving an estimate (5.7.4) can be chosen as

$$\rho = 2p^2, \quad \sigma = \frac{\log p}{\log 2}$$

and will be larger the closer ε is to 1.

5.9. Hölder continuous operators

An operator $T : E \to C(X)$ from an arbitrary Banach space E into the space $C(X)$ of continuous functions on a compact metric space X is called *Hölder continuous of type* α *with* $0 < \alpha \leqslant 1$ if, in analogy to (5.6.5),

$$\sup_{\delta > 0} \frac{\omega(T; \delta)}{\delta^\alpha} < \infty.$$

The value

$$|T|_\alpha = \sup_{\delta > 0} \frac{\omega(T; \delta)}{\delta^\alpha} \tag{5.9.1}$$

is related to the corresponding values $|f|_\alpha$ of the images $f = Tx$ of $x \in U_E$ in $C(X)$ by

$$|T|_\alpha = \sup_{\|x\| \leqslant 1} |Tx|_\alpha. \tag{5.9.2}$$

This is an immediate consequence of the definition (3.0.6) of the modulus of continuity $\omega(T; \delta)$ of an operator $T : E \to C(X)$ which enables us to write

$$|T|_\alpha = \sup_{\|x\| \leqslant 1} \sup_{\delta > 0} \frac{\omega(Tx; \delta)}{\delta^\alpha}$$

and thus to obtain (5.9.2) by Lemma 5.6.1. Apart from its similarity to the representation

$$\| T \| = \sup_{\| x \| \leqslant 1} \| Tx \| \tag{5.9.3}$$

for the operator norm $\| T \|$, the representation (5.9.2) for $|T|_\alpha$ is interesting in so far as it makes clear that a Hölder continuous operator $T : E \to C(X)$ of type α maps each $x \in U_E$, and hence also each $x \in E$, onto a Hölder continuous function $f = Tx$ of type α on X. Therefore T can also be considered as a linear operator $S : E \to C^\alpha(X)$ from E into the space $C^\alpha(X)$ of Hölder continuous functions of type α on X. This induced linear operator S with values in $C^\alpha(X)$ turns out to be continuous as well. Indeed, in view of (5.9.2) and (5.9.3) the expression

$$\| Sx \|_\alpha = \max (\| Tx \|, | Tx |_\alpha) \tag{5.9.4}$$

can be estimated by

$$\| Sx \|_\alpha \leqslant \max (\| T \|, | T |_\alpha) \| x \|.$$

This proves the continuity of the operator $S : E \to C^\alpha(X)$ and simultaneously

$$\| S \| \leqslant \max (\| T \|, | T |_\alpha). \tag{5.9.5}$$

On the other hand (5.9.4) implies

$$\| Tx \| \leqslant \| Sx \|_\alpha \quad \text{and} \quad | Tx |_\alpha \leqslant \| Sx \|_\alpha$$

and hence also

$$\| T \| \leqslant \| S \| \quad \text{and} \quad | T |_\alpha \leqslant \| S \|,$$

which finally provides the inequality

$$\max (\| T \|, | T |_\alpha) \leqslant \| S \|$$

converse to (5.9.5).

Summarizing the result of the above considerations we formulate the following proposition.

Proposition 5.9.1. *Any Hölder continuous operator $T : E \to C(X)$ of type α mapping an arbitrary Banach space E into a space $C(X)$ admits a factorization*

$$T = I_\alpha S \tag{5.9.6}$$

where $S \in L(E, C^\alpha(X))$ is an operator whose norm is given by

$$\| S \| = \max (\| T \|, | T |_\alpha) \tag{5.9.7}$$

and $I_\alpha : C^\alpha(X) \to C(X)$ is the identity map of $C^\alpha(X)$ into $C(X)$.

Conversely, the composition $T = I_\alpha S$ of an arbitary operator $S \in L(E, C^\alpha(X))$ with the identity map I_α of $C^\alpha(X)$ into $C(X)$ is always Hölder continuous of type α since

$$\omega(I_\alpha S; \delta) \leqslant \omega(I_\alpha; \delta) \| S \|,$$

by the multiplicativity (5.5.3) of the modulus of continuity, and since $\omega(I_\alpha; \delta) \leqslant \delta^\alpha$ by (5.6.7). It follows that

$$|T|_\alpha \leqslant \|S\|,$$

and

$$\|T\| \leqslant \|I_\alpha\| \|S\| \leqslant \|S\|$$

by (5.6.9). Thus again we have the inequality

$$\max(\|T\|, |T|_\alpha) \leqslant \|S\|.$$

But when considering the Hölder continuous operator $T = I_\alpha S$ as an operator from E into the space $C^\alpha(X)$ of Hölder continuous functions of type α on X, we are simply led back to the operator $S: E \to C^\alpha(X)$ that we started from. Therefore we even have equality between $\max(\|T\|, |T|_\alpha)$ and $\|S\|$ as stated in (5.9.7).

Altogether it has turned out that *the identity map* $I_\alpha: C^\alpha(X) \to C(X)$ is a particular Hölder continuous operator of type α which is a kind of *generating operator for the class* $L^\alpha(E, C(X))$ *of all Hölder continuous operators T of type* α from an arbitrary Banach space E into the space $C(X)$. The class $L^\alpha(E, C(X))$ is obviously a linear space consisting of compact operators only because

$$\lim_{\delta \to 0} \omega(T; \delta) = 0 \tag{5.5.9}$$

for every $T \in L^\alpha(E, C(X))$. The relation (5.9.7) between an operator $T \in L^\alpha(E, C(X))$ and the induced operator $S \in L(E, C^\alpha(X))$ is the motive for introducing the norm

$$\|T\|_\alpha = \max(\|T\|, |T|_\alpha) \tag{5.9.8}$$

on $L^\alpha(E, C(X))$. The linear space $L^\alpha(E, C(X))$ equipped with this norm is a Banach space. Moreover, if $T \in L^\alpha(E, C(X))$ is a Hölder continuous operator of type α defined on E, and $R \in L(Z, E)$ is an arbitrary operator with values in E, then the product TR is Hölder continuous of type α as well and its α-norm

$$\|TR\|_\alpha = \|SR\|$$

can be estimated by

$$\|TR\|_\alpha \leqslant \|T\|_\alpha \|R\| \tag{5.9.9}$$

according to the definition (5.9.8) of $\|T\|_\alpha$ and its relation (5.9.7) to the usual norm $\|S\|$ of the operator $S \in L(E, C^\alpha(X))$ induced by T. Hence the α-norm $\|T\|_\alpha$ of a Hölder continuous operator T of type α under multiplication from the right by an arbitrary linear and continuous operator R behaves in the same way as an ideal quasi-norm α connected with an operator ideal A (see section 1.6, (IQ3)(b)).

What can be said about *approximation and compactness properties of Hölder continuous operators of type* α? Theorem 5.6.1 combined with the definition (5.9.1) of $|T|_\alpha$ tells us that

$$a_{n+1}(T) \leqslant |T|_\alpha \varepsilon_n^\alpha(X), \quad n = 1, 2, 3, \ldots, \qquad (5.9.10)$$

for $T \in L^\alpha(E, C(X))$. If the compact metric space (X, d) consists of more than one point and possesses the additional property (5.7.4), the dyadic entropy numbers $e_n(T)$ of a Hölder continuous operator $T: E \to C(X)$ of type α provide an analogous estimate. This can be shown with the help of the factorization (5.9.6) of T by I_α and by applying Proposition 5.7.1. We need only write down

$$e_n(T) \leqslant \|S\| e_n(I_\alpha) \leqslant C \cdot \|T\|_\alpha \varepsilon_n^\alpha(X), \qquad (5.9.11)$$

$n = 1, 2, 3, \ldots$, by referring to (5.7.6), (5.9.6), and (5.9.7). In the case of the identity map $T = I_\alpha$ itself the dyadic entropy numbers $e_n(I_\alpha)$ in fact behave asymptotically like $\varepsilon_n^\alpha(X)$ under the above suppositions (see Proposition 5.7.1). The asymptotic behaviour of the approximation quantities

$$a_{n+1}(I_\alpha) = c_{n+1}(I_\alpha) \quad \text{and} \quad t_{n+1}(I_\alpha) = d_{n+1}(I_\alpha)$$

is determined by $\varepsilon_n^\alpha(X)$ without any additional assumptions on the compact metric space (X, d) (see Proposition 5.6.2). This makes it clear that the estimate from above by $|T|_\alpha \varepsilon_n^\alpha(X)$, even for the symmetrized approximation numbers $t_{n+1}(T) = d_{n+1}(T)$ of a Hölder continuous operator T of type α, is in general a best possible one. Under no circumstances can the decay of the Kolmogorov numbers $d_n(T)$ be faster than $(2n)^{-1/2} a_n(T)$ (see Proposition 2.4.6). If we subject the compact metric space (X, d) to a condition

$$\varepsilon_n(X) \leqslant \rho \cdot n^{-\sigma}, \quad n = 1, 2, 3, \ldots,$$

with certain constants $\rho, \sigma > 0$, and confine ourselves to Hölder continuous operators $T: H \to C(X)$ of type α defined on a Hilbert space H, we can actually produce the crucial power $n^{-1/2}$ of n as a factor improving a term which arises from the decay of the approximation numbers $a_n(T)$ (see Theorem 5.10.2). This will be done in the next section within a general framework of estimates for approximation and entropy quantities of operators $T \in L^\alpha(E, C(X))$ which rest on *a priori* estimates for finite-dimensional operators $S: E \to l_\infty^n$.

5.10. Application of local techniques to Hölder continuous operators

The method of employing estimates for approximation and entropy quantities of finite-dimensional operators to obtain corresponding estimates for infinite-dimensional operators will be referred to as a *local technique*.

We start with a lemma concerning the existence of certain *coupled systems of partitions of unity* $P_m = \{\varphi_{m,i}: 1 \leqslant i \leqslant N_m\}$, $1 \leqslant m \leqslant M$, of a compact metric space (X, d). The number N_m of functions $\varphi_{m,i}$ forming the partition of unity P_m will be limited from above by the minimal number $N(X; 2^{-(m+1)})$ of points in a $2^{-(m+1)}$-net of X (see section 1.1), and for $1 \leqslant m \leqslant M - 1$ the individual functions $\varphi_{m,i}$ of P_m will be obtained by dividing the functions $\varphi_{m+1,j}$ of P_{m+1} into groups and summing up the functions belonging to the same group.

Lemma 5.10.1. (see Carl, Heinrich and Kühn 1988). *Let (X, d) be a compact metric space and M an arbitrary natural number. Then for $m = 1, 2, \ldots, M$ there exist partitions of unity $P_m = \{\varphi_{m,i}: 1 \leqslant i \leqslant N_m\}$ and points $s_{m,k} \in X$, $1 \leqslant k \leqslant N_m$, with the properties*

$$N_m \leqslant N(X; 2^{-(m+1)}) \quad \text{for } 1 \leqslant m \leqslant M, \tag{5.10.1}$$

$$\varepsilon_1(\text{supp}(\varphi_{m,i})) \leqslant 2^{-m} \quad \text{for } 1 \leqslant i \leqslant N_m \text{ and } 1 \leqslant m \leqslant M, \tag{5.10.2}$$

$$\varphi_{m,i}(s_{m,k}) = \delta_{ik} \quad \text{for } 1 \leqslant i, k \leqslant N_m \text{ and } 1 \leqslant m \leqslant M, \tag{5.10.3}$$

$$\varphi_{m,i} \in \text{span}\{\varphi_{m+1,j}: 1 \leqslant j \leqslant N_{m+1}\} \tag{5.10.4}$$
$$\text{for } 1 \leqslant i \leqslant N_m \text{ and } 1 \leqslant m \leqslant M - 1.$$

Proof. The idea of the proof is a kind of 'backwards induction'. Therefore we begin with $m = M$ and consider a covering

$$X = \bigcup_{i=1}^{N_M} \mathring{U}(x_i; 2^{-M}) \tag{5.10.5}$$

of X by a minimal number $N_M = \mathring{N}(X; 2^{-M})$ of open balls of radius 2^{-M}. If we compare N_M with the minimal number $N(X; 2^{-(M+1)})$ of elements in a $2^{-(M+1)}$-net for X (see section 1.1.) we see that

$$N_M \leqslant N(X; 2^{-(M+1)}). \tag{5.10.1$'$}$$

For any partition of unity $P_M = \{\varphi_{M,i}: 1 \leqslant i \leqslant N_M\}$ subordinate to the open covering (5.10.5) in the sense that

$$\text{supp}(\varphi_{M,i}) \subseteq \mathring{U}(x_i; 2^{-M}) \quad \text{for } 1 \leqslant i \leqslant N_M$$

(see Lemma 5.2.1, (5.2.7)), the property

$$\varepsilon_1(\text{supp}(\varphi_{M,i}) \leqslant 2^{-M} \quad \text{for } 1 \leqslant i \leqslant N_M \tag{5.10.2$'$}$$

is obvious. Furthermore, since none of the sets $\mathring{U}(x_i; 2^{-M})$ of the covering (5.10.5) is superfluous we can find elements $s_{M,k} \in X$, $1 \leqslant k \leqslant N_M$, such that

$$s_{M,k} \in \mathring{U}(x_k; 2^{-M}) \text{ but } s_{M,k} \notin \mathring{U}(x_i; 2^{-M}) \quad \text{for } i \neq k$$

and, hence,

$$\varphi_{M,i}(s_{M,k}) = \delta_{ik} \quad \text{for } 1 \leqslant i, k \leqslant N_M. \tag{5.10.3$'$}$$

This settles the case $m = M$. Now let us assume that a partition of unity

$P_{m+1} = \{\varphi_{m+1,j} : 1 \leqslant j \leqslant N_{m+1}\}$ and a system of elements $s_{m+1,k} \in X$, $1 \leqslant k \leqslant N_{m+1}$, with the properties

$$N_{m+1} \leqslant N(X; 2^{-(m+2)}), \qquad (5.10.1)''$$

$$\varepsilon_1(\mathrm{supp}(\varphi_{m+1,j})) \leqslant 2^{-(m+1)} \quad \text{for } 1 \leqslant j \leqslant N_{m+1}, \qquad (5.10.2)''$$

$$\varphi_{m+1,j}(s_{m+1,k}) = \delta_{j,k} \quad \text{for } 1 \leqslant j,k \leqslant N_{m+1} \qquad (5.10.3)''$$

have been determined for some $m \leqslant M - 1$. The property (5.10.2)'' implies the existence of elements $y_j \in X$, $1 \leqslant j \leqslant N_{m+1}$, with

$$\mathrm{supp}(\varphi_{m+1,j}) \subseteq U(y_j; 2^{-(m+1)}) \qquad (5.10.6)$$

since, by the compactness of X, the first entropy number $\varepsilon_1(X_0)$ of a subset $X_0 \subseteq X$ is realized in the sense that X_0 is contained in an appropriate closed ball of radius $\varepsilon_1(X_0)$. On the other hand we can choose a minimal $2^{-(m+1)}$-net $z_1, z_2, \ldots, z_{\tilde{N}_m}$ for X so that

$$X = \bigcup_{i=1}^{\tilde{N}_m} U(z_i; 2^{-(m+1)}) \quad \text{with } \tilde{N}_m = N(X; 2^{-(m+1)}).$$

If we take mutually disjoint subsets

$$A_i \subseteq U(z_i; 2^{-(m+1)}), \quad 1 \leqslant i \leqslant \tilde{N}_m, \qquad (5.10.7)$$

of the closed balls $U(z_i; 2^{-(m+1)})$, which also cover X, we can prescribe a rule for summing up certain groups of functions $\varphi_{m+1,j} \in P_{m+1}$ as follows: given i, $1 \leqslant i \leqslant \tilde{N}_m$, check which of the elements $y_j \in X$ belong to A_i and then put

$$\varphi_{m,i} = \sum_{y_j \in A_i} \varphi_{m+1,j}. \qquad (5.10.8)$$

If A_i happens to contain none of the elements y_j, leave out the corresponding index i.

Having carried out the summation procedure (5.10.8) we have a system of

$$N_m \leqslant N(X; 2^{-(m+1)})$$

functions $\varphi_{m,i}$ which represent a partition of unity P_m. Indeed, any of the elements $y_j \in X$, $1 \leqslant j \leqslant N_{m+1}$, belongs to exactly one set A_i so that

$$\sum_{i=1}^{N_m} \varphi_{m,i} = \sum_{i=1}^{N_m} \sum_{y_j \in A_i} \varphi_{m+1,j} = \sum_{j=1}^{N_{m+1}} \varphi_{m+1,j} = 1.$$

Moreover, by (5.10.8) an element $x \in \mathrm{supp}(\varphi_{m,i})$ belongs to $\mathrm{supp}(\varphi_{m+1,j})$ for at least one j with $y_j \in A_i$. By (5.10.6) this implies

$$d(x, y_j) \leqslant 2^{-(m+1)}$$

for some j with $y_j \in A_i$. However, $y_j \in A_i$ implies

$$d(y_j, z_i) \leqslant 2^{-(m+1)}$$

by (5.10.7). All in all we get

$$d(x, z_i) \leqslant d(x, y_j) + d(y_j, z_i) \leqslant 2^{-m} \quad \text{for } x \in \mathrm{supp}(\varphi_{m,i})$$

and hence

$$\varepsilon_1(\mathrm{supp}(\varphi_{m,i})) \leqslant 2^{-m} \quad \text{for } 1 \leqslant i \leqslant N_m.$$

Finally we declare $s_{m,k}$ to be one of the elements $s_{m+1,j}$ with $y_j \in A_k$. Then

$$\varphi_{m,i}(s_{m,k}) = \sum_{y_l \in A_i} \varphi_{m+1,l}(s_{m+1,j}) = \sum_{y_l \in A_i} \delta_{lj} = \delta_{ik}$$

because $y_j \in A_i$ for $i = k$ but $y_j \notin A_i$ for $i \neq k$. This completes the proof of the properties (5.10.1), (5.10.2), and (5.10.3) for the partition of unity P_m constructed by using a partition of unity P_{m+1} with the analogous properties (5.10.1)″, (5.10.2)″, and (5.10.3)″. The property (5.10.4) of P_m is satisfied trivially in view of the definition (5.10.8) of the individual functions $\varphi_{m,i} \in P_m$.

The procedure ends up with $m = 1$. ∎

In the following we shall use Lemma 5.10.1 to decompose an arbitrary operator $T \in L(E, C(X))$ into a sum

$$T = \sum_{m=0}^{M-1} T_m + T_M \tag{5.10.9}$$

of finite rank operators $T_0, T_1, \ldots, T_{M-1}$ with special properties and a remainder T_M, all the operators $T_0, T_1, \ldots, T_M$ being subject to certain norm estimates. Compared with the statement

$$a_{n+1}(T) \leqslant \omega(T; \varepsilon_n(X)), \quad n = 1, 2, 3, \ldots \tag{3.0.8}$$

of Theorem 5.6.1 and its proof this may be regarded as a kind of *refined approximation procedure*.

Lemma 5.10.2. (cf. Carl, Heinrich and Kühn 1988). *Let (X, d) be a compact metric space, E a Banach space, and $T \in L(E, C(X))$. Then for each natural number M the operator T admits a decomposition*

$$T = \sum_{m=0}^{M-1} T_m + T_M \tag{5.10.9}$$

such that

$$T_m(E) \subseteq E_m, \tag{5.10.10}$$

where E_m is a finite-dimensional subspace of $C(X)$ of dimension $N_{m+1} \leqslant N(X; 2^{-(m+2)})$ isometrically isomorphic to $l_\infty^{N_{m+1}}$ for $m = 0, 1, 2, \ldots, M-1$,

$$\| T_M \| \leqslant \omega(T; 2^{-(M-1)}), \tag{5.10.11}$$

$$\| T_m \| \leqslant 2\omega(T; 2^{-(m-1)}) \quad \text{for } m = 1, 2, \ldots, M-1, \tag{5.10.12}$$

$$\| T_0 \| \leqslant \| T \|. \tag{5.10.13}$$

Proof. Let partitions of unity $P_m = \{\varphi_{m,i}: 1 \leqslant i \leqslant N_m\}$ and points $s_{m,k} \in X$, $1 \leqslant k \leqslant N_m$, for $m = 1, 2, \ldots, M$ be chosen according to Lemma 5.10.1. We then define finite rank operators $A_m \in L(E, C(X))$ by

$$A_m x = \sum_{i=1}^{N_m} (Tx)(s_{m,i}) \varphi_{m,i} \quad \text{for } m = 1, 2, \ldots, M \qquad (5.10.14)$$

and use A_m for approximating T. As in the proof of Theorem 5.6.1 property (5.2.3) of a partition of unity guarantees that

$$|(Tx)(t) - (A_m x)(t)| \leqslant \sum_{i=1}^{N_m} |(Tx)(t) - (Tx)(s_{m,i})| \varphi_i(t)$$

for all $t \in X$. Because the functions $\varphi_{m,i}, 1 \leqslant i \leqslant N_m$, have the additional property

$$\varepsilon_1(\text{supp}(\varphi_{m,i})) \leqslant 2^{-m} \quad \text{for } 1 \leqslant i \leqslant N_m, \qquad (5.10.2)$$

there is a collection of closed balls $U(t_i; 2^{-m})$, $1 \leqslant i \leqslant N_m$, such that

$$\text{supp}(\varphi_{m,i}) \subseteq U(t_i; 2^{-m}).$$

Accordingly, we have

$$|(Tx)(t) - (Tx)(s_{m,i})| \leqslant \omega(Tx; 2^{-(m-1)}) \quad \text{for } t \in U(t_i; 2^{-m}),$$

since

$$d(t, s_{m,i}) \leqslant d(t, t_i) + d(t_i, s_{m,i}) \leqslant 2^{-(m-1)} \quad \text{for } t \in U(t_i; 2^{-m}),$$

and

$$\varphi_i(t) = 0 \quad \text{for } t \notin U(t_i; 2^{-m}).$$

Thus we may conclude that

$$|(Tx)(t) - (A_m x)(t)| \leqslant \omega(Tx; 2^{-(m-1)}) \quad \text{for all } t \in X$$

and hence finally that

$$\|T - A_m\| \leqslant \omega(T; 2^{-(m-1)}) \quad \text{for } m = 1, 2, \ldots, M. \qquad (5.10.15)$$

The operators $T_0, T_1, \ldots, T_M$ are now introduced by

$$T_0 = A_1, T_m = A_{m+1} - A_m \quad \text{for } 1 \leqslant m \leqslant M - 1, T_M = T - A_M,$$

so that (5.10.9) is realized per definition. Furthermore, (5.10.15) implies

$$\|T_M\| \leqslant \omega(T; 2^{-(M-1)}) \qquad (5.10.11)$$

as well as

$$\begin{aligned} \|T_m\| &\leqslant \|T - A_{m+1}\| + \|T - A_m\| \\ &\leqslant \omega(T; 2^{-m}) + \omega(T; 2^{-(m-1)}) \\ &\leqslant 2\omega(T; 2^{-(m-1)}) \quad \text{for } 1 \leqslant m \leqslant M - 1, \end{aligned}$$

that is (5.10.12). The estimate (5.10.13) for $\|T_0\|$ immediately results from

$$\|T_0 x\| = \|A_1 x\| = \sup_{s \in X} \left| \sum_{i=1}^{N_1} (Tx)(s_{1,i}) \varphi_{1,i}(s) \right|$$

$$\leqslant \|Tx\| \sup_{s \in X} \sum_{i=1}^{N_1} \varphi_{1,i}(s) \leqslant \|T\| \, \|x\|.$$

To prove (5.10.10) we put

$$E_m = \text{span}\{\varphi_{m+1,j} : 1 \leqslant j \leqslant N_{m+1}\} \quad \text{for } 0 \leqslant m \leqslant M - 1.$$

Since

$$\varphi = \sum_{j=1}^{N_{m+1}} \xi_j \varphi_{m+1,j} \in E_m$$

takes the values

$$\varphi(s_{m+1,k}) = \xi_k$$

at the points $s_{m+1,k}$, $1 \leqslant k \leqslant N_{m+1}$ (see (5.10.3)), an isometry from E_m onto $l_\infty^{N_{m+1}}$ is given by

$$\varphi \to (\varphi(s_{m+1,1}), \varphi(s_{m+1,2}), \ldots, \varphi(s_{m+1,N_{m+1}})), \tag{5.10.4}$$

where $N_{m+1} \leqslant N(X, 2^{-(m+2)})$ by (5.10.1). Finally (5.10.4) guarantees that

$$A_m(E) \subseteq A_{m+1}(E) \quad \text{for } 1 \leqslant m \leqslant M - 1$$

and hence

$$T_m(E) \subseteq A_{m+1}(E) \subseteq E_m \quad \text{for } 0 \leqslant m \leqslant M - 1,$$

which completes the proof of (5.10.10) and thus also the proof of the lemma.

∎

Before developing the *local technique* for operators $T \in L^a(E, C(X))$ mentioned in the title of the present section we give an estimate for the dyadic entropy numbers $e_k(S)$ of finite rank operators $S : E \to F$ for natural numbers k exceeding a real number r with $r \geqslant \text{rank}(S)$.

Lemma 5.10.3. *Let $S \in L(E, F)$ be an operator between arbitrary Banach spaces E and F with $\text{rank}(S) \leqslant m$ and let $r \geqslant m$ and $\beta \geqslant 0$ be real numbers. Then*

$$e_k(S) \leqslant 16 \cdot \beta^\beta \cdot 2^{-k/r} \left(\frac{m}{r}\right)^\beta e_m(S) \quad \text{for } k \geqslant r. \tag{5.10.16}$$

Proof. We consider the canonical factorization

$$S = S_0 Q_N^E$$

of S by the quotient space E/N of E with respect to the null space N of S. Because of the surjectivity of the entropy numbers (see section 1.3, (ES)) we have

$$e_k(S) = e_k(S_0).$$

Writing the operator $S_0 : E/N \to F$ as

$$S_0 = S_0 I_{E/N}$$

with $I_{E/N}$ as the identity map on E/N and using the multiplicativity of the entropy numbers (see section 1.3, (DE3)) we obtain

$$e_k(S_0) \leqslant e_m(S_0) e_{k-m+1}(I_{E/N}) \quad \text{for } k \geqslant m.$$

Since $\dim(E/N) \leqslant m$ the entropy numbers $e_{k-m+1}(I_{E/N})$ are subject to the estimate

$$e_{k-m+1}(I_{E/N}) \leqslant 4 \cdot 2^{-(k-m)/m} = 8 \cdot 2^{-k/m}$$

(see (1.3.36), real vision). Hence it follows that

$$e_k(S) \leqslant 8 \cdot 2^{-k/m} e_m(S) \quad \text{for } k \geqslant m.$$

Combining this estimate with the elementary inequalities

$$2^{-k/m} \leqslant 2^{1-k/r-r/m} \quad \text{for } k \geqslant r \geqslant m$$

and

$$2^{-r/m} \leqslant \beta^\beta \left(\frac{m}{r}\right)^\beta \quad \text{for any } \beta \geqslant 0$$

we arrive at the desired result (5.10.16). ∎

From now on let $s_k(T)$ stand either for one of the approximation quantities $a_k(T), c_k(T), d_k(T), t_k(T)$ or for the dyadic entropy numbers $e_k(T)$. We shall show how certain *a priori* estimates for the $s_k(S)$ of operators $S: E \to l_\infty^m$ give rise to estimates of the approximation or entropy quantities $s_k(T)$ of a Hölder continuous operator $T: E \to C(X)$ of type α if the compact metric space (X, d) satisfies additional conditions. Since l_∞^n possesses the metric extension property we have

$$c_k(S) = a_k(S)$$

and, correspondingly,

$$t_k(S) = d_k(S)$$

because $t_k(S) = c_k(SQ_E)$ and hence $t_k(S) = a_k(SQ_E) = d_k(S)$. Therefore in the present context the meaning of $s_k(S)$ reduces either to one of the approximation quantities $a_k(S), d_k(S)$ or to the entropy quantities $e_k(S)$.

Theorem 5.10.1. *Let E be a Banach space such that there is a constant $\beta \geqslant 0$ with the property that for each $\varepsilon > 0$ there is a constant $c(\varepsilon) \geqslant 1$ granting uniform estimates*

$$s_k(S) \leqslant c(\varepsilon) \cdot \|S\| \cdot k^{-\beta} \left(\frac{n}{k}\right)^\varepsilon \tag{5.10.17}$$

for $n = 1, 2, \ldots, 1 \leqslant k \leqslant n$, and all operators $S: E \to l_\infty^n$. Furthermore, let (X, d) be a compact metric space such that

$$\varepsilon_n(X) \leqslant \rho \cdot n^{-\sigma}, \quad n = 1, 2, \ldots, \tag{5.10.18}$$

with appropriate constants $\rho, \sigma > 0$. Then the sequence $s_n(T)$ of a Hölder continuous operator $T \in L^\alpha(E, C(X))$ of type $\alpha, 0 < \alpha \leqslant 1$, is limited from above by

$$s_n(T) \leqslant C \cdot \|T\|_\alpha \cdot n^{-\beta - \alpha\sigma}, \quad n = 1, 2, \ldots, \tag{5.10.19}$$

where $C > 0$ is a constant depending on α, β, ρ, and σ.

Proof. Let us prove the assertion for the dyadic entropy numbers. Let $L \geqslant 1$ and $M \geqslant 1$ be arbitrary natural numbers. We decompose the operator $T \in L^\alpha(E, C(X))$ into a sum

$$T = \sum_{m=0}^{L-1} T_m + \sum_{m=L}^{L+M} T_m + T_{L+M+1} \tag{5.10.20}$$

according to Lemma 5.10.2. Hence we have

$$\| T_{L+M+1} \| \leqslant \omega(T; 2^{-L-M}) \leqslant \| T \|_\alpha \cdot 2^{-\alpha(L+M)}, \tag{5.10.11$'$}$$

$$\| T_m \| \leqslant 2\omega(T; 2^{-(m-1)}) \leqslant 2^{1+\alpha} \| T \|_\alpha \cdot 2^{-\alpha m} \tag{5.10.12$'$}$$

for $m = 1, 2, \ldots, L + M$,

$$\| T_0 \| \leqslant \| T \| \leqslant \| T \|_\alpha, \tag{5.10.13$'$}$$

by (5.10.11), (5.10.12), (5.10.13) and the assumption $T \in L^\alpha(E, C(X))$. The representation (5.10.20) can be employed to obtain the desired estimate (5.10.19). Indeed, using the additivity of the dyadic entropy numbers (DE2) we may conclude that

$$e_{n_0 + \cdots + n_{L+M} - L - M}(T) \leqslant \sum_{m=0}^{L-1} e_{n_m}(T_m) + \sum_{m=L}^{L+M} e_{n_m}(T_m) + \| T_{L+M+1} \| \tag{5.10.21}$$

for natural numbers n_m to be chosen later. Because of (5.10.11)$'$ the term $\| T_{L+M+1} \|$ will influence the right-hand side of (5.10.21) less the larger we choose M. So let us concentrate on the other two items $\sum_{m=0}^{L-1} e_{n_m}(T_m)$ and $\sum_{m=L}^{L+M} e_{n_m}(T_m)$ and try to estimate them by the natural number L independently of M. The operators T_m, $m = 0, 1, \ldots, L + M$, are of finite rank, and their ranges $T_m(E)$ are contained in N_{m+1}-dimensional subspaces E_m of $C(X)$ isometrically isomorphic to $l_\infty^{N_{m+1}}$. So let S_m denote the operator from E into $l_\infty^{N_{m+1}}$ induced by T_m and, correspondingly,

$$T_m = J_m S_m$$

with an isometric mapping $J_m : l_\infty^{N_{m+1}} \to C(X)$. Then

$$e_k(T_m) \leqslant e_k(S_m).$$

By the assumption (5.10.17) we have the estimate

$$e_k(S_m) \leqslant c(\varepsilon) \cdot k^{-\beta} \left(\frac{N_{m+1}}{k} \right)^\varepsilon \| S_m \|$$

for $1 \leqslant k \leqslant N_{m+1}$ and $m = 0, 1, \ldots, L + M$. Since $\| S_m \| = \| T_m \|$ we obtain

$$e_k(T_m) \leqslant 2^{1+\alpha} c(\varepsilon) \cdot \| T \|_\alpha \cdot 2^{-\alpha m} \cdot k^{-\beta} \left(\frac{N_{m+1}}{k} \right)^\varepsilon, \tag{5.10.22}$$

$1 \leqslant k \leqslant N_{m+1}$, $m = 0, 1, \ldots, L + M$, by (5.10.12)$'$ and (5.10.13)$'$. The dimension N_{m+1} is controlled by $N_{m+1} \leqslant N(X; 2^{-(m+2)})$ (see (5.10.10)) so that

$$\varepsilon_{N_{m+1}-1}(X) \geqslant 2^{-(m+2)}.$$

This implies

$$2^{-(m+2)} \leqslant \varepsilon_{N_{m+1}-1}(X) \leqslant \rho(N_{m+1}-1)^{-\sigma}$$

by the assumption (5.10.18). Thus

$$N_{m+1} \leqslant 1 + \rho^{1/\sigma}2^{(m+2)/\sigma} \leqslant (1 + \rho^{1/\sigma} \cdot 2^{2/\sigma}) \cdot 2^{m/\sigma}. \qquad (5.10.23)$$

We now estimate the values $e_{n_m}(T_m)$ for $0 \leqslant m \leqslant L-1$ and $L \leqslant m \leqslant L + M$ separately. For $0 \leqslant m \leqslant L-1$ we put

$$r_m = (1 + \rho^{1/\sigma} \cdot 2^{2/\sigma})2^{m/\sigma}$$

and choose a natural number K with

$$1 + \alpha + \frac{\beta}{\sigma} \leqslant K \leqslant 2 + \alpha + \frac{\beta}{\sigma}$$

as well as natural numbers n_m such that

$$K(L-m)r_m \leqslant n_m \leqslant 1 + K(L-m)r_m, \quad m = 0, \ldots, L-1. \qquad (5.10.24)$$

An upper bound for $e_{n_m}(T_m)$ then results from Lemma (5.10.3) with $r = r_m \geqslant N_{m+1}$ and with the constant $\beta \geqslant 0$ in the assumption (5.10.17), namely

$$e_{n_m}(T_m) \leqslant 16 \cdot \beta^\beta \cdot 2^{-n_m/r_m} \left(\frac{N_{m+1}}{r_m} \right)^\beta e_{N_{m+1}}(T_m).$$

The value $e_{N_{m+1}}(T_m)$ can be estimated by

$$e_{N_{m+1}}(T_m) \leqslant 2^{1+\alpha}c(\varepsilon) \cdot \| T \|_\alpha \cdot 2^{-\alpha m} \cdot N_{m+1}^{-\beta}$$

according to (5.10.22). This gives

$$e_{n_m}(T_m) \leqslant 16 \cdot c(\varepsilon) \beta^\beta 2^{1+\alpha} \| T \|_\alpha \cdot 2^{-n_m/r_m - \alpha m} r_m^{-\beta}.$$

Because of (5.10.24) we have

$$2^{-n_m/r_m} \leqslant 2^{-K(L-m)}.$$

If we take account of the above definition of r_m we get

$$e_{n_m}(T_m) \leqslant C_1(\varepsilon) \cdot \| T \|_\alpha \cdot 2^{-K(L-m)-(\alpha+\beta/\sigma)m}, \quad 0 \leqslant m \leqslant L-1,$$

where

$$C_1(\varepsilon) = 16 \cdot c(\varepsilon) \cdot \beta^\beta (1 + \rho^{1/\sigma} \cdot 2^{2/\sigma})^{-\beta} 2^{1+\alpha}.$$

The summation formula

$$\sum_{m=0}^{L-1} 2^{-K(L-m)-(\alpha+\beta/\sigma)m} = 2^{-KL} \frac{2^{(K-\alpha-\beta/\sigma)L} - 1}{2^{K-\alpha-\beta/\sigma} - 1}$$

leads us to the estimate

$$\sum_{m=0}^{L-1} e_{n_m}(T_m) \leqslant C_1(\varepsilon) \cdot \| T \|_\alpha \cdot 2^{-KL} \frac{2^{(K-\alpha-\beta/\sigma)L} - 1}{2^{K-\alpha-\beta/\sigma} - 1} \leqslant C_1(\varepsilon) \cdot \| T \|_\alpha \cdot 2^{-(\alpha+\beta/\sigma)L}$$

$$(5.10.25)$$

because $1 + \alpha + \beta/\sigma \leqslant K$.

It remains to limit the subscript $n_0 + \cdots + n_{L-1} - L$. From (5.10.24) we have

$$n_0 + \cdots + n_{L-1} - L \leqslant K \sum_{m=0}^{L-1} (L-m)r_m = (1 + \rho^{1/\sigma} \cdot 2^{2/\sigma})K \sum_{m=0}^{L-1} (L-m)2^{m/\sigma}.$$

By using the summation formula

$$\sum_{m=0}^{L-1} (L-m)2^{-m/\sigma} = \frac{2^{1/\sigma}}{(2^{1/\sigma}-1)^2}(2^{L/\sigma} + L \cdot 2^{-1/\sigma} - L - 1)$$

we verify that

$$n_0 + \cdots + n_{L-1} - L \leqslant C_2 \cdot 2^{L/\sigma}, \qquad (5.10.26)$$

where

$$C_2 = (1 + \rho^{1/\sigma} \cdot 2^{2/\sigma})2^{1/\sigma}(2^{1/\sigma}-1)^{-2}K.$$

Now we consider the sum $\sum_{m=L}^{L+M} e_{n_m}(T_m)$. Before choosing the subscripts n_m we give appropriate estimates for the $e_k(T_m)$ for $m = L, \ldots, L+M$ and all $k = 1, 2, 3, \ldots$ If $1 \leqslant k \leqslant N_{m+1}$ the estimate (5.10.22) applies. For $k \geqslant N_{m+1}$ we proceed in the same way as in the situation $0 \leqslant m \leqslant L-1$ with $k = n_m$ when $k = n_m \geqslant N_{m+1}$ appeared as a consequence of (5.10.24). We use Lemma 5.10.3 with $r = N_{m+1}$, namely

$$e_k(T_m) \leqslant 16 \cdot \beta^\beta \cdot 2^{-k/N_{m+1}} e_{N_{m+1}}(T_m) \quad \text{for } k \geqslant N_{m+1},$$

and then again use the estimate (5.10.22) for $e_{N_{m+1}}(T_m)$. The result this time reads as

$$e_k(T_m) \leqslant 16 \cdot c(\varepsilon) \cdot \beta^\beta \cdot 2^{1+\alpha} \cdot \| T \|_\alpha \cdot 2^{-k/N_{m+1} - \alpha m} \cdot N_{m+1}^{-\beta}, \quad k \geqslant N_{m+1}, \ldots$$

It can be given the form (5.10.22) by changing from $2^{-k/N_{m+1}}$ to

$$(\beta + \varepsilon)^{\beta+\varepsilon}\left(\frac{N_{m+1}}{k}\right)^{\beta+\varepsilon} \geqslant 2^{-k/N_{m+1}}.$$

Then we obtain

$$e_k(T_m) \leqslant 16c(\varepsilon) \cdot \beta^\beta(\beta+\varepsilon)^{\beta+\varepsilon} \cdot 2^{1+\alpha} \cdot \| T \|_\alpha \cdot 2^{-\alpha m} \cdot k^{-\beta}\left(\frac{N_{m+1}}{k}\right)^\varepsilon$$

in analogy to (5.10.22) for $k \geqslant N_{m+1}$. If we introduce the notation

$$C_3(\varepsilon) = 16c(\varepsilon) \cdot \beta^\beta(\beta+\varepsilon)^{\beta+\varepsilon} \cdot 2^{1+\alpha}$$

and take account of

$$16 \cdot \beta^\beta(\beta+\varepsilon)^{\beta+\varepsilon} \geqslant 16 \cdot e^{-2/e} \geqslant 1$$

we can state the validity of

$$e_k(T_m) \leqslant C_3(\varepsilon) \cdot \| T \|_\alpha \cdot 2^{-\alpha m} \cdot k^{-\beta}\left(\frac{N_{m+1}}{k}\right)^\varepsilon \qquad (5.10.22)'$$

for arbitrary $k = 1, 2, \ldots$ In this way we obtain

$$\sum_{m=L}^{L+M} e_{n_m}(T_m) \leqslant C_3(\varepsilon) \cdot \| T \|_\alpha \sum_{m=L}^{L+M} 2^{-\alpha m} n_m^{-\beta}\left(\frac{N_{m+1}}{n_m}\right)^\varepsilon.$$

Next we choose natural numbers n_m such that

$$2^{L/\sigma}(m - L + 1)^{-2} \leqslant n_m \leqslant 1 + 2^{L/\sigma}(m - L + 1)^{-2} \qquad (5.10.27)$$

for $L \leqslant m \leqslant L + M$. Inserting the estimates (5.10.27) for n_m and (5.10.23) for N_{m+1} we get

$$\sum_{m=L}^{L+M} 2^{-\alpha m} n_m^{-\beta} \left(\frac{N_{m+1}}{n_m} \right)^\varepsilon$$

$$\leqslant (1 + \rho^{1/\sigma} \cdot 2^{2/\sigma})^\varepsilon \sum_{m=L}^{L+M} 2^{(\varepsilon/\sigma - \alpha)m - (L/\sigma)(\beta + \varepsilon)}(m - L + 1)^{2(\beta + \varepsilon)}$$

$$= (1 + \rho^{1/\sigma} \cdot 2^{2/\sigma})^\varepsilon \cdot 2^{-(\beta/\sigma + \alpha)(L-1)} \sum_{m=L}^{L+M} 2^{(\varepsilon/\sigma - \alpha)(m-L+1)}(m - L + 1)^{2(\beta + \varepsilon)}$$

$$= (1 + \rho^{1/\sigma} \cdot 2^{2/\sigma})^\varepsilon \cdot 2^{\beta/\sigma + \alpha} \cdot 2^{-(\beta/\sigma + \alpha)L} \sum_{k=0}^{M} 2^{(\varepsilon/\sigma - \alpha)(k+1)}(k + 1)^{2(\beta + \varepsilon)}.$$

We then fix ε by putting $\varepsilon = \alpha\sigma/2$ and conclude that the sum on the right-hand side is uniformly bounded for all M, since

$$C_4 = \sum_{k=0}^{\infty} 2^{-(\alpha/2)(k+1)}(k + 1)^{2(\beta + \alpha\sigma/2)} < \infty.$$

Hence, we obtain

$$\sum_{m=L}^{L+M} e_{n_m}(T_m) \leqslant C_5 \cdot \| T \|_\alpha \cdot 2^{-(\beta/\sigma + \alpha)L} \qquad (5.10.28)$$

for $M = 1, 2, \ldots$, where

$$C_5 = (1 + \rho^{1/\sigma} \cdot 2^{2/\sigma})^{\alpha\sigma/2} \cdot 2^{\beta/\sigma + \alpha} \cdot C_3 \left(\frac{\alpha\sigma}{2} \right) \cdot C_4.$$

Again it remains to limit the subscript $n_L + \cdots + n_{L+M} - M$. From (5.10.27) we have

$$n_L + \cdots + n_{L+M} - M \leqslant 1 + 2^{L/\sigma} \cdot \sum_{m=L}^{L+M} (m - L + 1)^{-2}$$

$$\leqslant 1 + 2^{L/\sigma} \cdot \sum_{k=1}^{\infty} k^{-2} = 1 + 2^{L/\sigma} \frac{\pi^2}{6}$$

and thus

$$n_L + \cdots + n_{L+M} - M \leqslant 4 \cdot 2^{L/\sigma} \quad \text{for } M = 1, 2, \ldots \qquad (5.10.29)$$

Combining the estimate (5.10.25), for $\varepsilon = \alpha\sigma/2$, with (5.10.28) we arrive at

$$e_{n_0 + \cdots + n_{L+M} - L - M}(T) \leqslant C_6 \cdot \| T \|_\alpha \cdot 2^{-(\beta/\sigma + \sigma)L} + \| T_{L+M+1} \|,$$

by (5.10.21) where

$$C_6 = C_1 \left(\frac{\alpha\sigma}{2} \right) + C_5.$$

Moreover, the subscript $n_0 + \cdots + n_{L+M} - L - M$ is limited by

$$n_0 + \cdots + n_{L+M} - L - M \leqslant C_7 \cdot 2^{L/\sigma}$$

with

$$C_7 = C_2 + 4$$

independently of $M = 1, 2, \ldots$ by (5.10.26) and (5.10.29). Now let k_L, $L = 1, 2, \ldots$, be a non-decreasing sequence of natural numbers such that

$$C_7 \cdot 2^{L/\sigma} \leqslant k_L \leqslant (C_7 + 1) \cdot 2^{L/\sigma}.$$

Then we have

$$e_{k_L}(T) \leqslant e_{n_0 + \cdots + n_{L+M} - L - M}(T) \leqslant C_6 \cdot \| T \|_\alpha \cdot 2^{-(\beta/\sigma + \alpha)L} + \| T_{L+M+1} \|$$

for all $M = 1, 2, 3, \ldots$ But the estimate (5.10.11)$'$ tells us that $\| T_{L+M+1} \|$ becomes arbitrarily small when M is chosen sufficiently large. Hence it follows that

$$e_{K_L}(T) \leqslant C_6 \cdot \| T \|_\alpha \cdot 2^{-(\beta/\sigma + \alpha)L} \quad \text{for } L = 1, 2, 3, \ldots \qquad (5.10.30)$$

Given an arbitrary natural number $n \geqslant k_1$ we determine L such that

$$k_L \leqslant n \leqslant k_{L+1}.$$

Thus, from (5.10.30) we obtain

$$e_n(T) \leqslant e_{k_L}(T) \leqslant C_6 \cdot \| T \|_\alpha \cdot (2^{L/\sigma})^{-(\beta + \alpha\sigma)}$$

$$\leqslant 2^{\beta/\sigma + \alpha} \cdot C_6 \cdot \| T \|_\alpha (2^{(L+1)/\sigma})^{-(\beta + \alpha\sigma)}$$

$$\leqslant 2^{\beta/\sigma + \alpha} \cdot C_6 \cdot (C_7 + 1)^{\beta + \alpha\sigma} \cdot \| T \|_\alpha \cdot k_{L+1}^{-\beta - \alpha\sigma}$$

and therefore

$$e_n(T) \leqslant C \cdot \| T \|_\alpha \cdot n^{-\beta - \alpha\sigma} \quad \text{for } n \geqslant k_1 \qquad (5.10.31)$$

with the final constant

$$C = 2^{\beta/\sigma + \alpha} \cdot C_6 \cdot (C_7 + 1)^{\beta + \alpha\sigma}$$

depending on α, β, ρ and σ. Since $k_1 \leqslant (C_7 + 1) \cdot 2^{1/\sigma}$ and $C_6 \geqslant 1$ the inequality (5.10.31) also holds for $1 \leqslant n \leqslant k_1$. This completes the proof of the assertion (5.10.19) with $s_n(T) = e_n(T)$ in the real case. The complex case can be treated along the same lines.

Owing to the fact that $a_n(S) = d_n(S) = 0$ if $\text{rank}(S) < n$, the proof of (5.10.19) for $s_n(T) = a_n(T)$ and $s_n(T) = d_n(T)$ is simpler. ∎

Remark. Local estimates of the form

$$s_k(S) \leqslant c \cdot \| S \| \cdot k^{-\beta} \log^\gamma \left(\frac{n}{k} + 1 \right) \qquad (5.10.17)'$$

for $n = 1, 2, \ldots$, $1 \leqslant k \leqslant n$, and all operators $S : E \to l_\infty^n$ with constants $c \geqslant 1$ and $\beta, \gamma \geqslant 0$ can be brought under the conditions (5.10.17) of Theorem

5.10.1. Indeed, the elementary inequality

$$\log^{\gamma}\left(\frac{n}{k}+1\right) \leqslant \left(2\max\left(\frac{\gamma}{\varepsilon},1\right)\right)^{\gamma} \cdot \left(\frac{n}{k}\right)^{\varepsilon} \quad \text{for } \varepsilon > 0$$

implies (5.10.17) with

$$c(\varepsilon) = c \cdot \left(2\max\left(\frac{\gamma}{\varepsilon},1\right)\right)^{\gamma}.$$

Theorem 5.10.1 tells us that under additional assumptions on the compact metric space (X, d) the approximation or entropy quantities $s_n(T)$ in the case of a Hölder continuous oprator $T : E \to C(X)$ of type α tend to zero even faster than $n^{-\beta}$, where β is the exponent in the local conditions (5.10.17) or (5.10.17)′.

Before introducing the *local assumptions* (5.10.17) on the $s_k(S)$ for operators $S : E \to l_\infty^n$ we pointed out that it suffices to consider the approximation numbers $a_k(S)$, the Kolmogorov numbers $d_k(S)$, and the entropy numbers $e_k(S)$ in place of $s_k(S)$ (see also Theorem 5.3.2). The statement (5.10.19) of Theorem 5.10.1, under the corresponding local conditions (5.10.17) on the approximation or entropy quantities $s_k(S)$ and with the additional supposition (5.10.18) on the compact metric space (X, d), can also be expressed by

$$L^{\alpha}(E, C(X)) \subseteq L_{p,\infty}^{(s)}(E, C(X)) \quad \text{with } \frac{1}{p} = \beta + \alpha\sigma. \quad (5.10.32)$$

The strongest result of type (5.10.32) to be expected is the result

$$L^{\alpha}(E, C(X)) \subseteq L_{p,\infty}^{(a)}(E, C(X)) \quad \text{with } \frac{1}{p} = \beta + \alpha\sigma.$$

Indeed, by (2.2.19) we have

$$L_{p,\infty}^{(a)}(E, C(X)) \subseteq L_{p,\infty}^{(d)}(E, C(X))$$

and, furthermore,

$$L_{p,\infty}^{(d)}(E, C(X)) \subseteq L_{p,\infty}^{(e)}(E, C(X))$$

by (3.1.16) as a result of the Bernstein type inequalities (3.1.1). In particular, global estimates

$$s_n(T) \leqslant C \cdot \|T\|_{\alpha} \cdot n^{-\beta-\alpha\sigma}, \quad n = 1, 2, 3, \ldots, \quad (5.10.19)$$

for the approximation numbers $s_n(T) = a_n(T)$ or the Kolmogorov numbers $s_n(T) = d_n(T)$ of operators $T \in L^{\alpha}(E, C(X))$ imply global estimates

$$e_n(T) \leqslant \tilde{C} \cdot \|T\|_{\alpha} \cdot n^{-\beta-\alpha\sigma}$$

for the entropy numbers $e_n(T)$.

But what about Banach spaces E granting local properties (5.10.17) or (5.10.17)′ for approximation or entropy quantities $s_k(S)$? A result along these lines has been obtained by Carl and Pajor (1988). Since the proof lies beyond the scope of this book we present only the result itself.

Lemma 5.10.4. (see Carl and Pajor 1988). *If H is a Hilbert space then*

$$d_k(S) \leqslant c_0 \cdot \| S \| \cdot k^{-1/2} \log^{1/2} \left(\frac{n}{k} + 1 \right) \tag{5.10.33}$$

for $n = 1, 2, 3, \ldots, 1 \leqslant k \leqslant n$, and all operators $S: H \to l_\infty^n$ with a universal constant c_0.

On the basis of Theorem 5.10.1 and the comments after it we can now immediately write down *global estimates for the Kolmogorov numbers $d_n(T)$ and the entropy numbers $e_n(T)$* of Hölder continuous operators T defined on a Hilbert space.

Theorem 5.10.2. (see Carl, Heinrich and Kühn 1988). *Let (X, d) be a compact metric space such that*

$$\varepsilon_n(X) \leqslant \rho \cdot n^{-\sigma} \quad \text{for } n = 1, 2, 3, \ldots \tag{5.10.18}$$

with certain constats $\rho, \sigma > 0$ and let $T: H \to C(X)$ be a Hölder continuous operator of type α, $0 < \alpha \leqslant 1$, defined on a Hilbert space H. Then

$$d_n(T) \leqslant C \cdot \| T \|_\alpha \cdot n^{-1/2 - \alpha\sigma}, \quad n = 1, 2, 3, \ldots, \tag{5.10.34}$$

and

$$e_n(T) \leqslant \tilde{C} \cdot \| T \|_\alpha \cdot n^{-1/2 - \alpha\sigma}, \quad n = 1, 2, 3, \ldots, \tag{5.10.35}$$

with constants C and $\tilde{C}$ depending on α, ρ, and σ only.

We point out here that the local estimates (5.10.33) for the Kolmogorov numbers $d_k(S)$ of operators $S: H \to l_\infty^n$ can be used to derive corresponding local estimates or the entropy numbers $e_k(S)$ by means of the Bernstein type inequalities (3.1.1). This will be carried out in section 6.4 (see Lemma 6.4.1) for the purpose of estimating the volume of certain convex hulls in the n-dimensional euclidean space (see Proposition 6.4.1).

5.11. Integral operators

Our aim in this section and the next one is to apply the abstract results derived so far to integral operators with values in $C(X)$. This requires some preliminaries concerning measure theory and integration.

We recall that in an arbitrary topological space the smallest system $\mathscr{B}$ of sets that contains the open sets and is closed with respect to countable unions and to complementation is called *the Borel σ-algebra*, the elements of $\mathscr{B}$ being referred to as *Borel sets*. In the present context we operate with Borel subsets of a compact metric space (X, d), of the real line $\mathbb{R}$ as well as of the complex plane $\mathbb{C}$ as *measurable sets*. In accordance with that a real- or complex-valued function f defined on X is said to be *measurable* if the inverse image $f^{-1}(A)$ of any measurable subset $A \subseteq \mathbb{R}$ or $A \subseteq \mathbb{C}$, respectively, is a measurable subset of X. Obviously every continuous function $f \in C(X)$ is measurable. Moreover, the Borel σ-algebra $\mathscr{B}$ on X can be characterized as the smallest σ-algebra such that every continuous function f on X is measurable.

A *Borel measure* μ *on* X is a non-negative, countably additive set function defined on $\mathscr{B}$ with $\mu(\varnothing) = 0$ and $\mu(X) < \infty$. A subset $N \subset X$ is called a *μ-null set*, if $N \in \mathscr{B}$ and $\mu(N) = 0$. Two functions f and g on X are said to be *μ-equivalent*, if the set $\{s \in X : f(s) \neq g(s)\}$ is a μ-null set.

Given a Borel measure μ on X we make use of the notion of *an integrable function on* X *with respect to* μ. The notation $L_p(X, \mu)$, $1 \leqslant p < \infty$, as usual is used to stand for *the linear space of classes of μ-equivalent measurable functions f on X whose pth power $|f|^p$ is integrable with respect to μ.* The expression

$$\| f \|_p = \left(\int_X |f(t)|^p \, d\mu(t) \right)^{1/p} \tag{5.11.1}$$

then is independent of the choice of a representative f in a class of μ-equivalent functions and turns out to be a *norm on* $L_p(X, \mu)$. The space $L_p(X, \mu)$ is even a Banach space with respect to this norm.

Comparing spaces $L_p(X, \mu)$ with different subscripts p we recognize that

$$L_{p_2}(X, \mu) \subseteq L_{p_1}(X, \mu) \quad \text{for } p_1 \leqslant p_2. \tag{5.11.2}$$

and

$$\| f \|_{p_1} \leqslant \mu(X)^{1/p_1 - 1/p_2} \| f \|_{p_2} \quad \text{for } f \in L_{p_2}(X, \mu). \tag{5.11.3}$$

If we let $p = \infty$ we obtain *the space* $L_\infty(X, \mu)$ *of classes of μ-equivalent measurable functions f on X which are μ-essentially bounded.* This means that there is a μ-null set $N \subset X$ such that the restriction of f to $X \backslash N$ is bounded. Then the expression

$$\| f \|_\infty = \inf \{ \sup(|f(s)| : s \in X \backslash N) : \mu(N) = 0 \}$$

makes sense. It is called the *μ-essential supremum of* $|f|$ and, to differentiate it from the usual supremum

$$\| f \| = \sup \{ |f(s)| : s \in X \}, \tag{5..2.1}$$

it is denoted by

$$\| f \|_\infty = \mu - \operatorname{ess\,sup} \{ |f(s)| : s \in X \}.$$

The μ-essential supremum $\| f \|_\infty$ does not depend on the choice of a representative f in a class of μ-equivalent functions and, moreover, proves to be *a norm on* $L_\infty(X, \mu)$ according to which $L_\infty(X, \mu)$ becomes a Banach space. We have

$$L_\infty(X, \mu) \subseteq L_p(X, \mu) \quad \text{for all } p \geq 1$$

and

$$\|f\|_p \leq \mu^{1/p}(X) \cdot \|f\|_\infty \quad \text{for } f \in L_\infty(X, \mu).$$

The μ-essential supremum $\|f\|_\infty$ of a function $f \in C(X)$ obviously satisfies the inequality

$$\|f\|_\infty \leq \|f\|.$$

In the case of certain particular pairs (X, μ) of compact metric spaces X and Borel measures μ on X the μ-essential supremum $\|f\|_\infty$ of a continuous function $f \neq 0$ on X may actually vanish. This means that $\|f\|_\infty$, like $\|f\|_p$, is *a semi-norm on* $C(X)$. Indeed, when introducing the spaces $L_p(X, \mu)$ and $L_\infty(X, \mu)$ we considered classes of μ-equivalent functions f for the very purpose of obtaining norms instead of semi-norms. Nevertheless we shall refer to the elements of $L_p(X, \mu)$ and $L_\infty(X, \mu)$ as if they were functions f since no confusion can arise.

The integral operators

$$(T_{K,\mu}f)(s) = \int_X K(s, t)f(t) \, d\mu(t) \tag{5.11.4}$$

to be studied in this section and in the next one will be considered on the space $L_p(X, \mu)$ for $1 \leq p < \infty$ and on the space $C(X)$. Since they are meant to have their values in the space $C(X)$ in any case we assume *the generating 'kernel'* $K(s, t)$ to be continuous on the product space $X \times X$. Note that the space $X \times X$ equipped with the metric

$$d((s_1, t_1), (s_2, t_2)) = \max(d(s_1, t_1), d(s_2, t_2))$$

is again compact. Therefore the above assumption on $K(s, t)$ guarantees that the norm

$$\|K\| = \sup\{|K(s, t)| : s, t \in X\}$$

is finite. It also ensures the existence of the integral $\int_X K(s, t)f(t)d\mu(t)$ for every $f \in L_p(X, \mu)$, $1 \leq p < \infty$, and all $s \in X$, as well as the continuity of the function $T_{K,\mu}f$ defined by (5.11.4) on X. But, what is more, the operator $T_{K,\mu} : L_p(X, \mu) \to C(X)$ will even turn out to be compact.

Proposition 5.11.1. *Let (X, d) be a compact metric space, μ a Borel measure on X, and $K \in C(X \times X)$. Then the integral operator $T_{K,\mu} : L_p(X, \mu) \to C(X)$, $1 \leq p < \infty$, defined by (5.11.4) is compact.*

Proof. Because of (5.11.3) the identity map $I_p : L_p(X, \mu) \to L_1(X, \mu)$ is continuous. Hence it suffices to consider $T_{K,\mu}$ as an operator on $L_1(X, \mu)$.

The fact that $T_{K,\mu}$ actually has its values in $C(X)$ is proved simultaneously with the compactness of $T_{K,\mu}$. Indeed, both the continuity of a function f on X and the compactness of an operator T with values in $C(X)$ can be verified with the help of the corresponding modulus of continuity. For a function $f \in L_1(X, \mu)$ with $\|f\|_1 \leqslant 1$ we have

$$\omega(T_{K,\mu}f; \delta) = \sup_{d(s_1, s_2) \leqslant \delta} \left| \int_X (K(s_1, t) - K(s_2, t)) f(t) \, d\mu(t) \right|$$

$$\leqslant \sup_{d(s_1, s_2) \leqslant \delta} \|K(s_1, \cdot) - K(s_2, \cdot)\|_\infty \leqslant \sup_{d(s_1, s_2) \leqslant \delta} \|K(s_1, \cdot) - K(s_2, \cdot)\|$$

and hence

$$\omega(T_{K,\mu}f; \delta) \leqslant \varepsilon \quad \text{for } \delta \leqslant \delta_\varepsilon$$

with a sufficiently small $\delta_\varepsilon > 0$ because of the continuity of K on $X \times X$. Thus the inequality

$$\omega(T_{K,\mu}; \delta) \leqslant \varepsilon$$

also holds for $\delta \leqslant \delta_\varepsilon$. But this amounts to the continuity of the function $T_{K,\mu}f$ on X and the compactness of the operator $T_{K,\mu}: L_1(X, \mu) \to C(X)$ (see sections 5.4 and 5.5. Proposition 5.5.1). ∎

In order to be able to express *the degree of compactness of the integral operator* $T_{K,\mu}: L_p(X, \mu) \to C(X)$ or $T_{K,\mu}: C(X) \to C(X)$ in terms of its kernel $K \in C(X \times X)$ we shall introduce *a modulus of continuity* $\Omega_p(K; \delta)$, $1 \leqslant p \leqslant \infty$, *directly related to the kernel* K. The definition of $\Omega_p(K; \delta)$, in contrast to the usual definition of $\omega(K; \delta)$ for K as an element of $C(X \times X)$, refers to the semi-norms

$$\|K(s, \cdot)\|_p = \left(\int_X |K(s, t)|^p \, d\mu(t) \right)^{1/p} \quad \text{for } 1 \leqslant p < \infty$$

and

$$\|K(s, \cdot)\|_\infty = \mu - \operatorname{ess\,sup}\{|K(s, t)| : t \in X\} \quad \text{for } p = \infty$$

on $C(X)$, depending on a fixed element $s \in X$. We put

$$\Omega_p(K; \delta) = \sup\{\|K(s_1, \cdot) - K(s_2, \cdot)\|_p : s_1, s_2 \in X, d(s_1, s_2) \leqslant \delta\} \quad (5.11.5)$$

for $1 \leqslant p \leqslant \infty$ and arbitrary $\delta \geqslant 0$. Furthermore, for the purpose of representing the operator norm of the integral operator $T_{K,\mu}$ generated by the kernel $K \in C(X \times X)$ we use the notation

$$\|\|K\|\|_p = \sup_{s \in X} \|K(s, \cdot)\|_p \quad \text{for } 1 \leqslant p \leqslant \infty. \quad (5.11.6)$$

Let us emphasize that $\|\|\cdot\|\|_p$ is also in general only a semi-norm on $C(X \times X)$.

Proposition 5.11.2. *Let (X, d) be a compact metric space, μ a Borel measure on X, and $K \in C(X \times X)$. Then, for the integral operator (5.11.4) generated by K we have*

$$\| T_{K,\mu} \| = \| K \|_{p'} \tag{5.11.7}$$

and

$$\omega(T_{K,\mu}; \delta) = \Omega_{p'}(K; \delta) \tag{5.11.8}$$

with $1/p + 1/p' = 1$, $1 \leqslant p \leqslant \infty$, if $T_{K,\mu}: L_p(X, \mu) \to C(X)$; and we have

$$\| T_{K,\mu} \| = \| K \|_1 \tag{5.11.9}$$

and

$$\omega(T_{K,\mu}; \delta) = \Omega_1(K; \delta) \tag{5.11.10}$$

if $T_{K,\mu}: C(X) \to C(X)$.

Proof. We first consider the case $T_{K,\mu}: L_p(X, \mu) \to C(X)$ with $1 < p < \infty$. Then for arbitrary $f, g \in L_p(X, \mu)$ Hölder's inequality gives

$$|(T_{K,\mu}f)(s)| \leqslant \int_X |K(s, t)| \, |f(t)| \, d\mu(t) \leqslant \| K(s, \cdot) \|_{p'} \cdot \| f \|_p \tag{5.11.11}$$

and

$$|(T_{K,\mu}g)(s_1) - (T_{K,\mu}g)(s_2)| \leqslant \| K(s_1, \cdot) - K(s_2, \cdot) \|_{p'} \cdot \| g \|_p \tag{5.11.12}$$

which imply

$$\| T_{K,\mu} \| \leqslant \| K \|_{p'} \quad \text{and} \quad \omega(T_{K,\mu}; \delta) \leqslant \Omega_{p'}(K; \delta)$$

from the definitions of $\| K \|_{p'}$ and $\Omega_{p'}(K; \delta)$, respectively. For the proof of the converse inequalities we may assume that $\| K \|_{p'} > 0$ and $\Omega_{p'}(K; \delta) > 0$. The compactness of X and the continuity of the function $\| K(s, \cdot) \|_{p'}$ on X imply the existence of a point $s_0 \in X$ with

$$\| K \|_{p'} = \sup_{s \in X} \| K(s, \cdot) \|_{p'} = \| K(s_0, \cdot) \|_{p'} > 0 \tag{5.11.13}$$

and, similarly, the existence of two points $s_1^{(0)}, s_2^{(0)}$ with $d(s_1^{(0)}, s_2^{(0)}) \leqslant \delta$ such that

$$\Omega_{p'}(K; \delta) = \| K(s_1^{(0)}, \cdot) - K(s_2^{(0)}, \cdot) \|_{p'} > 0. \tag{5.11.14}$$

We now consider the inequalities (5.11.11) and (5.11.12) with $s = s_0$ and $(s_1, s_2) = (s_1^{(0)}, s_2^{(0)})$, respectively. They turn into equalities if the functions $f, g \in L_p(X, \mu)$ are chosen in an appropriate way. Namely, with

$$f(t) = \begin{cases} e^{-i \arg(K(s_0, t))} \cdot | K(s_0, t) |^{p'-1} & \text{For } K(s_0, t) \neq 0 \\ 0 & \text{for } K(s_0, t) = 0 \end{cases}$$

and

$$g(t) = \begin{cases} e^{-i \arg(K(s_1^{(0)}, t) - K(s_2^{(0)}, t))} \cdot | K(s_1^{(0)}, t) - K(s_2^{(0)}, t) |^{p'-1} \\ \qquad \text{for } K(s_1^{(0)}, t) - K(s_2^{(0)}, t) \neq 0 \\ 0 \qquad \text{for } K(s_1^{(0)}, t) - K(s_2^{(0)}, t) = 0 \end{cases}$$

we obtain

$$\|f\|_p = \left(\int_X |K(s_0,t)|^{p'} d\mu(t)\right)^{1/p} = \|K(s_0,\cdot)\|_{p'}^{p'-1}$$

and

$$\|g\|_p = \|K(s_1^{(0)},\cdot) - K(s_2^{(0)},\cdot)\|_{p'}^{p'-1},$$

respectively, and thus

$$|(T_{K,\mu}f)(s_0)| = \int_X |K(s_0,t)|^{p'} d\mu(t) = \|K(s_0,\cdot)\|_{p'} \cdot \|f\|_p$$

and

$$|T_{K,\mu}g)(s_1^{(0)}) - (T_{K,\mu}g)(s_2^{(0)})| = \int_X |K(s_1^{(0)},t) - K(s_2^{(0)},t)|^{p'} d\mu(t)$$
$$= \|K(s_1^{(0)},\cdot) - K(s_2^{(0)},\cdot)\|_{p'} \cdot \|g\|_p.$$

Since (5.11.13) and (5.11.14) also guarantee $\|f\|_p > 0$ and $\|g\|_p > 0$ for the functions f and g under consideration, we can take

$$f_0 = \frac{f}{\|f\|_p} \quad \text{and} \quad g_0 = \frac{g}{\|g\|_p}$$

and hence arrive at

$$|(T_{K,\mu}f_0)(s_0)| = \|K(s_0,\cdot)\|_{p'}$$

and

$$|(T_{K,\mu}g_0)(s_1^{(0)}) - (T_{K,\mu}g_0)(s_2^{(0)})| = \|K(s_1^{(0)},\cdot) - K(s_2^{(0)},\cdot)\|_{p'}.$$

These equalities lead us to the desired conclusions

$$\|\|K\|\|_{p'} \leqslant \|T_{K,\mu}\| \quad \text{and} \quad \Omega_{p'}(K;\delta) \leqslant \omega(T_{K,\mu};\delta)$$

which completes the proof of (5.11.7) and (5.11.8) for $1 < p < \infty$.

The case $p = 1$ is treated via the inequalities

$$|(T_{K,\mu}f)(s)| \leqslant \|K(s,\cdot)\|_\infty \cdot \|f\|_1$$

and

$$|T_{K,\mu}g)(s_1) - (T_{K,\mu}g)(s_2)| \leqslant \|K(s_1,\cdot) - K(s_2,\cdot)\|_\infty \cdot \|g\|_1$$

analogous to (5.11.11) and (5.11.12). It follows that

$$\|T_{K,\mu}\| \leqslant \|\|K\|\|_\infty \quad \text{and} \quad \omega(T_{K,\mu};\delta) \leqslant \Omega_\infty(K;\delta).$$

For the proof of the converse estimates we again may assume that $\|\|K\|\|_\infty > 0$ and $\Omega_\infty(K;\delta) > 0$, respectively. So let ε be given with $0 < \varepsilon < \|\|K\|\|_\infty$. Then there exists a point $s_0 \in X$ such that

$$\|\|K\|\|_\infty - \varepsilon < \|K(s_0,\cdot)\|_\infty.$$

According to the definition of the μ-essential supremum $\|K(s_0,\cdot)\|_\infty$ the Borel set

$$X_\varepsilon = \{t \in X : |K(s_0,t)| > \|\|K\|\|_\infty - \varepsilon\}$$

is of positive measure μ. Similarly, given η with $0 < \eta < \Omega_\infty(K; \delta)$, there exists a pair of points $s_1^{(0)}, s_2^{(0)} \in X$ with $d(s_1^{(0)}, s_2^{(0)}) \leqslant \delta$ such that

$$\Omega_\infty(K; \delta) - \eta < \| K(s_1^{(0)}, \cdot) - K(s_2^{(0)}, \cdot) \|_\infty.$$

Obviously the Borel set

$$Y_\eta = \{ t \in X : | K(s_1^{(0)}, t) - K(s_2^{(0)}, t) | > \Omega_\infty(K; \delta) - \eta \}$$

also has positive measure $\mu(Y_\eta)$. For what follows we fix an arbitrary number $p > 1$ and, on the basis of the definitions of the sets X_ε and Y_η, conclude that

$$\mu^{1/p'}(X_\varepsilon)(\|| K \||_\infty - \varepsilon) \leqslant \left(\int_X | K(s_0, t) |^{p'} \, d\mu(t) \right)^{1/p'} = \| K(s_0, \cdot) \|_{p'}$$

$$\leqslant \|| K \||_{p'}$$

and

$$\mu^{1/p'}(Y_\eta)(\Omega_\infty(K; \delta) - \eta) \leqslant \left(\int_X | K(s_1^{(0)}, t) - K(s_2^{(0)}, t) |^{p'} \, d\mu(t) \right)^{1/p'}$$

$$= \| K(s_1^{(0)}, \cdot) - K(s_2^{(0)}, \cdot) \|_{p'} \leqslant \Omega_{p'}(K; \delta),$$

respectively. Now we can make use of the results (5.11.7) and (5.11.8) which have already been proved for $p > 1$ and thus obtain

$$\mu^{1/p'}(X_\varepsilon)(\|| K \||_\infty - \varepsilon) \leqslant \| T_{K,\mu} : L_p(X, \mu) \to C(X) \|$$

and

$$\mu^{1/p'}(Y_\eta)(\Omega_\infty(K; \delta) - \eta) \leqslant \omega(T_{K,\mu} : L_p(X, \mu) \to C(X); \delta).$$

The operator $T_{K,\mu} : L_p(X, \mu) \to C(X)$, however, can be factorized over $L_1(X, \mu)$ by using the identity map $I_p : L_p(X, \mu) \to L_1(X, \mu)$. This factorization gives rise to the estimates

$$\| T_{K,\mu} : L_p(X, \mu) \to C(X) \| \leqslant \| I_p \| \cdot \| T_{K,\mu} : L_1(X, \mu) \to C(X) \|$$

and

$$\omega(T_{K,\mu} : L_p(X, \mu) \to C(X); \delta) \leqslant \| I_p \| \cdot \omega(T_{K,\mu} : L_1(X, \mu) \to C(X); \delta)$$

(see (5.5.3)). From (5.11.3) we have

$$\| I_p \| \leqslant \mu^{1/p'}(X).$$

From the resulting inequalities

$$\mu^{1/p'}(X_\varepsilon)(\|| K \||_\infty - \varepsilon) \leqslant \mu^{1/p'}(X) \cdot \| T_{K,\mu} : L_1(X, \mu) \to C(X) \|$$

and

$$\mu^{1/p'}(Y_\eta)(\Omega_\infty(K; \delta) - \eta) \leqslant \mu^{1/p'}(X) \cdot \omega(T_{K,\mu} : L_1(X, \mu) \to C(X); \delta)$$

we obtain

$$\|| K \||_\infty - \varepsilon \leqslant \| T_{K,\mu} : L_1(X, \mu) \to C(X) \|$$

and

$$\Omega_\infty(K; \delta) - \eta \leqslant \omega(T_{K,\mu} : L_1(X, \mu) \to C(X); \delta),$$

respectively, by taking the limit $p \to 1$. Since $\varepsilon > 0$ and $\eta > 0$ can be chosen arbitrarily small we finally arrive at the desired inequalities

$$\||K\||_\infty \leqslant \| T_{K,\mu} : L_1(X, \mu) \to C(X) \|$$

and

$$\Omega_\infty(K; \delta) \leqslant \omega(T_{K,\mu} : L_1(X, \mu) \to C(X); \delta).$$

This completes the proof of (5.11.7) and (5.11.8) for $p = 1$.

It remains to consider $T_{k,\mu}$ as an operator from $C(X)$ into $C(X)$. In a similar way to the case $T_{K,\mu} : L_p \to C(X), 1 < p < \infty$, we may conclude that

$$\| T_{K,\mu} \| \leqslant \||K\||_1 \tag{5.11.15}$$

and

$$\omega(T_{K,\mu}; \delta) \leqslant \Omega_1(K; \delta) \tag{5.11.16}$$

and, furthermore, determine a point $s_0 \in X$ with

$$\||K\||_1 = \| K(s_0, \cdot) \|_1 \tag{5.11.17}$$

as well as a pair of points $s_1^{(0)}, s_2^{(0)} \in X$ with $d(s_1^{(0)}, s_2^{(0)}) \leqslant \delta$ such that

$$\Omega_1(K; \delta) = \| K(s_1^{(0)}, \cdot) - K(s_2^{(0)}, \cdot) \|_1. \tag{5.11.18}$$

The converses of inequalities (5.11.15) and (5.11.16) will be derived by estimating $|(T_{K,\mu}f)(s_0)|$ and $|(T_{K,\mu}g)(s_1^{(0)}) - (T_{K,\mu}g)(s_2^{(0)})|$, respectively, from below for appropriate functions $f, g \in C(X)$. For arbitrary $\varepsilon > 0$ we put

$$f_\varepsilon(t) = \frac{\overline{K(s_0, t)}}{\varepsilon + |K(s_0, t)|}$$

and

$$g_\varepsilon(t) = \frac{\overline{K(s_1^{(0)}, t) - K(s_2^{(0)}, t)}}{\varepsilon + |K(s_1^{(0)}, t) - K(s_2^{(0)}, t)|}.$$

These functions belong to $C(X)$ and their norms satisfy the inequalities $\| f_\varepsilon \| \leqslant 1$ and $\| g_\varepsilon \| \leqslant 1$. We have

$$\| T_{K,\mu}f_\varepsilon \| \geqslant |(T_{K,\mu}f_\varepsilon)(s_0)| \tag{5.11.19}$$

as well as

$$\omega(T_{K,\mu}; \delta) \geqslant \omega(T_{K,\mu}g_\varepsilon; \delta) \geqslant |(T_{K,\mu}g_\varepsilon)(s_1^{(0)}) - (T_{K,\mu}g_\varepsilon)(s_2^{(0)})|. \tag{5.11.20}$$

The values $|(T_{K,\mu}f_\varepsilon)(s_0)|$ and $|(T_{K,\mu}g_\varepsilon)(s_1^{(0)}) - (T_{K,\mu}g_\varepsilon)(s_2^{(0)})|$ can easily be calculated and estimated from below, namely

$$|(T_{K,\mu}f_\varepsilon)(s_0)| = \int_X \frac{|K(s_0, t)|^2}{\varepsilon + |K(s_0, t)|} \, d\mu(t)$$

$$= \| K(s_0, \cdot) \|_1 - \varepsilon \cdot \int_X \frac{|K(s_0, t)|}{\varepsilon + |K(s_0, t)|} \, d\mu(t)$$

$$\geqslant \| K(s_0, \cdot) \|_1 - \varepsilon \cdot \mu(X) \tag{5.11.21}$$

and

$$|(T_{K,\mu}g_\varepsilon)(s_1^{(0)}) - (T_{K,\mu}g_\varepsilon)(s_2^{(0)})| = \int_X \frac{|K(s_1^{(0)},t) - K(s_2^{(0)},t)|^2}{\varepsilon + |K(s_1^{(0)},t) - K(s_2^{(0)},t)|} \, d\mu(t)$$
$$\geq \| K(s_1^{(0)},\cdot) - K(s_2^{(0)},\cdot) \|_1 - \varepsilon \cdot \mu(X),$$

(5.11.22)

respectively. Combining (5.11.19), (5.11.21), (5.11.17) and (5.11.20), (5.11.22), (5.11.18) we get

$$\| T_{K,\mu} \| \geq \|\!\| K \|\!\|_1 - \varepsilon \cdot \mu(X)$$

and

$$\omega(T_{K,\mu};\delta) \geq \Omega_1(K;\delta) - \varepsilon \cdot \mu(X),$$

respectively, which amount to

$$\| T_{K,\mu} \| \geq \|\!\| K \|\!\|_1 \text{ and } \omega(T_{K,\mu};\delta) \geq \Omega_1(K;\delta).$$

This completes the proof of (5.11.9) and (5.11.10). ∎

We are now in a position to carry over without any difficulties the results of sections 5.6 and 5.8 to integral operators. We explicitly restate Theorem 5.6.1 and Proposition 5.8.4.

Proposition 5.11.3. *Let (X,d) be a compact metric space, μ a Borel measure on X, and $K \in C(X \times X)$. Then for the integral operator (5.11.4) generated by K we have*

$$a_{n+1}(T_{K,\mu}) \leq \Omega_{p'}(K;\varepsilon_n(X)) \quad \text{for } n = 1,2,3,\dots \qquad (5.11.23)$$

with $1/p + 1/p' = 1$, $1 \leq p < \infty$, if $T_{K,\mu}:L_p(X,\mu) \to C(X)$; and

$$a_{n+1}(T_{K,\mu}) \leq \Omega_1(K;\varepsilon_n(X)) \quad \text{for } n = 1,2,3,\dots \qquad (5.11.24)$$

if $T_{K,\mu}$ is considered as an operator in $C(X)$.

Proposition 5.11.4. *Let X be a compact and convex subset of a normed space with more than one point, μ a Borel measure on X, and $K \in C(X \times X)$ such that the induced integral operator $T_{K,\mu}:L_p(X,\mu) \to C(X)$ or $T_{K,\mu}:C(X) \to C(X)$ has the property $\text{rank}(T_{K,\mu}) > 1$. Then*

$$e_n(T_{K,\mu}) \leq C_0 \cdot \frac{\|\!\| K \|\!\|_{p'}}{\Omega_{p'}(K;\varepsilon_1(X))} \cdot \Omega_{p'}(K;\varepsilon_n(X)) \quad \text{for } n = 1,2,3,\dots$$

with $1/p + 1/p' = 1$, $1 \leq p < \infty$, if $T_{K,\mu}:L_p(X,\mu) \to C(X)$; and

$$e_n(T_{K,\mu}) \leq C_0 \cdot \frac{\|\!\| K \|\!\|_1}{\Omega_1(K;\varepsilon_1(X))} \cdot \Omega_1(K;\varepsilon_n(X)) \quad \text{for } n = 1,2,3,\dots$$

if $T_{K,\mu}:C(X) \to C(X)$, where C_0 is a universal constant.

Remark. In the case of the Lebesgue measure on the Borel subsets of a closed interval $[a,b]$ the supposition $\text{rank}(T_{K,\mu}) > 1$ amounts to the supposition that $K \in C(X \times X)$ should not be representable as a product $K(s,t) = g(s) \cdot h(t)$ of two functions $g, h \in C(X)$.

5.12. Hölder continuous integral operators

For an integral operator $T_{K,\mu}$ generated by a continuous kernel $K \in C(X \times X)$ the condition

$$|T|_\alpha = \sup_{\delta>0} \frac{\omega(T;\delta)}{\delta^\alpha} < \infty \qquad (5.9.1)$$

of Hölder continuity, $0 < \alpha \leqslant 1$, says that

$$|T_{K,\mu}|_\alpha = \sup_{\delta>0} \frac{\Omega_{p'}(K;\delta)}{\delta^\alpha} < \infty \qquad (5.12.1)$$

with $1/p + 1/p' = 1$, $1 \leqslant p < \infty$, if $T_{K,\mu} : L_p(X,\mu) \to C(X)$ is regarded as an operator on $L_p(X,\mu)$; and

$$|T_{K,\mu}|_\alpha = \sup_{\delta>0} \frac{\Omega_1(K;\delta)}{\delta^\alpha} < \infty, \qquad (5.12.2)$$

if $T_{K,\mu}$ is regarded as an operator in $C(X)$ (see Proposition 5.11.2). Hence it is obvious to work with the quantities

$$|K|_{p,\alpha} = \sup_{\delta>0} \frac{\Omega_p(K;\delta)}{\delta^\alpha} \qquad (5.12.3)$$

directly derived from the kernel K, where $1 \leqslant p \leqslant \infty$ and $0 < \alpha \leqslant 1$, and to introduce Hölder classes $C_p^\alpha(X \times X)$ of continuous kernels K from the very beginning by demanding

$$|K|_{p,\alpha} < \infty.$$

In analogy to the ordering (5.11.2) of the $L_p(X,\mu)$ spaces we have

$$C_{p_2}^\alpha(X \times X) \subseteq C_{p_1}^\alpha(X \times X) \quad \text{for } p_1 \leqslant p_2$$

as well as

$$|K|_{p_1,\alpha} \leqslant \mu(X)^{1/p_1 - 1/p_2} \cdot |K|_{p_2,\alpha} \quad \text{if } K \in C_{p_2}^\alpha(X \times X).$$

The expression

$$\|K\|_{p,\alpha} = \max(\|K\|_p, |K|_{p,\alpha}) \qquad (5.12.4)$$

represents a semi-norm on $C_p^\alpha(X \times X)$ subject to the same estimates as $|K|_{p,\alpha}$, namely

$$\|K\|_{p_1,\alpha} \leqslant \mu(X)^{1/p_1 - 1/p_2} \cdot \|K\|_{p_2,\alpha} \quad \text{for } p_1 \leqslant p_2 \qquad (5.12.5)$$

and $K \in C_{p_2}^\alpha(X \times X)$.

If we consider a kernel $K \in C_{p'}^\alpha(X \times X)$, where $1/p + 1/p' = 1$ with $1 \leqslant p \leqslant \infty$, and the operator $T_{K,\mu} : L_p(X,\mu) \to C(X)$ for $1 \leqslant p < \infty$ or

$T_{K,\mu}: C(X) \to C(X)$ in the case of $p = \infty$, generated by K, we recognize that

$$\||K\||_{p',\alpha} = \max(\|T_{K,\mu}\|, |T_{K,\mu}|_\alpha) \qquad (5.12.6)$$

in view of (5.12.3) and (5.12.1) or (5.12.2); and (5.11.7) or (5.11.9), respectively. The right-hand side of (5.12.6), however, coincides with the Hölder norm $\|T_{K,\mu}\|_\alpha$ of $T_{K,\mu}$ as a Hölder continuous operator of type α (see (5.9.8)). This means that

$$\||K\||_{p',\alpha} = \|T_{K,\mu}\|_\alpha \qquad (5.12.7)$$

and

$$|K|_{p',\alpha} = |T_{K,\mu}|_\alpha,$$

where $T_{K,\mu}$ is thought to act on $L_p(X, \mu)$ for $1 \leqslant p < \infty$ and on $C(X)$ for $p = \infty$, $1/p + 1/p' = 1$.

Having related the quantities $|T_{K,\mu}|_\alpha$ and $\|T_{K,\mu}\|_\alpha$ to the generating kernel $K \in C_{p'}^\alpha(X \times X)$ we can immediately restate the claims of Proposition 5.11.3 and corresponding claims concerning the dyadic entropy numbers $e_n(T_{K,\mu})$ of $T_{K,\mu}$. For $K \in C_{p'}^\alpha(X \times X)$ both the estimates (5.11.23) and (5.11.24) amount to

$$a_{n+1}(T_{K,\mu}) \leqslant |K|_{p',\alpha} \cdot \varepsilon_n^\alpha(X) \quad \text{for } n = 1, 2, 3, \ldots \qquad (5.12.8)$$

with $1/p + 1/p' = 1$ and $1 \leqslant p \leqslant \infty$, where $T_{K,\mu}: L_p(X, \mu) \to C(X)$ for $1 \leqslant p < \infty$ and $T_{K,\mu}: C(X) \to C(X)$ for $p = \infty$. If the compact metric space (X, d) consists of more than one point and possesses the additional property

$$\left(\frac{k}{n}\right)^\sigma \varepsilon_k(X) \leqslant \rho \varepsilon_n(X) \qquad (5.7.4)$$

for $n = 1, 2, 3, \ldots$ and $k = 1, 2, \ldots, n$ the dyadic entropy numbers $e_n(T_{K,\mu})$ of an integral operator $T_{K,\mu}$ generated by a kernel $K \in C_{p'}^\alpha(X \times X)$ with $1/p + 1/p' = 1$ and $1 \leqslant p \leqslant \infty$ allow an analogous estimate, namely

$$e_n(T_{K,\mu}) \leqslant C \cdot \||K\||_{p',\alpha} \cdot \varepsilon_n^\alpha(X) \quad \text{for } n = 1, 2, 3, \ldots \qquad (5.12.9)$$

(see (5.9.11)). Note that C is the same constant as in the estimate

$$e_n(I_\alpha) \leqslant C \cdot \varepsilon_n^\alpha(X)$$

of the dyadic entropy numbers $e_n(I_\alpha)$ of the identity map $I_\alpha: C^\alpha(X) \to C(X)$ (see Proposition 5.7.1). Furthermore let us recall that the supposition (5.7.4) implies

$$\frac{\varepsilon_1(X)}{\rho} \cdot n^{-\sigma} \leqslant \varepsilon_n(X) \quad \text{for } n = 1, 2, 3, \ldots \qquad (5.7.5)$$

so that the decay of the sequence $\varepsilon_n(X)$ is at most of order $n^{-\sigma}$ if X consists of more than one point.

On the other hand, when applying local techniques to Hölder continuous operators $T \in L^\alpha(E, C(X))$ in section 5.10 we assumed

$$\varepsilon_n(X) \leqslant \rho \cdot n^{-\sigma} \quad \text{for } n = 1, 2, 3, \ldots \qquad (5.10.18)$$

instead of (5.7.4), which means that the sequence $\varepsilon_n(X)$ decreases at least as $n^{-\sigma}$ (see Theorem 5.10.1). The local assumptions of Theorem 5.10.1 are satisfied for a Hilbert space H as the domain of T and for the Kolmogorov numbers d_k in place of s_k, as stated in Lemma 5.10.4. Hence the global conclusions (5.10.34) and (5.10.35) of Theorem 5.10.2 can be formulated for integral operators $T_{K,\mu}: L_2(X,\mu) \to C(X)$ generated by a kernel $K \in C_2^\alpha(X \times X)$, namely

$$d_n(T_{K,\mu}) \leqslant C \cdot \interleave K \interleave_{2,\alpha} \cdot n^{-1/2 - \alpha\sigma}, \quad n = 1, 2, 3, \ldots, \qquad (5.12.10)$$

and

$$e_n(T_{K,\mu}) \leqslant \tilde{C} \cdot \interleave K \interleave_{2,\alpha} \cdot n^{-1/2 - \alpha\sigma}, \quad n = 1, 2, 3, \ldots, \qquad (5.12.11)$$

the compact metric space (X,d) being subject to the condition (5.10.18) with $\rho, \sigma > 0$. Moreover, estimates of type (5.12.10) and (5.12.11) for the Kolmogorov numbers $d_n(T_{K,\mu})$ and the dyadic entropy numbers $e_n(T_{K,\mu})$, respectively, are also valid if $T_{K,\mu}$ is considered as an operator from $L_p(X,\mu)$ with $p \geqslant 2$ into $C(X)$ or from $C(X)$ into $C(X)$, provided that the generating kernel K belongs to $C_2^\alpha(X \times X)$. This is due to the possibility of factorizing $T_{K,\mu}: L_p(X,\mu) \to C(X)$ with $p \geqslant 2$ and $T_{K,\mu}: C(X) \to C(X)$ over $L_2(X,\mu)$ by using the identity map $I_{p,2}: L_p(X,\mu) \to L_2(X,\mu)$ for $2 \leqslant p < \infty$ and $I_{\infty,2}: C(X) \to L_2(X,\mu)$ for $p = \infty$, respectively. Since

$$\| I_{p,2} \| \leqslant \mu(X)^{1/2 - 1/p} \quad \text{for } 2 \leqslant p < \infty$$

by (5.11.3) and

$$\| I_{\infty,2} \| \leqslant \mu(X)^{1/2}$$

we obtain

$$d_n(T_{K,\mu}: L_p(X,\mu) \to C(X)) \leqslant C \cdot \mu(X)^{1/2 - 1/p} \interleave K \interleave_{2,\alpha} \cdot n^{-1/2 - \alpha\sigma}, \qquad (5.12.12)$$

$n = 1, 2, 3, \ldots,$ and

$$d_n(T_{K,\mu}: C(X) \to C(X)) \leqslant C \cdot \mu(X)^{1/2} \interleave K \interleave_{2,\alpha} \cdot n^{-1/2 - \alpha\sigma}, \qquad (5.12.13)$$

$n = 1, 2, 3, \ldots,$ as well as

$$e_n(T_{K,\mu}: L_p(X,\mu) \to C(X)) \leqslant \tilde{C} \cdot \mu(X)^{1/2 - 1/p} \interleave K \interleave_{2,\alpha} \cdot n^{-1/2 - \alpha\sigma}, \qquad (5.12.14)$$

$n = 1, 2, 3, \ldots,$ and

$$e_n(T_{K,\mu}: C(X) \to C(X)) \leqslant \tilde{C} \cdot \mu(X)^{1/2} \interleave K \interleave_{2,\alpha} \cdot n^{-1/2 - \alpha\sigma}, \qquad (5.12.15)$$

$n = 1, 2, 3, \ldots$

So far it has made no difference whether the spaces $L_p(X,\mu)$ and $C(X)$ were regarded as real or complex Banach spaces. When investigating the sequence of eigenvalues $\lambda_1(T_{K,\mu}), \lambda_2(T_{K,\mu}), \ldots, \lambda_n(T_{K,\mu}), \ldots$, however, one has to think of $T_{K,\mu}$ as an operator in a complex Banach space, in this case the space $C(X)$ of complex-valued continuous functions on X. According to the general estimate (4.2.16) we have

$$|\lambda_n(T_{K,\mu})| \leqslant \sqrt{2} e_n(T_{K,\mu}), \quad n = 1, 2, 3, \ldots,$$

so that (5.12.15) implies

$$|\lambda_n(T_{K,\mu})| \leqslant \sqrt{2}\tilde{C} \cdot \mu(X)^{1/2} \cdot |\!|\!| K |\!|\!|_{2,\alpha} \cdot n^{-1/2-\alpha\sigma}, \qquad (5.12.16)$$

$n = 1, 2, 3, \ldots$ Hence under the assumption that μ is a Borel measure on X and K a kernel of the class $C_2^\alpha(X \times X)$ the Kolmogorov numbers $d_n(T_{K,\mu})$, the entropy numbers $e_n(T_{K,\mu})$, and the eigenvalues $\lambda_n(T_{K,\mu})$ of the integral operator $T_{K,\mu}: C(X) \to C(X)$ all decrease of order $O(n^{-1/2-\alpha\sigma})$.

An argument in the converse direction has been carried out by Heinrich and Kühn (1985). They supposed

$$|\lambda_n(T_{K,\mu})| = O(n^{-1/2-\alpha\sigma}) \qquad (5.12.17)$$

to be true for all integral operators $T_{K,\mu}: C(X) \to C(X)$ defined by any Borel measure μ on X and any kernel K belonging to a subclass $C^{\alpha,0}(X \times X)$ of $C_2^\alpha(X \times X)$. This subclass $C^{\alpha,0}(X \times X)$ is the class of functions $K \in C(X \times X)$ which are Hölder continuous of type α in the first variable uniformly with respect to the second variable. An exact characterization of the class $C^{\alpha,0}(X \times X)$ can be given by using the modulus of continuity

$$\Omega(K;\delta) = \sup\{\|K(s_1,\cdot) - K(s_2,\cdot)\| : s_1, s_2 \in X, d(s_1,s_2) \leqslant \delta\} \qquad (5.12.18)$$

which, in contrast with

$$\Omega_\infty(K;\delta) = \sup\{\|K(s_1,\cdot) - K(s_2,\cdot)\|_\infty : s_1, s_2 \in X, d(s_1,s_2) \leqslant \delta\}$$

(see (5.11.5)), refers to the usual norm

$$\|K(s_1,\cdot) - K(s_2,\cdot)\| = \sup\{|K(s_1,t) - K(s_2,t)| : t \in X\}$$

of $K(s_1,t) - K(s_2,t)$ as a continuous function of $t \in X$. The definition of $C^{\alpha,0}(X \times X)$ states that $K \in C^{\alpha,0}(X \times X)$ if the expression

$$|K|_\alpha = \sup_{\delta > 0} \frac{\Omega(K;\delta)}{\delta^\alpha} \qquad (5.12.19)$$

is finite. Observe that $C^{\alpha,0}(X \times X)$ turns out to be a Banach space under the norm

$$\|K\|_\alpha = \max(\|K\|, |K|_\alpha), \qquad (5.12.20)$$

contained in the class $C_\infty^\alpha(X \times X)$ of functions $K \in C(X \times X)$ and hence also in the class $C_2^\alpha(X \times X)$, as stated above. If (5.12.17) holds for all Borel measures μ on X and all kernels $K \in C^{\alpha,0}(X \times X)$ with $T_{K,\mu}: C(X) \to C(X)$ as the corresponding integral operators, then the entropy numbers $\varepsilon_n(X)$ of X necessarily decrease of order $O(n^{-\sigma})$, which amounts to (5.10.18). A proof of this converse implication can be found in Heinrich and Kühn (1985) as already mentioned. Carl, Heinrich and Kühn (1988) started from the supposition

$$d_n(T_{K,\mu}) = O(n^{-1/2-\alpha\sigma})$$

for all Borel measures μ on X and all kernels $K \in C^{\alpha,0}(X \times X)$ when deriving

(5.10.18) for the entropy numbers $\varepsilon_n(X)$ of X. In the complex case this does not require any additional work except reference to (5.12.17). On the other hand, the real case can easily be reduced to the complex one.

Apart from Kolmogorov numbers, entropy numbers, and eigenvalues the approximation numbers $a_n(T_{K,\mu})$ of the integral operators $T_{K,\mu}: C(X) \rightarrow C(X)$ in question also reflect the entropy behaviour of the underlying compact metric space (X, d) to a certain degree. One can extract from the paper by Carl, Heinrich and Kühn (1988) that a decrease of the approximation numbers $a_n(T_{K,\mu})$ of order

$$a_n(T_{K,\mu}) = O(n^{-\alpha\sigma})$$

for all integral operators $T_{K,\mu}: C(X) \rightarrow C(X)$ induced by any Borel measure μ on X and any kernel $K \in C^{\alpha,0}(X \times X)$ implies

$$\varepsilon_n(X) = O(n^{-\sigma+\varepsilon})$$

for arbitrary $\varepsilon > 0$. Note that in view of (5.12.8) the estimate

$$a_{n+1}(T_{K,\mu}) \leqslant \rho^\alpha |K|_{1,\alpha} \cdot n^{-\alpha\sigma}, \quad n = 1, 2, 3, \ldots,$$

applies to all integral operators $T_{K,\mu}: C(X) \rightarrow C(X)$ generated by a kernel $K \in C_1^\alpha(X \times X)$ so long as

$$\varepsilon_n(X) \leqslant \rho \cdot n^{-\sigma} \quad \text{for } n = 1, 2, 3, \ldots \tag{5.10.18}$$

5.13. Operators defined by abstract kernels

In analogy to $C(X)$ we introduce *the set $C_Z(X)$ of all continuous Z-valued functions K on a compact metric space (X, d), where Z is an arbitrary Banach space.* It is obvious that $C_Z(X)$ becomes a Banach space under the *norm*

$$\| K \| = \sup_{s \in X} \| K(s) \|_Z. \tag{5.13.1}$$

In a similar way to that of scalar-valued bounded functions f on X, the *modulus of continuity*

$$\Omega_Z(K; \delta) = \sup \{ \| K(s) - K(t) \|_Z : s, t \in X, d(s, t) \leqslant \delta \} \tag{5.13.2}$$

makes sense for arbitrary bounded Z-valued functions K on X, the limit relation

$$\lim_{\delta \rightarrow +0} \Omega_Z(K; \delta) = 0 \tag{5.13.3}$$

again being *a characteristic property of functions $K \in C_Z(X)$*. Moreover, the modulus of continuity $\Omega_Z(K; \delta)$ gives rise to *classes C_Z^α of Hölder continuous Z-valued functions of type α on X, $0 < \alpha \leqslant 1$. $C_Z^\alpha(X)$ is meant to be the subclass of functions $K \in C_Z(X)$ for which the expression

$$|K|_{Z,\alpha} = \sup_{\delta > 0} \frac{\Omega_Z(K; \delta)}{\delta^\alpha} \tag{5.13.4}$$

is finite. The class $C_Z^\alpha(X)$ turns out to be a Banach space with respect to the norm

$$\|K\|_{Z,\alpha} = \max\{\|K\|, |K|_{Z,\alpha}\}. \tag{5.13.5}$$

In the situation where $Z = E'$ is the dual of the Banach space E the elements $K \in C_{E'}(X)$ can be used to generate operators $T_K: E \to C(X)$ according to the rule

$$(T_K x)(s) = \langle x, K(s) \rangle. \tag{5.13.6}$$

These operators are linear and also continuous since

$$\|T_K\| = \sup_{\|x\| \leqslant 1} \sup_{s \in X} |\langle x, K(s) \rangle| = \sup_{s \in X} \|K(s)\|_{E'} < \infty$$

and hence

$$\|T_K\| = \|K\|. \tag{5.13.7}$$

If we calculate the modulus of continuity $\omega(T_K; \delta)$ we see that the operators $T_K: E \to C(X)$ generated by an *abstract kernel* $K \in C_{E'}(X)$ in the sense of (5.13.6) are even compact. In fact, on the basis of the general definition (3.0.6) we obtain

$$\omega(T_K; \delta) = \sup_{\|x\| \leqslant 1} \omega(T_K x; \delta)$$

$$= \sup_{\|x\| \leqslant 1} \sup\{|\langle x, K(s) - K(t) \rangle| : s, t \in X, d(s,t) \leqslant \delta\}$$

$$= \sup\{\|K(s) - K(t)\|_{E'} : s, t \in X, d(s,t) \leqslant \delta\}$$

and thus finally

$$\omega(T_K; \delta) = \Omega_{E'}(K; \delta). \tag{5.13.8}$$

From (5.13.3) we therefore have

$$\lim_{\delta \to +0} \omega(T_K; \delta) = 0$$

which proves the compactness of T_K for arbitrary $K \in C_{E'}(X)$.

On the other hand, any compact operator $T: E \to C(X)$ can be generated by an 'abstract kernel' $K \in C_{E'}(X)$ in the sense of (5.13.6). Indeed, given $T \in K(E, C(X))$ we put

$$K(s) = T'\delta_s, \tag{5.13.9}$$

where δ_s is the well-known Dirac functional

$$\langle f, \delta_s \rangle = f(s)$$

on $C(X)$. The function K defined by (5.13.9) is a bounded E'-valued function on X since

$$\|K\| = \sup_{s \in X} \|K(s)\|_{E'} = \sup_{s \in X} \sup_{\|x\| \leqslant 1} |\langle x, T'\delta_s \rangle|$$

$$= \sup_{\|x\| \leqslant 1} \sup_{s \in X} |(Tx)(s)|$$

reduces to

$$\|K\| = \|T\|. \tag{5.13.10}$$

To show the continuity of K we compute the norm $\|K(s) - K(t)\|_{E'}$ appearing in the modulus of continuity $\Omega_{E'}(K; \delta)$, that is

$$\|K(s) - K(t)\|_{E'} = \sup_{\|x\| \leq 1} |\langle x, T'\delta_s - T'\delta_t \rangle|$$
$$= \sup_{\|x\| \leq 1} |(Tx)(s) - (Tx)(t)|.$$

In this way we recognize that

$$\Omega_{E'}(K; \delta) = \omega(T; \delta) \tag{5.13.11}$$

and, by the compactness of T, we may conclude that

$$\lim_{\delta \to +0} \Omega_{E'}(K; \delta) = 0$$

which amounts to $K \in C_{E'}(X)$. Moreover, by (5.13.6) and (5.13.9) the operator T_K generated by K coincides with the original operator T since

$$(T_K x)(s) = \langle x, T'\delta_s \rangle = \langle Tx, \delta_s \rangle = (Tx)(s).$$

The essence of the above considerations is summarized in the following proposition.

Proposition 5.13.1. *Let (X, d) be a compact metric space and E an arbitrary Banach space. Then the* map $\phi: K(E, C(X)) \to C_{E'}(X)$ *defined by*

$$\phi(T)(s) = T'\delta_s \quad \text{for } s \in X \tag{5.13.12}$$

is a metric isomorphism from $K(E, C(X))$ onto $C_{E'}(X)$ as well as a metric isomorphism from the subclass $L^\alpha(E, C(X))$ of $K(E, C(X))$ onto the subclass $C_{E'}^\alpha(X)$ of $C_{E'}(X)$.

Proof. The map ϕ is obviously linear. Therefore the first part of the assertion is clear by what has already been pointed out. Furthermore, because of (5.13.4) and (5.13.11) the image $K = \phi(T)$ of $T \in L^\alpha(E, C(X))$ in $C_{E'}(X)$ satisfies the condition

$$|K|_{E',\alpha} = |T|_\alpha < \infty. \tag{5.13.13}$$

If we combine (5.13.10) and (5.13.13) we obtain

$$\|K\|_{E',\alpha} = \|T\|_\alpha$$

which proves that ϕ is a metric injection of $L^\alpha(E, C(X))$ into $C_{E'}^\alpha(X)$. On the other hand, the equality (5.13.8) tells us that the inverse image $\phi^{-1}(K)$ of any kernel $K \in C_{E'}^\alpha(X)$ in $K(E, C(X))$ is an operator T_K of type L^α. Thus ϕ even turns out to be a metric isomorphism from $L^\alpha(E, C(X))$ onto $C_{E'}^\alpha(X)$. ∎

The *operators defined by abstract kernels* mentioned in the title of this section coincide with the *compact operators* $T: E \to C(X)$. The integral operators $T_{K,\mu}:L_p(X,\mu)\rightarrow C(X)$ or $T_{K,\mu}:C(X)\rightarrow C(X)$ generated by a continuous kernel $K\in C(X\times X)$ in the proper sense can be refound in this general framework. One has only to interpret the kernel $K\in C(X\times X)$ as an $L'_p(X,\mu)$-valued or $C(X)'$-valued function on X whose norm $\|K\|$ is given by

$$\|K\|=\sup_{s\in X}\|K(s,\cdot)\|_{p'},\quad 1/p+1/p'=1,\qquad(5.13.14)$$

if $T_{K,\mu}$ is considered as an operator on $L_p(X,\mu)$ with $1\leqslant p<\infty$, and by

$$\|K\|=\sup_{s\in X}\|K(s,\cdot)\|_1\qquad(5.13.15)$$

for $T_{K,\mu}$ as an operator in $C(X)$. In section 5.11 we used the abbreviation $\|K\|_{p'}$, $1\leqslant p\leqslant\infty$, for the expressions on the right-hand sides of (5.13.14) and (5.13.15). In the same context we introduced a modulus of continuity

$$\Omega_p(K;\delta)=\sup\{\|K(s_1,\cdot)-K(s_2,\cdot)\|_p:s_1,s_2\in X,d(s_1,s_2)\leqslant\delta\}\qquad(5.11.5)$$

for $1\leqslant p\leqslant\infty$ directly related to the kernel $K\in C(X\times X)$. We now see that

$$\Omega_p(K;\delta)=\Omega_{L_p(X,\mu)}(K;\delta)$$

is just the modulus of continuity $\Omega_Z(K;\delta)$ defined by (5.13.2) with $Z=L_p(X,\mu)$, the function $K\in C(X\times X)$ being regarded as an $L_p(X,\mu)$-valued function $K(s,\cdot)$ on X in the first variable. These remarks bring to light that Propositions 5.11.2, 5.11.3, and 5.11.4 can be formulated under the more general supposition that $K\in C_{L'_p(X,\mu)}$ or $K\in C_{C(X)'}(X)$ is a generalized kernel generating a compact operator $T_{K,\mu}:L_p(X,\mu)\rightarrow C(X)$ or $T_K:C(X)\rightarrow C(X)$, respectively, by (5.13.6). In this particular situation the procedure (5.13.6) of course automatically provides an integral operator. For an arbitrary Banach space E and an E'-valued kernel $K\in C_{E'}(X)$ the equalities (5.13.7) and (5.13.8) take the place of (5.11.7) or (5.11.9), and (5.11.8) or (5.11.10), respectively, for the induced operator $T:E\rightarrow C(X)$.

Integral operators $T_{K,\mu}:L_p(X,\mu)\rightarrow C(X)$ for $1\leqslant p<\infty$ and $T_{K,\mu}:C(X)\rightarrow C(X)$ for $p=\infty$ generated by a kernel K of a Hölder class $C_p^\alpha(X\times X)$ with $1/p+1/p'=1$ were studied in section 5.12 by using the quantities

$$|K|_{p',\alpha}=\sup_{\delta>0}\frac{\Omega_{p'}(K;\delta)}{\delta^\alpha}\qquad(5.12.3)'$$

and

$$\|K\|_{p',\alpha}=\max(\|K\|_{p'},|K|_{p',\alpha}),\qquad(5.12.4)'$$

which can now give way to the quantities (5.13.4) and (5.13.5) with $Z=L'_p(X,\mu)$. All the claims of section 5.12, under the corresponding assumptions on the compact metric space (X,d), remain valid if we take generalized kernels $K\in C^\alpha_{L'_p(X,\mu)}(X)$ or $K\in C^\alpha_{C(X)'}(X)$ instead of kernels $K\in C_p^\alpha(X\times X)$ with $1\leqslant p\leqslant\infty$ and $1/p+1/p'=1$.

6

Operator theoretical methods in the local theory of Banach spaces

As pointed out at the beginning of section 2.7 the term *local*, when applied to an operator $T: E \to F$, refers to the behaviour of T on the finite-dimensional subspaces of E. Similarly, the *local theory of Banach spaces* concerns finite-dimensional subspaces of a Banach space E (see section 5.2). In this final chapter we shall see how local properties of operators can be used for studying local properties of Banach spaces. Certain classes of operators defined by local properties, namely the so-called *absolutely 2-summing operators*, which are well-known since Pietsch's 1967 paper on absolutely p-summing operators, will serve us for estimating *norms of projections onto finite-dimensional subspaces* of a Banach space E in section 6.1. The corresponding results were formulated without proof in section 2.4 (see Lemmas 2.4.1 and 2.4.2) and were subsequently used to estimate the extension constants $p_n(E, F)$ and the lifting constants $q_n(E, F)$ (see (2.4.10) and (2.4.27), respectively). These *geometrical parameters* were introduced for the purposes of finding estimates for the approximation numbers $a_n(T)$ of an arbitrary operator $T: E \to F$ from above by the Gelfand numbers $c_n(T)$ and the Kolmogorov numbers $d_n(T)$ (see Propositions 2.4.1 and 2.4.4, respectively). But since the extension constants $p_n(E, F)$ and the lifting constants $q_n(E, F)$ depend on the pair of Banach spaces (E, F), it is generally inconvenient to work with them. In section 6.2 we shall use *projection constants* related to a single Banach space E instead. In this way the investigations of section 6.2 can be considered as a continuation of the investigations of section 2.4. Their proper culminations, however, are the estimates for the projection constants from below. Roughly speaking, these estimates reflect the situation of *badly complemented finite-dimensional subspaces* of a given Banach space F. We in particular apply them to $F = l_\infty^n$, using the local estimates for the Kolmogorov numbers $d_k(S)$ of operators $S: H \to l_\infty^n$ dealt with in section 5.10 (see Lemma 5.10.4). By similar associations of ideas we obtain estimates from below for the *Banach–Mazur distance* $d(M, l_2^k)$ between a k-dimensional subspace M of l_∞^n and the k-dimensional euclidean space l_2^k in section 6.3.

The estimation of the *volumes of certain convex bodies* in section 6.4 also rests on the local estimates (5.10.33) for the Kolmogorov numbers $d_k(S)$ of operators $S:H \to l_\infty^n$. In fact, corresponding local estimates for the entropy numbers $e_k(S)$ of operators $S:l_1^n \to H$ are derived on the basis of (5.10.33). Finally, the inequality

$$g_n(S) \leqslant 2e_n(S)$$

between the nth entropy modulus $g_n(S)$ and the nth entropy number $e_n(S)$ combined with the relation between volume ratios and entropy moduli provides the volume estimates for the convex bodies in question.

The proof for the famous *Santaló inequality* (see Santaló 1949; Saint Raymond 1980/81) *and its inverse* (see Bourgain and Milman 1987) that we give in section 6.5 is on the same lines. *Volume ratios*, which are the subject of Santaló's inequality, are estimated by means of entropy numbers, the behaviour of the latter this time being derived from the behaviour of the Kolmogorov numbers $d_k(S)$ and of the Gelfand numbers $c_k(S^{-1})$ of certain isomorphisms $S:l_2^n \to E$ and $S^{-1}:E \to l_2^n$, respectively. The tool again is *the inequality* (3.1.1) *of Bernstein type*. The crucial point of the considerations in section 6.5, however, is a recent and deep result of Pisier (1989, 1988), which is formulated and interpreted geometrically, but not proved.

Sections 6.6 and 6.7 deal with *a weakened version of Grothendieck's theorem*, which is called *the little Grothendieck theorem*. It says that every operator $T:l_1 \to l_2$ is absolutely 2-summing. Though this statement concerns operators, it expresses properties of the Banach spaces l_1 and l_2 involved. The little Grothendieck theorem has to come under the local theory of Banach spaces, because it is a consequence of the local properties of l_1 and l_2. The proof we present rests on the estimates for the entropy numbers $e_k(S)$ of operators $S:l_1^n \to l_2^m$ already employed in section 6.4 and thus excellently fits into the framework of this chapter.

6.1. Norms of projections

The problem connected with projections onto finite-dimensional subspaces M of a Banach space E consists in *finding projections with a small norm*. This is obviously a question concerning the geometry of the underlying Banace space E, since the norm $\|P\|$ of a projection P has a definite geometrical meaning. However, our approach to a solution of this question is an operator theoretical one. It rests on the notion of an absolutely 2-summing operator.

An *operator* $T:E \to F$ between arbitrary Banach spaces E and F is called *absolutely p-summing* with $1 \leqslant p < \infty$, if there exists a constant $\rho \geqslant 0$ such that

$$\left(\sum_{i=1}^{n} \| Tx_i \|^p \right)^{1/p} \leqslant \rho \cdot \sup_{\|a\| \leqslant 1} \left(\sum_{i=1}^{n} |\langle x_i, a \rangle|^p \right)^{1/p} \tag{6.1.1}$$

for any finite system of elements $x_1, x_2, \ldots, x_n \in E$. Let $\pi_p(T)$ denote the *smallest possible constant* guaranteeing the inequality (6.1.1). Then we can claim that the class P_p of all absolutely p-summing operators forms an *injective operator ideal*, the quantity π_p representing an *injective ideal norm* on P_p according to which P_p is complete (see Pietsch 1967 and section 1.6). Note that the phrase 'absolutely p-summing operator' refers to the fact that T maps any weakly p-summable sequence $x_i \in E$ in the sense that $\sum_{i=1}^{\infty} |\langle x_i, a \rangle|^p < \infty$ for all $a \in E'$ onto an absolutely p-summable sequence $Tx_i \in F$ in the sense that $\sum_{i=1}^{\infty} \| Tx_i \|^p < \infty$. For $1 \leqslant p \leqslant q < \infty$ we have

$$P_p(E,F) \subseteq P_q(E,F) \quad \text{and} \quad \pi_q(T) \leqslant \pi_p(T) \tag{6.1.2}$$

for $T \in P_p(E,F)$. Moreover, the norm $\pi_p(T)$ of any absolutely p-summing operator T satisfies the inequality

$$\| T \| \leqslant \pi_p(T).$$

Among the operators of type P_p the *absolutely 2-summing operators* are of major interest to us. They may be regarded as a *generalization of the well-known Hilbert–Schmidt operators to Banach spaces*. We formulate an important factorization property of absolutely 2-summing operators due to Pietsch 1967 (see also Pietsch 1978) without proof.

Factorization theorem for absolutely 2-summing operators. *Every operator $T \in P_2(E,F)$ admits a factorization involving an absolutely 2-summing operator $S: l_\infty(\Gamma) \to H$, where $l_\infty(\Gamma)$ is the Banach space of bounded number families over an appropriate index set Γ (see section 2.3) and H a Hilbert space, in detail*

$$T = YSX \tag{6.1.3}$$

with $X \in L(E, l_\infty(\Gamma))$, $S \in P_2(l_\infty(\Gamma), H)$, and $Y \in L(H,F)$ such that

$$\pi_2(T) = \| Y \| \pi_2(S) \| X \|. \tag{6.1.4}$$

On the basis of the factorization theorem for absolutely 2-summing operators we prove a supplementary result for finite rank operators.

Lemma 6.1.1. *An operator $T: E \to F$ of rank n between arbitrary Banach spaces E and F can be factorized as*

$$T = VU \tag{6.1.5}$$

with $U \in L(E, l_2^n)$ and $V \in L(l_2^n, F)$ such that

$$\pi_2(T) = \| V \| \cdot \pi_2(U). \tag{6.1.6}$$

232 *Operator theoretical methods*

Proof. Let $F_0 = R(T)$ be the range of the rank n operator T so that

$$T = I^F_{F_0} T_0$$

with the operator $T_0 : E \to F_0$ induced by T and the natural embedding $I^F_{F_0} : F_0 \to F$ of F_0 into F. Because of the injectivity of the absolutely 2-summing norm π_2 we have

$$\pi_2(T) = \pi_2(T_0). \qquad (6.1.7)$$

Moreover, the above theorem ensures a factorization

$$T_0 = Y_0 S_0 X_0 \qquad (6.1.3)'$$

of T_0 involving an absolutely 2-summing operator $S_0 : l_\infty(\Gamma) \to H$ with values in a Hilbert space H such that

$$\pi_2(T_0) = \| Y_0 \| \pi_2(S_0) \| X_0 \|. \qquad (6.1.4)'$$

The quotient space H/N of H with respect to the kernel $N = N(Y_0)$ of the operator $Y_0 : H \to F_0$ is isometric to the n-dimensional euclidean space l^n_2 since $\dim(H/N) = \dim(F_0) = n$. Hence the canonical factorization

$$Y_0 = Y_1 Q^H_N$$

of Y_0, inserted in $(6.1.3)'$, leads us to a factorization

$$T = I^F_{F_0} Y_0 S_0 X_0 = (I^F_{F_0} Y_1)(Q^H_N S_0) X_0 = YSX_0$$

of the original operator T with

$$S = Q^H_N S_0 \quad \text{and} \quad Y = I^F_{F_0} Y_1.$$

Putting

$$U = SX_0 \quad \text{and} \quad V = Y$$

we arrive at (6.1.5) with $U \in L(E, l^n_2)$, $V \in L(l^n_2, F)$ and see that

$$\pi_2(T) \leqslant \| V \| \cdot \pi_2(U).$$

On the other hand, we obtain

$$\| V \| \cdot \pi_2(U) = \| Y_0 \| \cdot \pi_2(SX_0) \leqslant \| Y_0 \| \pi_2(S) \| X_0 \|$$
$$\leqslant \| Y_0 \| \pi_2(S_0) \| X_0 \| = \pi_2(T)$$

by $(6.1.4)'$ and $(6.1.7)$. This proves (6.1.6). ∎

The following result, concerning *the absolutely 2-summing norm of identity maps on finite-dimensional Banach spaces*, was discovered by Garling and Gordon (1971). They used a famous 1948 theorem of John on ellipsoids of maximal volume inscribed into a convex body (see also Lewis 1979; Pełczyński 1980). The proof we present is due to Kwapień (unpublished).

Lemma 6.1.2. *The absolutely 2-summing norm of the identity map I_E of an*

n-dimensional Banach space E is given by

$$\pi_2(I_E) = \sqrt{n}. \qquad (6.1.8)$$

Proof. First we prove the assertion for the n-dimensional Hilbert space $H = l_2^n$. Let $e_1, e_2, \ldots, e_n$ be the unit vector basis in l_2^n and $x_1, x_2, \ldots, x_k$ arbitrary elements in l_2^n. Then

$$\sum_{i=1}^{k} \|x_i\|^2 = \sum_{i=1}^{k} \sum_{j=1}^{n} |(x_i, e_j)|^2 = \sum_{j=1}^{n} \sum_{i=1}^{k} |(x_i, e_j)|^2$$

$$\leqslant n \cdot \sup_{\|a\| \leqslant 1} \sum_{i=1}^{k} |\langle x_i, a \rangle|^2.$$

Hence it follows that

$$\pi_2^2(I_H) \leqslant n$$

by the definition of the absolutely 2-summing norm. On the other hand, we have

$$n = \sum_{j=1}^{n} \|e_j\|^2 \leqslant \pi_2^2(I_H) \cdot \sup_{\|a\| \leqslant 1} \sum_{j=1}^{n} |\langle e_j, a \rangle|^2$$

$$= \pi_2^2(I_H).$$

This proves (6.1.8) for $E = H = l_2^n$.

Now let E be an arbitrary n-dimensional Banach space. Given elements $x_1, x_2, \ldots, x_k \in E$ we define an operator A from l_2^k into E by

$$Ae_i = x_i, \quad 1 \leqslant i \leqslant k. \qquad (6.1.9)$$

Then we obtain

$$\sum_{i=1}^{k} \|x_i\|^2 \leqslant \pi_2^2(A) \cdot \sup_{\|a\| \leqslant 1} \sum_{i=1}^{k} |\langle e_i, a \rangle|^2 = \pi_2^2(A). \qquad (6.1.10)$$

To estimate $\pi_2(A)$ from above we use the canonical factorization

$$A = A_0 Q$$

of A through the quotient space $H = l_2^k / N(A)$ of l_2^k with respect to the kernel $N(A)$ of A concluding that

$$\pi_2(A) \leqslant \|A_0\| \cdot \pi_2(Q).$$

The operator $A_0 : H \to E$ induced by A on the quotient space H of l_2^k has the same norm $\|A_0\| = \|A\|$ as A itself. Moreover, the definition (6.1.9) of A implies

$$\|A\| = \sup_{\|b\| \leqslant 1} \left(\sum_{i=1}^{k} |\langle x_i, b \rangle|^2 \right)^{1/2}, \qquad (6.1.11)$$

as can easily be checked. The absolutely 2-summing norm $\pi_2(Q)$ of the canonical surjection $Q : l_2^k \to H$ can be estimated by

$$\pi_2(Q) \leqslant \pi_2(I_H) \cdot \|Q\| = \pi_2(I_H).$$

But the quotient space $H = l_2^k/N(A)$ is a Hilbert space of dimension $\dim(H) \leqslant \dim(E) = n$. Therefore we have

$$\pi_2(I_H) \leqslant \sqrt{n}$$

by what has been proved in the first step. Thus we arrive at

$$\pi_2(A) \leqslant \sqrt{n} \cdot \sup_{\|b\| \leqslant 1} \left(\sum_{i=1}^{k} |\langle x_i, b \rangle|^2 \right)^{1/2}.$$

Combining this result with the inequality (6.1.10) we recognize that

$$n_2(I_E) \leqslant \sqrt{n}.$$

For the proof of the converse inequality we remind the reader of Lemma 6.1.1. It guarantees a factorization

$$I_E = VU \tag{6.1.5$'$}$$

of the identity map I_E of E by l_2^m such that

$$\pi_2(I_E) = \|V\| \cdot \pi_2(U). \tag{6.1.6$'$}$$

The product UV obviously represents the identity map

$$I_H = UV$$

of $H = l_2^m$. Hence it follows that

$$\sqrt{n} = n_2(I_H) \leqslant \pi_2(U) \cdot \|V\| = \pi_2(I_E)$$

by the first step. This completes the proof of (6.1.8). ∎

Lemma 6.1.2 and the *extension property of absolutely 2-summing operators*, to be dealt with in the following lemma, will provide the desired claims for norms of finite rank projections.

Lemma 6.1.3. *Let $M \subseteq E$ be a subspace of the Banach space E and $T: M \to F$ an absolutely 2-summing operator defined on M. Then T can be extended to an absolutely 2-summing operator $\tilde{T}: E \to F$ on E such that*

$$\pi_2(\tilde{T}) = \pi_2(T). \tag{6.1.12}$$

Proof. According to the factorization theorem for absolutely 2-summing operators the operator T can be factorized as

$$T = YSX$$

with $S \in P_2(l_\infty(\Gamma), H)$, $Y \in L(H, F)$, and $X \in L(M, l_\infty(\Gamma))$, where

$$\pi_2(T) = \|Y\| \pi_2(S) \|X\|.$$

Since the Banach space $l_\infty(\Gamma)$ possesses the metric extension property (see section 2.3), the operator X can be extended to an operator $\tilde{X}: E \to l_\infty(\Gamma)$ with the same norm $\|\tilde{X}\| = \|X\|$. The product $T = YS\tilde{X}$ then represents an extension of T to E which is also absolutely 2-summing, the norm $\pi_2(\tilde{T})$ being subject to the estimate

$$\pi_2(\tilde{T}) \leqslant \| Y \| \pi_2(S) \| \tilde{X} \| = \pi_2(T).$$

On the other hand, we have

$$\pi_2(T) \leqslant \pi_2(\tilde{T}),$$

because $T = \tilde{T} I_M^E$ is the restriction of $\tilde{T}$ to M. This completes the proof of (6.1.12). ∎

We now are in a position to prove the existence of a projection P with $\| P \| \leqslant \sqrt{n}$ onto every n-dimensional subspace $M \subseteq E$ of an arbitrary Banach space E. This result, which has already been stated in Lemma 2.4.1, is due to Kadec and Snobar (1971). A stronger version proved by Garling and Gordon (1971) states that there is a projection P with $\pi_2(P) = \sqrt{n}$ onto every n-dimensional subspace $M \subseteq E$. We shall derive this stronger version by using

$$\pi_2(I_M) = \sqrt{n} \qquad (6.1.8)'$$

(see Lemma 6.1.2) thus following an unpublished idea of Kwapień.

Theorem 6.1.1. *Let E be an arbitrary Banach space and $M \subseteq E$ an n-dimensional subspace. Then there exists a projection P from E onto M with $\pi_2(P) = \sqrt{n}$ so that*

$$\| P \| \leqslant \sqrt{n}.$$

Moreover, P can be factorized as

$$P = VU \qquad (6.1.5)''$$

with $U \in L(E, l_2^n)$ and $V \in L(l_2^n, E)$ such that

$$\| P \| \leqslant \| V \| \cdot \| U \| \leqslant \sqrt{n}.$$

Proof. By Lemma 6.1.3 the identity map I_M of M can be extended to an operator $\tilde{I}_M$ from E onto M such that

$$\pi_2(\tilde{I}_M) = \pi_2(I_M)$$

and hence

$$\pi_2(\tilde{I}_M) = \sqrt{n}$$

from (6.1.8)'. The operator

$$P = I_M^E \tilde{I}_M$$

is the desired projection. In fact, from

$$I_M = \tilde{I}_M I_M^E$$

we get $P^2 = P$. Because of the injectivity of the absolutely 2-summing norm π_2 we have

$$\pi_2(P) = \pi_2(\tilde{I}_M) = \sqrt{n}.$$

Finally, the assertion concerning the factorization (6.1.5)″ is an immediate consequence of Lemma 6.1.1 and of $\| U \| \leqslant \pi_2(U)$. ∎

We remark that König and Lewis (1988) proved the existence of a projection P from E onto an arbitrary n-dimensional subspace $M \subseteq E$ with $n > 1$ whose norm satisfies the strict inequality

$$\| P \| < \sqrt{n}$$

(see König 1986).

Next we want to prove a counterpart to Theorem 6.1.1 concerning the absolutely 2-summing norms of the duals of projections with a prescribed null space of finite codimension. For this purpose we introduce so-called *2-nuclear representations for finite rank operators*. These representations characterize *a subclass of the class P_2* of absolutely 2-summing operators, the class N_2 of so-called 2-nuclear operators.

An *operator* T between arbitrary Banach space E and F is said to be *2-nuclear if T can be represented as*

$$Tx = \sum_{i=1}^{\infty} \langle x, a_i \rangle y_i, \tag{6.1.13}$$

with $a_i \in E'$ and $y_i \in F$ such that

$$\sum_{i=1}^{\infty} \| a_i \|^2 < \infty \quad \text{and} \quad \sup_{\|b\| \leqslant 1} \sum_{i=1}^{\infty} |\langle y_i, b \rangle|^2 < \infty \tag{6.1.14}$$

(see Persson and Pietsch 1969; Pietsch 1978, 1987).

The condition (6.1.14) obviously guarantees the convergence of the series (6.1.13) for each $x \in E$. Considering all possible 2-nuclear representations (6.1.13) of a 2-nuclear operator $T : E \to F$ and putting

$$\nu_2(T) = \inf \left\{ \left(\sum_{i=1}^{\infty} \| a_i \|^2 \right)^{1/2} \sup_{\|b\| \leqslant 1} \left(\sum_{i=1}^{\infty} |\langle y_i, b \rangle|^2 \right)^{1/2} \right\}$$

we obtain a norm on the linear space $N_2(E, F)$ of 2-nuclear operators from E into F according to which $N_2(E, F)$ is a Banach space. The *norm $\nu_2(T)$* is referred to as *the 2-nuclear norm of the operator $T \in N_2(E, F)$*. Altogether the class N_2 of 2-nuclear operators, equipped with the 2-nuclear norm ν_2, turns out to be *a complete normed operator ideal* (see section 1.6).

It can easily be verified that every 2-nuclear operator $T : E \to F$ is absolutely 2-summing. Moreover,

$$\pi_2(T) = v_2(T), \quad \text{for } T \in N_2(E, F). \tag{6.1.15}$$

The verification of this equality, however, requires more effort (see Pietsch 1987).

Applying (6.1.15) to finite rank operators we can prove the counterpart to Theorem 6.1.1 mentioned above. The claim of the corresponding theorem (Garling and Gordon 1971; Pietsch 1987), like the claim of Theorem 6.1.1, is stronger than the claim of the lemma established in section 2.4 without proof (see Lemma 2.4.2).

Theorem 6.1.2. *Let E be an arbitrary Banach space and $N \subseteq E$ a subspace of codimension n. Then for a given $\varepsilon > 0$ there exists a projection $P \in L(E, E)$ with*

$$N(P) = N \tag{6.1.16}$$

and

$$\pi_2(P') \leqslant (1 + \varepsilon)\sqrt{n}. \tag{6.1.17}$$

Moreover, P can be factorized as

$$P = VU \tag{6.1.18}$$

with $U \in L(E, l_2^n)$ and $V \in L(l_2^n, E)$ such that

$$\|P\| \leqslant \|V\| \cdot \|U\| \leqslant (1 + \varepsilon)\sqrt{n}. \tag{6.1.19}$$

Proof. Since $\dim(E/N) = n$ the dual $(E/N)'$ is also of dimension n so that

$$v_2(I_{(E/N)'}) = \pi_2(I_{(E/N)'}) = \sqrt{n}$$

by (6.1.15) and Lemma 6.1.2. Recalling the definition of the 2-nuclear norm and identifying the bidual $(E/N)''$ of E/N with E/N we can find a representation

$$I_{(E/N)'} \hat{a} = \sum_{i=1}^{\infty} \langle \hat{a}, \hat{x}_i \rangle \hat{a}_i \tag{6.1.20}$$

of the identity map of the space $(E/N)'$ with elements $\hat{x}_i \in E/N$ and functionals $\hat{a}_i \in (E/N)'$ such that

$$\left(\sum_{i=1}^{\infty} \|\hat{x}_i\|^2 \right)^{1/2} \leqslant \sqrt{1 + \varepsilon} \cdot \sqrt{n} \tag{6.1.21}$$

and

$$\sup_{\|\hat{x}\| \leqslant 1} \left(\sum_{i=1}^{\infty} |\langle \hat{x}, \hat{a}_i \rangle|^2 \right)^{1/2} \leqslant 1. \tag{6.1.22}$$

Since $I_{(E/N)'} = I'_{E/N}$ the 2-nuclear representation (6.1.20) of $I_{(E/N)'}$ gives rise to a representation

$$I_{(E/N)} \hat{x} = \sum_{i=1}^{\infty} \langle \hat{x}, \hat{a}_i \rangle \hat{x}_i$$

of the identity map $I_{E/N}$. To obtain a projection $P \in L(E, E)$ with the desired properties we first construct a factorization of $I_{E/N}$ over E and for this purpose determine elements $x_i \in E$ with

$$Q_N^E x_i = \hat{x}_i \quad \text{and} \quad \|x_i\| \leqslant \sqrt{1 + \varepsilon} \cdot \|\hat{x}_i\|.$$

Putting

$$S\hat{x} = \sum_{i=1}^{\infty} \langle \hat{x}, \hat{a}_i \rangle x_i \qquad (6.1.23)$$

we obtain

$$I_{E/N} = Q_N^E S.$$

Hence, the operator

$$P = S Q_N^E \qquad (6.1.24)$$

represents a projection in E with $N(P) = N$. Moreover, taking into consideration that

$$P' = (Q_N^E)' S',$$

where

$$S'a = \sum_{i=1}^{\infty} \langle a, K_E x_i \rangle \hat{a}_i,$$

we see that

$$v_2(P') \leqslant v_2(S') \leqslant \left(\sum_{i=1}^{\infty} \|x_i\|^2 \right)^{1/2} \cdot \sup_{\|\hat{x}\| \leqslant 1} \left(\sum_{i=1}^{\infty} |\langle \hat{x}, \hat{a}_i \rangle|^2 \right)^{1/2} \leqslant (1 + \varepsilon)\sqrt{n}.$$

Because of (6.1.15) this amounts to (6.1.17). Finally, we have

$$Px = \sum_{i=1}^{\infty} \langle x, a_i \rangle x_i$$

with

$$a_i = (Q_N^E)' \hat{a}_i$$

from (6.1.24) and (6.1.23). This representation of P by an infinite series suggests a factorization

$$P = BA \qquad (6.1.25)$$

of P over l_2 involving the operators

$$Ax = \sum_{i=1}^{\infty} \langle x, a_i \rangle e_i \quad \text{from } E \text{ into } l_2$$

and

$$Bz = \sum_{i=1}^{\infty} (z, e_i) x_i \quad \text{from } l_2 \text{ into } E.$$

Note that the series defining the operator $A:E \to l_2$ converges for each $x \in E$, since

$$\left(\sum_{i=1}^{\infty} |\langle x, a_i \rangle|^2 \right)^{1/2} = \left(\sum_{i=1}^{\infty} |\langle Q_N^E x, \hat{a}_i \rangle|^2 \right)^{1/2} \leqslant \|Q_N^E x\| \leqslant \|x\|$$

by (6.1.22), and that, correspondingly,

$$\|A\| \leqslant 1.$$

On the other hand, we have

$$\|Bz\| \leqslant \sum_{i=1}^{\infty} |(z, e_i)| \cdot \|x_i\| \leqslant \|z\| \left(\sum_{i=1}^{\infty} \|x_i\|^2 \right)^{1/2}$$
$$\leqslant (1 + \varepsilon)\sqrt{n} \cdot \|z\|$$

and hence

$$\|B\| \leqslant (1 + \varepsilon)\sqrt{n}.$$

Therefore the factorization (6.1.25) is subject to the norm estimate

$$\|P\| \leqslant \|B\| \cdot \|A\| \leqslant (1 + \varepsilon)\sqrt{n}.$$

The desired factorization (6.1.18) results from (6.1.25) by inserting the orthogonal projection $P_n:l_2 \to l_2^n$ from l_2 onto the inverse image $B^{-1}(M)$ of $R(P) = M$ in l_2 which, of course, is isometrically isomorphic to l_2^n. With

$$U = P_n A \quad \text{and} \quad V = BI_n,$$

where $I_n:l_2^n \to l_2$ denotes the natural embedding of l_2^n into l_2, we obtain (6.1.18) and the corresponding norm estimate (6.1.19). ∎

6.2. Projection constants

In section 2.4 we introduced *the relative projection constant*

$$\lambda(M, E) = \inf \{ \|P\| : P \in L(E, E) \text{ projection with } R(P) = M \}$$

of a subspace $M \subseteq E$ with respect to the Banach space E. Letting M vary in the class of all k-dimensional subspaces M of E we now define *the kth projection constant $\lambda_k(E)$ of the Banach space E* itself by

$$\lambda_k(E) = \sup \{ \lambda(M, E) : M \subseteq E \text{ with } \dim(M) = k \}.$$

The value $\lambda_k(E)$, roughly speaking, describes the smallest possible norm of a projection onto a 'badly' complemented k-dimensional subspace of E. By Theorem 6.1.1 we have

$$\lambda_k(E) \leqslant \sqrt{k} \tag{6.2.1}$$

for any E. However, if E is itself finite-dimensional, the estimate (6.2.1) can be strengthened.

Proposition 6.2.1. *Let E be an arbitrary n-dimensional Banach space. Then*

$$\lambda_k(E) \leqslant \min(\sqrt{k}, 1 + \sqrt{n-k}) \quad \text{for } k = 1, 2, 3, \ldots, n. \qquad (6.2.2)$$

Proof. Any subspace $M \subseteq E$ with $\dim(M) = k$ is of codimension $\text{codim}(M) = n - k$. Hence, according to Theorem 6.1.2, there exists a projection $P \in L(E, E)$ with $N(P) = M$ and

$$\|P\| \leqslant (1 + \varepsilon)\sqrt{n-k}.$$

The operator

$$P_M = I_E - P$$

then represents a projection with the range $R(P_M) = M$, its norm satisfying the estimate

$$\|P_M\| \leqslant 1 + (1 + \varepsilon)\sqrt{n-k}.$$

This implies

$$\lambda(M, E) \leqslant 1 + \sqrt{n-k}$$

and thus also

$$\lambda_k(E) \leqslant 1 + \sqrt{n-k}. \qquad (6.2.3)$$

Combining (6.2.3) and (6.2.1) we obtain (6.2.2). ∎

Now we proceed to develop estimates for $\lambda_k(E)$ from below. For this purpose we recall the definition

$$q_{k+1}(Z, E) = \sup\{q(Z, E; N) : N \subseteq E, \dim(N) = k\} \qquad (2.4.26)'$$

of *the (k + 1)th lifting constant $q_{k+1}(Z, E)$ of the pair of Banach spaces* (Z, E) which is based on the definition (2.4.16) of the quantity $q(Z, E; N)$. What we actually need is the estimate

$$q(Z, E; N) \leqslant \inf\{\|P\| : P \in L(E, E) \text{ is a projection with } N(P) = N\} \qquad (2.4.23)'$$

derived in section 2.4. Indeed, (2.4.23)' can be extended to

$$q(Z, E; N) \leqslant 1 + \inf\{\|P_N\| : P_N \in L(E, E) \text{ is a projection with } R(P_N) = N\}$$

$$= 1 + \lambda(N, E), \qquad (6.2.4)$$

since

$$I_E = P + P_N$$

expresses a one-to-one correspondence between projections $P \in L(E, E)$ and $P_N \in L(E, E)$ with $N(P) = N$ and $R(P_N) = N$, respectively, where

$$\|P\| \leqslant 1 + \|P_N\|.$$

Taking the supremum with respect to all k-dimensional subspaces $N \subseteq E$

on both sides of the inequality (6.2.4) we obtain

$$q_{k+1}(Z, E) \leqslant 1 + \lambda_k(E) \quad \text{for } k = 1, 2, 3, \ldots$$

and arbitrary Banach spaces Z.

The variability of the Banach space Z provides the main opportunity to obtain satisfactory estimates of $\lambda_k(E)$ from below. On the other hand, since the lifting constants $q_{k+1}(Z, E)$ depend on a pair of Banach spaces (Z, E), it is difficult to work with them in general, as already emphasized at the beginning of the chapter. For this reason we use Proposition 2.4.4 and replace $q_{k+1}(Z, E)$ by the quotient

$$\frac{a_{k+1}(S)}{d_{k+1}(S)} \leqslant q_{k+1}(Z, E)$$

of the approximation numbers $a_{k+1}(S)$ and the Kolmogorov numbers $d_{k+1}(S)$ of an operator $S: Z \to E$ from an arbitrary Banach space Z into E with $\operatorname{rank}(S) > k$. The inequality

$$\frac{a_{k+1}(S)}{d_{k+1}(S)} \leqslant 1 + \lambda_k(E) \tag{6.2.5}$$

then opens up the possibility of estimating $\lambda_k(E)$ from below on the basis of the behaviour of approximation and Kolmogorov numbers of appropriate operators $S: Z \to E$ with $\operatorname{rank}(S) > k$. In the case of the Banach space $E = l_\infty^n$ we make use of universal estimates

$$d_k(S) \leqslant c_0 \|S\| \cdot k^{-1/2} \log^{1/2} \left(\frac{n}{k} + 1 \right) \tag{5.10.33}$$

for $n = 1, 2, 3, \ldots, 1 \leqslant k \leqslant n$, for the Kolmogorov numbers $d_k(S)$ of arbitrary operators $S: H \to l_\infty^n$ from above (see Lemma 5.10.4). Hence it remains to find estimates for the approximation numbers $a_k(S)$ of appropriate operators $S: H \to l_\infty^n$ from below. If we confine ourselves to $H = l_2^n$ and use isomorphisms $S: l_2^n \to l_\infty^n$ with special properties, we actually succeed in getting the desired estimates for the approximation numbers $a_k(S)$. Since the result does not depend on the particular Banach space l_∞^n, we derive it for an arbitrary n-dimensional Banach space E.

Lemma 6.2.1. *Let E be an arbitrary n-dimensional Banach space. Then there exists an isomorphism $S: l_2^n \to E$ with $\|S\| = 1$ and*

$$a_{k+1}(S) \geqslant \left(1 - \frac{k}{n} \right)^{1/2} \quad \text{for } k = 0, 1, 2, \ldots, n. \tag{6.2.6}$$

Proof. By Lemma 6.1.1 there exists an isomorphism $S: l_2^n \to E$ with

$$\pi_2(I_E) = \|S\| \cdot \pi_2(S^{-1}).$$

Without loss of generality we may assume that

$$\|S\| = 1 \quad \text{and} \quad \pi_2(S^{-1}) = \pi_2(I_E)$$

which, by (6.1.8) (see Lemma 6.1.2), amounts to

$$\pi_2(S^{-1}) = \sqrt{n}. \tag{6.2.7}$$

This guarantees the validity of (6.2.6) for $k = 0$. To prove (6.2.6) for the isomorphism S under consideration and $k > 0$ we take the Gelfand numbers $c_{k+1}(S) \leqslant a_{k+1}(S)$. The representation formula (2.3.7)' for $c_{k+1}(S)$ states that for any $\varepsilon > 0$ there is a subspace $M \subseteq l_2^n$ with $\text{codim}(M) = k$ such that

$$\| S I_M^{l_2^n} \| < c_{k+1}(S) + \varepsilon. \tag{6.2.8}$$

To estimate $\| S I_M^{l_2^n} \|$ from below we consider the factorization

$$I_M^{l_2^n} = S^{-1}(S I_M^{l_2^n})$$

of the natural embedding $I_M^{l_2^n} : M \to l_2^n$ concluding that

$$\pi_2(I_M^{l_2^n}) \leqslant \pi_2(S^{-1}) \| S I_M^{l_2^n} \|. \tag{6.2.9}$$

Because of the injectivity of the absolutely 2-summing norm π_2 we have

$$\pi_2(I_M^{l_2^n}) = \pi_2(I_M)$$

and thus

$$\pi_2(I_M^{l_2^n}) = \sqrt{n-k}$$

by Lemma 6.1.2. If we make use of (6.2.7) and (6.2.8), the inequality (6.2.9) turns into

$$\sqrt{n-k} \leqslant \sqrt{n} \cdot (c_{k+1}(S) + \varepsilon).$$

In this way we finally arrive at

$$\left(1 - \frac{k}{n}\right)^{1/2} \leqslant c_{k+1}(S) \leqslant a_{k+1}(S),$$

since $\varepsilon > 0$ can be chosen arbitrarily small.   ∎

Lemma 6.2.1 is the key for the desired estimate of $\lambda_k(l_\infty^n)$ from below.

Proposition 6.2.2. *The kth projection constant of the Banach space l_∞^n satisfies the inequality*

$$\lambda_k(l_\infty^n) \geqslant \frac{1}{2c_0} \cdot \left(1 - \frac{k}{n}\right)^{1/2} k^{1/2} \log^{-1/2}\left(\frac{n}{k} + 1\right) \tag{6.2.10}$$

for $n = 1, 2, 3, \ldots$ and $1 \leqslant k \leqslant n$, where c_0 is the universal constant entering the estimate (5.10.33).

Proof. Let $S : l_2^n \to l_\infty^n$ be an isomorphism subject to the claims of

Lemma 6.2.1. Combining (6.2.6) with (5.10.33) and taking account of $\|S\| = 1$ we get

$$\frac{a_{k+1}(S)}{d_{k+1}(S)} \geq \frac{1}{c_0} \cdot \left(1 - \frac{k}{n}\right)^{1/2} k^{1/2} \log^{-1/2}\left(\frac{n}{k} + 1\right)$$

for $n = 1, 2, 3, \ldots$ and $1 \leq k \leq n$. The proof of (6.2.10) is completed by referring to (6.2.5) and to $1 \leq \lambda_k(E)$. ∎

We remark that the behaviour of the kth projection constant $\lambda_k(l_\infty^n)$ for values of k which are large compared with the dimension n of l_∞^n, say $k \geq n/2$, is determined by $\sqrt{n-k}$. Indeed, with $k/n \geq 1/2$ the right-hand side of (6.2.10) can be estimated from below by

$$\frac{1}{2c_0} \cdot \left(1 - \frac{k}{n}\right)^{1/2} k^{1/2} \log^{-1/2}\left(\frac{n}{k} + 1\right) \geq \frac{1}{2\sqrt{2}c_0} \cdot \sqrt{n-k} \cdot (\log 3)^{-1/2}$$

$$= \frac{1}{2c_0}(\log 9)^{-1/2} \cdot \sqrt{n-k},$$

while (6.2.2) gives $1 + \sqrt{n-k}$ as an upper bound for $\lambda_k(l_\infty^n)$. Altogether we have

$$\frac{1}{2c_0} \cdot (\log 9)^{-1/2} \cdot \sqrt{n-k} \leq \lambda_k(l_\infty^n) \leq 1 + \sqrt{n-k} \quad \text{for } \frac{n}{2} \leq k \leq n.$$

If the dimension $n = 2k$ is twice the dimension k of the subspaces under consideration, we see that

$$\frac{1}{2c_0} \cdot (\log 9)^{-1/2} \cdot \sqrt{k} \leq \lambda_k(l_\infty^{2k}) \leq \sqrt{k}, \quad k = 1, 2, 3, \ldots, \qquad (6.2.11)$$

because in this situation the term $\sqrt{k}$ under the minimum on the right-hand side of (6.2.2) is good enough for the upper estimate of $\lambda_k(l_\infty^n)$. The left-hand part of the estimate (6.2.11) is interesting in so far as it implies a corresponding estimate

$$\lambda_k(l_\infty) \geq \frac{1}{2c_0} \cdot (\log 9)^{-1/2} \cdot \sqrt{k}$$

of the kth projection constant $\lambda_k(l_\infty)$ of the infinite-dimensional Banach space l_∞ from below (see Figiel and Johnson 1980; König 1986).

As already mentioned at the beginning of this section, the definition of the kth projection constant $\lambda_k(E)$ refers to 'badly' complemented k-dimensional subspaces of a Banach space E. In general, however, a Banach space E also possesses 'well' complemented subspaces. For instance, the sequence spaces l_p, $1 \leq p \leq \infty$, have a norm 1 projection onto the k-dimensional subspaces l_p^k consisting of the elements

$x = (\xi_1, \xi_2, \ldots, \xi_k, 0, 0, \ldots)$. It is therefore all the more surprising that there are Banach spaces with the property that all subspaces of a certain dimension $k > 1$ are badly complemented. Szarek (1983) constructed an infinite sequence of Banach spaces E_{2k} with $\dim(E_{2k}) = 2k$ such that

$$\|P\| > C \cdot \sqrt{k} \tag{6.2.12}$$

for any rank k projection $P \in L(E_{2k}, E_{2k})$ with a universal constant $C > 0$. An example of an infinite-dimensional Banach space E such that for arbitrary $k > 1$ all rank k projections $P \in L(E, E)$ satisfy the inequality (6.2.12) was discovered by Pisier (1983).

6.3. Banach–Mazur distances $d(M, l_2^k)$ of subspaces $M \subseteq l_\infty^n$

Having investigated badly complemented k-dimensional subspaces of l_∞^n we now study the Banach–Mazur distance $d(M, l_2^k)$ between an arbitrary k-dimensional subspace $M \subseteq l_\infty^n$ and l_2^k, once more using the estimates (5.10.33) for the Kolomogorov numbers $d_k(S)$ of operators $S: H \to l_\infty^n$.

Let us recall the definition of the Banach–Mazur distance

$$d(E, F) = \inf\{\|T\| \cdot \|T^{-1}\| : T \in L(E, F) \text{ is an isomorphism}\}$$

of two isomorphic Banach spaces E and F. If we want to estimate the Banach–Mazur distance $d(E, l_2^k)$ between an arbitrary k-dimensional Banach space E and the k-dimensional euclidean space l_2^k from above, we may use the isomorphism $S: l_2^k \to E$, subject of Lemma 6.2.1, and conclude that

$$\|S\| \cdot \|S^{-1}\| \leqslant \|S\| \cdot \pi_2(S^{-1}) = \sqrt{k}.$$

This inequality implies

$$d(E, l_2^k) \leqslant \sqrt{k}. \tag{6.3.1}$$

To estimate $d(E, l_2^k)$ from below, however, we have to permit arbitrary isomorphisms $T: l_2^k \to E$. In the following we shall consider a k-dimensional subspace $M \subseteq l_\infty^n$, as mentioned above, and isomorphisms $T: l_2^k \to M$.

Proposition 6.3.1. *Let $M \subseteq l_\infty^n$ be an arbitrary k-dimensional subspace of l_∞^n. Then*

$$d(M, l_2^k) \geqslant \frac{1}{c_0} \cdot k^{1/2} \log^{-1/2}\left(\frac{n}{k} + 1\right) \tag{6.3.2}$$

for $n = 1, 2, 3, \ldots$ and $1 \leqslant k \leqslant n$, where c_0 is the universal constant in the estimate (5.10.33).

Proof. Starting from an isomorphism $T: l_2^k \to M$ we take the operator $S = I_M^{l_\infty^n} T$ with values in l_∞^n and apply the estimate (5.10.33). Unfortunately, the Kolmogorov numbers $d_k(S)$ are not injective. But the symmetrized approximation numbers $t_k(S) \leqslant d_k(S)$ are (see section 2.6), namely

$$t_k(T) = t_k(I_M^{l_\infty^n} T) \leqslant d_k(I_M^{l_\infty^n} T) = d_k(S).$$

Hence we obtain

$$t_k(T) \leqslant c_0 \| T \| \cdot k^{-1/2} \log^{1/2} \left(\frac{n}{k} + 1 \right) \tag{6.3.3}$$

from (5.10.33) since $\| I_M^{l_\infty^n} T \| = \| T \|$. On the other hand, the norm determining property (T5) and the version (T3)(a) of the multiplicativity of the symmetrized approximation numbers give rise to the estimate

$$1 = t_k(I_M) \leqslant t_k(T) \cdot \| T^{-1} \|. \tag{6.3.4}$$

Combining (6.3.3) and (6.3.4) we arrive at

$$1 \leqslant c_0 \cdot \| T \| \cdot \| T^{-1} \| \cdot k^{-1/2} \log^{1/2} \left(\frac{n}{k} + 1 \right)$$

which implies

$$\frac{1}{c_0} \cdot k^{1/2} \log^{-1/2} \left(\frac{n}{k} + 1 \right) \leqslant d(M, l_2^k)$$

and thus completes the proof of (6.3.2). ∎

If the dimension k of the subspace $M \subseteq l_\infty^n$ is large in the sense that $k \geqslant n/2$, the right-hand side of (6.3.2) can be estimated from below by $(1/c_0) \cdot (\log 3)^{-1/2} \sqrt{k}$. By (6.3.1) we then have

$$\frac{1}{c_0} \cdot (\log 3)^{-1/2} \cdot \sqrt{k} \leqslant d(M, l_2^k) \leqslant \sqrt{k}.$$

Hence the Banach–Mazur distance $d(M, l_2^k)$ between l_2^k and a k-dimensional subspace $M \subseteq l_\infty^n$ with $k \geqslant n/2$ shows an obvious analogy to the asymptotic behaviour (6.2.11) of the kth projection constant $\lambda_k(l_\infty^{2k})$ of the space l_∞^{2k}. We quote without proof the result that

$$d(l_\infty^n, l_2^n) = \sqrt{n} \quad \text{for } n = 1, 2, 3, \dots$$

(see Tomczak and Jaegermann 1988).

6.4. Volumes of convex hulls of finite sets

In this section we shall deal with *the n-dimensional euclidean volume of the convex hull of an arbitrary finite set of elements* $x_1, x_2, \dots, x_N \in l_2^n$. Recall that the convex hull $\mathrm{conv}(x_1, x_2, \dots, x_N)$ of the elements $x_1, x_2, \dots, x_N$ is defined as the smallest convex set containing these elements. It can be described by

$$\text{conv}(x_1, x_2, \ldots, x_N)$$

$$= \left\{ x \in l_2^n : x = \sum_{i=1}^{N} \lambda_i x_i \text{ with } \lambda_i \geq 0 \text{ for } 1 \leq i \leq N \text{ and } \sum_{i=1}^{N} \lambda_i = 1 \right\}$$

This description of $\text{conv}(x_1, x_2, \ldots, x_N)$ suggests the introduction of an operator $S: l_1^N \to l_2^n$ by prescribing the values

$$Se_i = x_i, \quad 1 \leq i \leq N, \tag{6.4.1}$$

on the unit vector basis $e_1, e_2, \ldots, e_N$ of l_1^N. Obviously

$$\|S\| = \max_{1 \leq i \leq N} \|x_i\|. \tag{6.4.2}$$

Finally, the definition of S is justified by the inclusion

$$\text{conv}(x_1, x_2, \ldots, x_N) \subseteq S(U_1^N),$$

which will be the starting point for the estimation of the n-dimensional euclidean volume $\text{vol}_n(\text{conv}(x_1, x_2, \ldots, x_N))$ that follows. Indeed, applying the estimate (1.2.3) of the volume ratio $(\text{vol}_m(M)/\text{vol}_m(U_E))^{1/m}$ to the subset $M = S(U_1^N)$ of $E = l_2^n$ we see that

$$\text{vol}_n^{1/n}(S(U_1^N)) \leq g_n(S) \cdot \text{vol}_n^{1/n}(U_2^n)$$

from the definition (1.2.4) of the nth entropy modulus $g_n(S) = g_n(S(U_1^N))$. Moreover, we have

$$g_n(S) \leq 2e_n(S)),$$

so that an estimation of $\text{vol}_n^{1/n}(S(U_1^N))$ reduces to an estimation of $e_n(S)$ for $S: l_1^N \to l_2^n$. This again can be managed on the basis of the estimate

$$d_k(S) \leq c_0 \cdot \|S\| \cdot k^{-1/2} \log^{1/2}\left(\frac{n}{k} + 1\right), \quad 1 \leq k \leq n, \tag{5.10.33}$$

claimed for the Kolmogorov numbers $d_k(S)$ of operators $S: H \to l_\infty^n$ in Lemma 5.10.4.

Lemma 6.4.1. *Let $S: H \to l_\infty^n$ or $S: l_1^n \to H$ be an operator defined on a Hilbert space or having its values in a Hilbert space H, respectively, and let l_∞^n or l_1^n be the range or the domain of S, respectively. Then*

$$e_k(S) \leq c \cdot \|S\| \cdot k^{-1/2} \log^{1/2}\left(\frac{n}{k} + 1\right) \tag{6.4.3}$$

for $n = 1, 2, 3, \ldots$ and $1 \leq k \leq n$ where c is a universal constant.

Proof. We employ the Bernstein type inequality (3.1.13) with $p = 1$ stated in the Supplement at the end of section 3.1, namely

$$ke_k(S) \leq \sup_{1 \leq j \leq k} je_j(S) \leq 2c_1 \cdot \sup_{1 \leq j \leq k} jt_j(S).$$

In the case $S: H \to l_\infty^n$ we can refer to $t_j(S) \leqslant d_j(S)$ and the estimate (5.10.33) for $d_j(S)$, that is to say

$$ke_k(S) \leqslant 2c_1 c_0 \cdot \|S\| \cdot \sup_{1 \leqslant j \leqslant k} \left(j \log \left(\frac{n}{j} + 1 \right) \right)^{1/2}.$$

Since the function $f(x) = x \log(n/x + 1)$ is monotonously increasing for $x \geqslant 1$ we arrive at

$$e_k(S) \leqslant 2c_1 c_0 \cdot \|S\| \cdot k^{-1/2} \log^{1/2} \left(\frac{n}{k} + 1 \right). \tag{6.4.4}$$

On the other hand, in the case $S: l_1^n \to H$ we refer to $t_j(S) \leqslant c_j(S)$ and to $c_j(S) = d_j(S')$ (see Proposition 2.5.5, (2.5.35)). Since the operator $S': H \to l_\infty^n$ acts from H into l_∞^n, we can this time apply the estimate (5.10.33) with S' in place of S. Because $\|S'\| = \|S\|$ the final result again is the inequality (6.4.4). This completes the proof of (6.4.3) for both the cases $S: H \to l_\infty^n$ and $S: l_1^n \to H$ with the universal constant $c = 2c_1 c_0$.   ∎

Let us point out that Lemma 6.4.1 may also be proved directly by probabilistic arguments due to Maurey (Pisier 1980/81; Carl 1985).

In the following we shall make use of the estimate (6.4.3) for the nth entropy number $e_n(S)$ of the operator $S: l_1^N \to l_2^n$ defined by (6.4.1).

Proposition 6.4.1. *Let $x_1, x_2, \ldots, x_N$ be an arbitrary finite system of elements in the n-dimensional euclidean space l_2^n. Then*

$$\mathrm{vol}_n^{1/n}(\mathrm{conv}(x_1, x_2, \ldots, x_N)) \leqslant C \cdot \max_{1 \leqslant i \leqslant N} \|x_i\| \cdot n^{-1} \log^{1/2} \left(\frac{N}{n} + 1 \right)$$

$$\text{for } n, N = 1, 2, 3, \ldots \tag{6.4.5}$$

where C is a universal constant.

Proof. The considerations carried out at the beginning of this section make it clear that

$$\mathrm{vol}_n^{1/n}(\mathrm{conv}(x_1, x_2, \ldots, x_N)) \leqslant 2e_n(S) \, \mathrm{vol}_n^{1/n}(U_2^n),$$

where $S: l_1^N \to l_2^n$ is the operator defined by (6.4.1). The norm of S is determined by (6.4.2). Hence the estimate (6.4.3) with N in place of n, and n in place of k, gives

$$\mathrm{vol}_n^{1/n}(\mathrm{conv}(x_1, x_2, \ldots, x_N))$$

$$\leqslant 2c \cdot \max_{1 \leqslant i \leqslant N} \|x_i\| \, n^{-1/2} \log^{1/2} \left(\frac{N}{n} + 1 \right) \cdot \mathrm{vol}_n^{1/n}(U_2^n).$$

The volume $\mathrm{vol}_n(U_2^n)$ of the unit ball U_2^n in the n-dimensional euclidean space is given by

$$\text{vol}_n (U_2^n) = \frac{\pi^{n/2}}{\Gamma\left(\dfrac{n}{2} + 1\right)},$$

where Γ denotes the well-known Gamma-function. The functional equation

$$\Gamma\left(\frac{n}{2} + 1\right) = \frac{n}{2} \cdot \Gamma\left(\frac{n}{2}\right)$$

and the estimate

$$\Gamma\left(\frac{n}{2}\right) \geqslant \left(\frac{4\pi}{n}\right)^{1/2} \cdot \left(\frac{n}{2e}\right)^{n/2}$$

imply

$$\text{vol}_n(U_2^n) = \frac{2 \cdot \pi^{n/2}}{n \cdot \Gamma\left(\dfrac{n}{2}\right)} \leqslant \frac{(2e\pi)^{n/2} n^{1/2}}{\pi^{1/2} n^{n/2} \cdot n} \leqslant \left(\frac{2e\pi}{n}\right)^{n/2}$$

and thus

$$\text{vol}_n^{1/n}(U_2^n) \leqslant (2e\pi)^{1/2} \cdot n^{-1/2}.$$

In this way we obtain the desired result

$$\text{vol}_n^{1/n}(\text{conv}(x_1, x_2, \ldots, x_N)) \leqslant C \cdot \max_{1 \leqslant i \leqslant N} \|x_i\| \cdot n^{-1} \log^{1/2}\left(\frac{N}{n} + 1\right) \quad (6.4.5)$$

with $C = 2c \cdot (2e\pi)^{1/2}$. ∎

For further information on volumes of convex hulls we refer the reader to Carl and Pajor (1988).

6.5. On a theorem of Pisier

Pisier (1988, 1989) proved a deep theorem concerning the existence of certain isomorphisms from l_2^n onto an arbitrary n-dimensional Banach space. In its full form this theorem states the following:

Given a real number $\alpha > 1/2$ there is a constant $b(\alpha)$ such that for any n-dimensional Banach space E there exists on isomorphism $S : l_2^n \to E$ from l_2^n onto E with

$$d_k(S) < b(\alpha) \cdot \left(\frac{n}{k}\right)^{\alpha} \quad (6.5.1)$$

and

$$c_k(S^{-1}) < b(\alpha) \cdot \left(\frac{n}{k}\right)^{\alpha} \tag{6.5.2}$$

for $1 \leqslant k \leqslant n$. The constant $b(\alpha)$ is of order $(\alpha - \frac{1}{2})^{-1/2} \cdot O(1)$ for $\alpha \to \frac{1}{2}$.

Let us emphasize that the statement of Pisier's theorem is not true for $\alpha < \frac{1}{2}$. The case $\alpha = \frac{1}{2}$ is still open (see Pisier 1988, 1989).

The proof of Pisier's theorem uses delicate ingredients which are beyond the scope of this book. We therefore confine ourselves to a *geometrical interpretation* and some conclusions. The geometry operates in the n-dimensional real space $\mathbb{R}^n$. It does not refer to a norm on $\mathbb{R}^n$. However, the notion of a compact subset $X \subset \mathbb{R}^n$ as well as the notion of an interior point of a set $M \subset \mathbb{R}^n$ will be basic for the considerations that follow. We emphasize that these notions are actually independent of the particular norm used on the linear space $\mathbb{R}^n$.

A subset $M \subseteq \mathbb{R}^n$ is said to be *symmetric* if $x \in M$ implies $-x \in M$. A convex, compact, and symmetric subset $U \subset \mathbb{R}^n$ with the origin as an interior point is called *a ball*. A ball U of $\mathbb{R}^n$ gives rise to a *norm*

$$\|x\|_U = \inf\{\lambda > 0 : x \in \lambda U\}$$

on $\mathbb{R}^n$ according to which U turns out to be the closed unit ball, namely

$$U = \{x \in \mathbb{R}^n : \|x\|_U \leqslant 1\}.$$

On the other hand, if the space $\mathbb{R}^n$ is equipped with a norm $\|\cdot\|$ from the start and if E stands for the n-dimensional Banach space $(\mathbb{R}^n, \|\cdot\|)$, the closed unit ball U_E is a ball in the above sense such that

$$\|x\|_{U_E} = \|x\|.$$

Hence there is *a one-to-one correspondence between norms over $\mathbb{R}^n$ and balls in $\mathbb{R}^n$.*

If a is a linear functional over $\mathbb{R}^n$ we write $\langle x, a \rangle$ for the value of a on the element $x \in \mathbb{R}^n$ as we are used to doing within the Banach space setting. The linear space of all linear functionals over $\mathbb{R}^n$ can be identified with $\mathbb{R}^n$. Since the *polar set M^0 of a subset* $M \subseteq \mathbb{R}^n$ is defined as a certain set of linear functionals over $\mathbb{R}^n$ it likewise appears as a subset of $\mathbb{R}^n$, namely

$$M^0 = \{a \in \mathbb{R}^n : \langle x, a \rangle \leqslant 1 \text{ for all } x \in M\}.$$

In particular, the polar set $(U_2^n)^0$ of the closed unit ball U_2^n of l_2^n coincides with U_2^n, that is to say

$$(U_2^n)^0 = U_2^n. \tag{6.5.3}$$

Quite generally the polar set U^0 of a ball U is again a ball, the norm corresponding to U^0 being given by

$$\|a\|_{U^0} = \sup_{x \in U} |\langle x, a \rangle|. \tag{6.5.4}$$

The representation formula (6.5.4) expresses the fact that the norm induced by the ball U^0 coincides with the norm $\|a\|$ of the functional a over the Banach space $E = (\mathbb{R}^n, \|\cdot\|_U)$. In other words, *the polar set U^0 of a ball* is nothing other than *the unit ball $U_{E'}$ of the dual space E' of $E = (\mathbb{R}^n, \|\cdot\|_U)$.*

An *ellipsoid* $\mathscr{E}$ is meant to be a subset of $\mathbb{R}^n$ of the form

$$\mathscr{E} = S(U_2^n), \tag{6.5.5}$$

where $S : \mathbb{R}^n \to \mathbb{R}^n$ is a linear isomorphism. It can easily be checked that the norm corresponding to the 'ball' $\mathscr{E}$ is determined by

$$\|x\|_{\mathscr{E}} = \|S^{-1}x\|_{U_2^n}.$$

Furthermore, the polar set $\mathscr{E}^0$ of $\mathscr{E}$ can be described by

$$\mathscr{E}^0 = S'^{-1}(U_2^n) \tag{6.5.6}$$

and thus turns out to be an ellipsoid as well. The corresponding norm reads as

$$\|a\|_{\mathscr{E}^0} = \|S'a\|_{U_2^n}.$$

Now we are ready to present the geometrical interpretation of Pisier's theorem mentioned above. For the sake of convenience we use the abbreviation

$$\delta_{k,n}(\alpha) = b(\alpha) \cdot \left(\frac{n}{k}\right)^{\alpha},$$

taking the real number $\alpha > \frac{1}{2}$ as fixed. If we remind the reader of the definition (2.2.4) of the Kolmogorov numbers and of the representation (2.3.7) for the Gelfand numbers (see Proposition 2.3.2) we can reformulate *Pisier's theorem* as follows:

For any ball $U \subseteq \mathbb{R}^n$ there exists an ellipsoid $\mathscr{E} \subseteq \mathbb{R}^n$ such that for every k with $1 \leqslant k \leqslant n$ there are subspaces N_1 and M_1 in $\mathbb{R}^n$ with $\dim(N_1) < k$ and $\operatorname{codim}(M_1) < k$, respectively, guaranteeing the inclusions

$$\mathscr{E} \subset N_1 + \delta_{k,n}(\alpha) \cdot U \tag{6.5.7}$$

and

$$U \cap M_1 \subset \delta_{k,n}(\alpha) \cdot (\mathscr{E} \cap M_1), \tag{6.5.8}$$

respectively.

The inclusion (6.5.7) with $\mathscr{E} = S(U_2^n)$ and $U = U_E$ obviously corresponds to (6.5.1). The inclusion (6.5.8) expresses the norm estimate

$$\|S^{-1}I_{M_1}^E\| < \delta_{k,n}(\alpha) \tag{6.5.9}$$

valid in view of (6.5.2) for the restriction $S^{-1}I_{M_1}^E$ of S^{-1} to an appropriate subspace $M_1 \subset E$ with $\text{codim}(M_1) < k$. Indeed, (6.5.9) amounts to

$$S^{-1}(U \cap M_1) \subset \delta_{k,n}(\alpha) \cdot U_2^n$$

and hence to

$$U \cap M_1 \subset \delta_{k,n}(\alpha) \cdot S(U_2^n). \tag{6.5.10}$$

The subset $S(U_2^n)$ of E is the ellipsoid $\mathscr{E}$ in question. The left-hand side of (6.5.10) is contained in the subspace $M_1 \subset E$. Therefore (6.5.10) is equivalent to a corresponding inclusion with the intersection $S(U_2^n) \cap M_1$ instead of $S(U_2^n)$ on the right-hand side, namely

$$U \cap M_1 \subset \delta_{k,n}(\alpha) \cdot (\mathscr{E} \cap M_1). \tag{6.5.8}$$

The *duality relations*

$$d_k(S) = c_k(S') \quad \text{and} \quad c_k(S^{-1}) = d_k(S'^{-1})$$

(see Propositions 2.5.6 and 2.5.5) finally enable us to establish inclusions which concern *the polar sets U^0 and $\mathscr{E}^0$ of the ball $U \subset \mathbb{R}^n$ and of the ellipsoid $\mathscr{E} \subset \mathbb{R}^n$* assigned to U by Pisier's theorem. Let us rewrite (6.5.2) and (6.5.1) as

$$d_k(S'^{-1}) < \delta_{k,n}(\alpha) \tag{6.5.11}$$

and

$$c_k(S') < \delta_{k,n}(\alpha), \tag{6.5.12}$$

respectively, for $1 \leqslant k \leqslant n$. Note that $S'^{-1} : l_2^n \to E'$ is an isomorphism from l_2^n onto the dual space E' of E with $S'^{-1}(U_2^n) = \mathscr{E}^0$ (see (6.5.6)) and that the unit ball $U_{E'}$ of E' coincides with the polar set U^0 of the ball $U = U_E$ that we started from. Hence the ellipsoid $\mathscr{E}^0$ is related to the ball U^0 in the same way as the ellipsoid $\mathscr{E}$ to the ball U. This means that *there are subspaces N_2 and M_2 in $\mathbb{R}^2$ with $\dim(N_2) < k$ and $\text{codim}(M_2) < k$, respectively, guaranteeing the inclusions*

$$\mathscr{E}^0 \subset N_2 + \delta_{k,n}(\alpha) \cdot U^0 \tag{6.5.13}$$

and

$$U^0 \cap M_2 \subset \delta_{k,n}(\alpha) \cdot (\mathscr{E}^0 \cap M_2), \tag{6.5.14}$$

which correspond to the inequalities (6.5.11) and (6.5.12), respectively.

The inequality (3.1.13) of Bernstein type, stated in the Supplement at the end of section 3.1, can be used to derive a conclusion from Pisier's theorem which involves the entropy numbers $e_k(S)$, $e_k(S^{-1})$, $e_k(S')$, and $e_k(S'^{-1})$ of the isomorphisms $S : l_2^n \to E$, $S^{-1} : E \to l_2^n$, $S' : E' \to l_2^n$, and $S'^{-1} : l_2^n \to E'$. Let us think of T as of one of the operators S, S^{-1}, S', or S'^{-1}. If we replace the Kolmogorov numbers and the Gelfand numbers on the left-hand sides of (6.5.1), (6.5.2), (6.5.11) and (6.5.12) by the symmetrized approximation numbers $t_k(T)$, we may write

$$t_k(T) < b(\alpha) \cdot \left(\frac{n}{k}\right)^\alpha, \quad 1 \leqslant k \leqslant n, \tag{6.5.15}$$

as a consequence of the four inequalities (6.5.1), (6.5.2), (6.5.11), and (6.5.12) in a unfied form. The Bernstein type inequality (3.1.13) with $p = 1/\alpha$ then gives

$$\sup_{1 \leq k \leq n} k^\alpha e_k(T) < 2c_{1/\alpha} b(\alpha) \cdot \sup_{1 \leq k \leq n} k^\alpha \left(\frac{n}{k}\right)^\alpha = 2c_{1/\alpha} b(\alpha) \cdot n^\alpha,$$

which implies

$$e_k(T) < 2c_{1/\alpha} b(\alpha) \cdot \left(\frac{n}{k}\right)^\alpha \quad \text{for } 1 \leq k \leq n.$$

If we introduce the abbreviation

$$\eta_{k,n}(\alpha) = 2c_{1/\alpha} b(\alpha) \cdot \left(\frac{n}{k}\right)^\alpha,$$

we can formulate the above *conclusion from Pisier's theorem* as follows:

For any n-dimensional Banach space E there exists an isomorphism $S: l_2^n \to E$ *such that*

$$\max \{e_k(S), e_k(S^{-1}), e_k(S'), e_k(S'^{-1})\} < \eta_{k,n}(\alpha) \tag{6.5.16}$$

for $1 \leq k \leq n$. *For* $\alpha = 1$ *and* $k = n$ *in particular we obtain*

$$\max \{e_n(S), e_n(S^{-1}), e_n(S'), e_n(S'^{-1})\} < C \tag{6.5.17}$$

where C is the universal constant

$$C = 2c_1 b(1), \tag{6.5.18}$$

Since

$$d_k(S) = c_k(S^{-1}) = c_k(S') = d_k(S'^{-1}) = 0 \quad \text{for } k > n,$$

and because $b(\alpha) \cdot (n/k)^\alpha > 0$, the estimates (6.5.15) and the above considerations apply also to the case $k > n$. Therefore even the estimates (6.5.16) remain true for $k > n$. However, they are not interesting in this region, because the estimates

$$e_k(T) \leq e_n(T) \cdot e_{k-n+1}(I_{l_2^n}) \leq 4 \cdot 2^{-(k-n)/n} e_n(T)$$
$$= 8 \cdot 2^{-k/n} e_n(T), \quad k = n+1, n+2, \dots,$$

give better insight into the asymptotic entropy behaviour of the operators S, S^{-1}, S', and S'^{-1} (see (1.3.36)).

The *geometrical version* of the analytical conclusion (6.5.16) from Pisier's theorem in detail goes as follows:

For any ball $U \subset \mathbb{R}^n$ *there exists an ellipsoid* $\mathscr{E} \subset \mathbb{R}^n$ *such that for every* k *with* $1 \leq k \leq n$ *there are elements* x_i, y_i, a_i *and* b_i *in* $\mathbb{R}^n$, $1 \leq i \leq 2^{k-1}$ *producing coverings*

$$\mathscr{E} \subseteq \bigcup_{i=1}^{2^{k-1}} \{x_i + \eta_{k,n}(\alpha) \cdot U\} \quad and \quad U \subseteq \bigcup_{i=1}^{2^{k-1}} \{y_i + \eta_{k,n}(\alpha) \cdot \mathscr{E}\},$$

$$\mathscr{E}^0 \subseteq \bigcup_{i=1}^{2^{k-1}} \{a_i + \eta_{k+n}(\alpha) \cdot U^0\} \quad and \quad U^0 \subseteq \bigcup_{i=1}^{2^{k-1}} \{b_i + \eta_{k,n}(\alpha) \cdot \mathscr{E}^0\}.$$

The geometrical interpretation of inequality (6.5.17) for the *case* $\alpha = 1$ *and* $k = n$ says that *for any ball* $U \subset \mathbb{R}^n$ *there exist an ellipsoid* $\mathscr{E} \subset \mathbb{R}^n$ *and elements* x_i, y_i, a_i, b_i *in* $\mathbb{R}^n$, $1 \leqslant i \leqslant 2^{n-1}$, *such that*

$$\mathscr{E} \subseteq \bigcup_{i=1}^{2^{n-1}} \{x_i + C \cdot U\} \quad and \quad U \subseteq \bigcup_{i=1}^{2^{n-1}} \{y_i + C \cdot \mathscr{E}\}, \tag{6.5.19}$$

$$\mathscr{E}^0 \subseteq \bigcup_{i=1}^{2^{n-1}} \{a_i + C \cdot U^0\} \quad and \quad U^0 \subseteq \bigcup_{i=1}^{2^{n-1}} \{b_i + C \cdot \mathscr{E}^0\} \tag{6.5.20}$$

where C *is the universal constant introduced in* (6.5.18).

By a comparison of volumes we get

$$\mathrm{vol}_n(\mathscr{E}) \leqslant 2^{n-1} C^n \cdot \mathrm{vol}_n(U) \tag{6.5.21}$$

and

$$\mathrm{vol}_n(U) \leqslant 2^{n-1} C^n \cdot \mathrm{vol}_n(\mathscr{E}) \tag{6.5.22}$$

from (6.5.19), as well as

$$\mathrm{vol}_n(\mathscr{E}^0) \leqslant 2^{n-1} C^n \cdot \mathrm{vol}_n(U^0) \tag{6.5.23}$$

and

$$\mathrm{vol}_n(U^0) \leqslant 2^{n-1} C^n \cdot \mathrm{vol}_n(\mathscr{E}^0) \tag{6.5.24}$$

from (6.5.20). Realizing that

$$\mathscr{E} = S(U_2^n) \quad and \quad \mathscr{E}^0 = S'^{-1}(U_2^n)$$

(see (6.5.5) and (6.5.6)) and, correspondingly,

$$\mathrm{vol}_n(\mathscr{E}) \cdot \mathrm{vol}_n(\mathscr{E}^0) = |\det(S)| \cdot |\det(S'^{-1})| \cdot \mathrm{vol}_n^2(U_2^n)$$
$$= \mathrm{vol}_n^2(U_2^n),$$

we see that

$$\mathrm{vol}_n(U) \cdot \mathrm{vol}_n(U^0) \leqslant 2^{2n-2} C^{2n} \cdot \mathrm{vol}_n(\mathscr{E}) \cdot \mathrm{vol}_n(\mathscr{E}^0) = 2^{2n-2} C^{2n} \cdot \mathrm{vol}_n^2(U_2^n),$$

which provides

$$\left(\frac{\mathrm{vol}_n(U) \cdot \mathrm{vol}_n(U^0)}{\mathrm{vol}_n(U_2^n)^2} \right)^{1/n} \leqslant 4C^2. \tag{6.5.25}$$

This is a *weakened version of the so-called 'Santaló inequality'*. Santaló himself proved the estimate

$$\left(\frac{\mathrm{vol}_n(U) \cdot \mathrm{vol}_n(U^0)}{\mathrm{vol}_n(U_2^n)^2} \right)^{1/n} \leqslant 1$$

(see Santaló 1949; Saint Raymond 1980/81) which, in view of (6.5.3), turns out to be clear-cut.

Even the *inverse Santaló inequality* established by Bourgain and Milman (1987) is an immediate consequence of (6.5.21) and (6.5.23). Indeed, starting from

$$\mathrm{vol}_n^2(U_2^n) = \mathrm{vol}_n(\mathscr{E}) \cdot \mathrm{vol}_n(\mathscr{E}^0) \leqslant 2^{2n-2} C^{2n} \cdot \mathrm{vol}_n(U) \cdot \mathrm{vol}_n(U^0)$$

we obtain an estimate for the volume ratio $[(\mathrm{vol}_n(U)\cdot\mathrm{vol}_n(U^0)/\mathrm{vol}_n^2(U_2^n)]^{1/n}$ from below, namely

$$\frac{1}{4C^2}\leqslant\left(\frac{\mathrm{vol}_n(U)\cdot\mathrm{vol}_n(U^0)}{\mathrm{vol}_n^2(U_2^n)}\right)^{1/n}, \qquad (6.5.26)$$

the so-called 'inverse Santaló inequality'.

Pisier in his 1989 paper and his 1988 book derived both the estimates (6.5.25) and (6.5.26) for the volume ratio $[\mathrm{vol}_n(U)\cdot\mathrm{vol}_n(U^0)/\mathrm{vol}_n^2(U_2^n)]^{1/n}$ and, in addition, improved previous results of Milman (1985 and 1986), of Bourgain and Milman (1987), and König and Milman (1987). These improvements concern the so-called 'inverse Brunn–Minkowski inequality' and Milman's 'quotient of a subspace theorem'.

6.6. On absolutely 1-summing and 2-summing operators

Grothendieck (1956) published a fascinating paper entitled 'Résumé de la théorie métrique des produits tensoriel topologiques'. The highlight of this 'Résumé' is a result that Grothendieck called *the fundamental theorem of the metric theory of tensor products* and which is now called *Grothendieck's theorem* or sometimes *Grothendieck's inequality*. It states that every operator from l_1 into l_2 is absolutely 1-summing and that, correspondingly,

$$\pi_1(T)\leqslant c_G\cdot\|T\| \quad \text{for } T\in L(l_1,l_2) \qquad (6.6.1)$$

where c_G is a universal constant (cf. Lindenstrauss and Pełczyński 1968; Pisier 1986).

When introducing the absolutely p-summing operators in section 6.1 we observed that

$$P_p(E,F)\subseteq P_q(E,F) \quad \text{and} \quad \pi_q(T)\leqslant\pi_p(T) \qquad (6.1.2)$$

for $T\in P_p(E,F)$ and $1\leqslant p\leqslant q<\infty$. Hence Grothendieck's inequality (6.6.1) implies

$$\pi_p(T)\leqslant c_G\cdot\|T\| \quad \text{for } T\in L(l_1,l_2) \text{ and } 1\leqslant p<\infty.$$

In the next section we shall show that

$$\pi_2(T)\leqslant c_K\cdot\|T\| \quad \text{for } T\in L(l_1,l_2) \qquad (6.6.2)$$

where c_K is another universal constant appearing in *Khinchine's inequality* (see (6.6.3)). The inequality (6.6.2) is the assertion of the so-called *little Grothendieck theorem*. The proof we shall present rests on the local estimates

$$e_k(S)\leqslant c\cdot\|S\|\cdot k^{-1/2}\log^{1/2}\left(\frac{n}{k}+1\right), \qquad (6.4.3)$$

$n=1,2,3,\ldots,1\leqslant k\leqslant n$, for the entropy numbers $e_k(S)$ of operators

$S:l_1^n \to H$, proved in Lemma 6.4.1 by means of the local estimates

$$d_k(S) \leqslant c_0 \cdot \| S \| \cdot k^{-1/2} \log^{1/2} \left(\frac{n}{k} + 1 \right) \qquad (5.10.33)$$

for the Kolmogorov numbers $d_k(S)$ of operators $S:H \to l_\infty^n$, $n = 1, 2, 3, \ldots$, $1 \leqslant k \leqslant n$. However, the estimates (5.10.33) were established in Lemma 5.10.4 without proof and in fact a proof would be beyond the scope of this book. The proof given by Carl and Pajor (1988) actually makes use of the little Grothendieck theorem, so that applying the result (5.10.33) in the proof of (6.6.2) would amount to a statement of the equivalence of the two claims (6.6.2) and (5.10.33). For this reason we shall avoid the estimates (5.10.33) and employ the estimates (6.4.3) instead, which, as already mentioned, may be proved directly by probabilistic arguments due to Maurey. These arguments do not involve the Grothendieck theorem (see Pisier 1980/81; Carl 1985). What we intend is simply to give an approach to the little Grothendieck theorem via local estimates of certain compactness quantities of operators. They will be combined with striking tensor product techniques and thus lead us to the desired result (6.6.2).

Besides the estimates (6.4.3) we need the fact that the identity map $J:l_1 \to l_2$ is absolutely 1-summing. This, of course, is contained in the general Grothendieck theorem (6.6.1). But it can be verified separately by using *Khinchine's inequality* (cf. Pietsch 1978).

Proposition 6.6.1. *The identity map J from l_1 into l_2 is absolutely 1-summing.*

Proof. Let $\mathscr{E}_n$ denote the set of n-dimensional vectors $e = (\varepsilon_1, \varepsilon_2, \ldots, \varepsilon_n)$ with components $\varepsilon_i = \pm 1$ and let $\| x \|_2$ denote the euclidean norm of an arbitrary n-dimensional vector $x \in \mathbb{K}^n$, where $\mathbb{K}$ is either the field $\mathbb{R}$ of real numbers or the field $\mathbb{C}$ of complex numbers. Then Khinchine's inequality says that

$$\| x \|_2 \leqslant \frac{c_K}{2^n} \sum_{e \in \mathscr{E}_n} |\langle x, e \rangle| \quad \text{for all } x \in \mathbb{K}^n, \qquad (6.6.3)$$

where c_K is a universal constant. We first show that this constant c_K is an upper bound for the absolutely 1-summing norm $\pi_2(J_n)$ of the identity map $J_n:l_1^n \to l_2^n$ for all $n = 1, 2, 3, \ldots$ Indeed, given $x_1, x_2, \ldots x_k \in l_1^n$ we obtain

$$\sum_{i=1}^k \| J_n x_i \| \leqslant \frac{c_K}{2^n} \sum_{e \in \mathscr{E}_n} \sum_{i=1}^k |\langle x_i, e \rangle| \qquad (6.6.4)$$

by (6.6.3), the n-dimensional vector $e = (\varepsilon_1, \varepsilon_2, \ldots, \varepsilon_n)$ in this connection being considered as an element of l_∞^n. But since $\| e \|_\infty = 1$ for all $e \in \mathscr{E}_n$ we may extend the estimate (6.6.4) to

$$\sum_{i=1}^{k} \|J_n x_i\| \leqslant c_K \cdot \sup_{b \in U_\infty^n} \sum_{i=1}^{k} |\langle x_i, b \rangle|, \qquad (6.6.5)$$

which amounts to

$$\pi_1(J_n) \leqslant c_K \quad \text{for } n = 1, 2, 3, \ldots \qquad (6.6.6)$$

Now we show that (6.6.6) is the *local version* of

$$\pi_1(J) \leqslant c_K. \qquad (6.6.7)$$

For this purpose we choose arbitrary elements $x_1, x_2, \ldots, x_k \in l_1$ and map them onto elements $P_n x_1, P_n x_2, \ldots, P_n x_k \in l_1^n$ by using the natural projection $P_n : l_1 \to l_1^n$. Then (6.6.5) gives

$$\sum_{i=1}^{k} \|J_n P_n x_i\| \leqslant c_K \cdot \sup_{b \in U_\infty^n} \sum_{i=1}^{k} |\langle P_n x_i, b \rangle| = c_K \cdot \sup_{b \in U_\infty^n} \sum_{i=1}^{k} |\langle x_i, P_n' b \rangle|$$

$$\leqslant c_K \cdot \sup_{a \in U_\infty} \sum_{i=1}^{k} |\langle x_i, a \rangle|$$

for $n = 1, 2, 3, \ldots$ since $\|P_n\| = \|P_n'\| = 1$. The proof of (6.6.7) is completed by letting n tend to infinity and observing that

$$\lim_{n \to \infty} \|J_n P_n x_i\| = \|J x_i\|. \qquad \blacksquare$$

Next we study the class of absolutely 2-summing operators over l_2.

Proposition 6.6.2. *Over the Banach space l_2 the ideal $\boldsymbol{P}_2$ of absolutely 2-summing operators coincides with the approximation ideal $\boldsymbol{L}_{2,2}^{(a)}$, that is to say*

$$\boldsymbol{P}_2(l_2, l_2) = \boldsymbol{L}_{2,2}^{(a)}(l_2, l_2).$$

Moreover, we have

$$\pi_2(T) = \lambda_{2,2}^{(a)}(T) \quad \text{for } T \in \boldsymbol{P}_2(l_2, l_2) \qquad (6.6.8)$$

and, simultaneously,

$$\pi_2(T) = \left(\sum_{n=1}^{\infty} \|T e_n\|^2 \right)^{1/2}, \qquad (6.6.9)$$

where $e_1, e_2, e_3, \ldots,$ is the unit vector basis in l_2.

Proof. Let $T : l_2 \to l_2$ be an absolutely 2-summing operator so that T maps every weakly 2-summable sequence $x_i \in l_2$ onto an absolutely 2-summable sequence $T x_i \in l_2$ (see section 6.1). Since the sequence e_i of the unit vectors in l_2 is weakly 2-summable we in particular have

$$\sum_{i=1}^{\infty} \|T e_i\|^2 < \infty \qquad (6.6.10)$$

and hence, given $\varepsilon > 0$, can determine n_ε such that

$$\left(\sum_{i=n_\varepsilon+1}^{\infty} \| Te_i \|^2 \right)^{1/2} \leqslant \varepsilon.$$

This is the key to the proof of the compactness of T. We start from the representation

$$x = \sum_{i=1}^{\infty} (x, e_i) e_i$$

of the elements $x \in l_2$, apply the operator T, and then split the infinite series for Tx into two parts:

$$Tx = \sum_{i=1}^{n_\varepsilon} (x, e_i) Te_i + \sum_{i=n_\varepsilon+1}^{\infty} (x, e_i) Te_i.$$

The finite rank operator $\sum_{i=1}^{n_\varepsilon} (x, e_i) Te_i$ can be estimated by

$$\left\| \sum_{i=1}^{n_\varepsilon} (x, e_i) Te_i \right\| \leqslant \sup_{1 \leqslant i \leqslant n_\varepsilon} |\langle x, a_i \rangle|$$

with appropriate elements $a_i \in l_2$, $1 \leqslant i \leqslant n_\varepsilon$, while the remainder gives rise to a term

$$\left\| \sum_{i=n_\varepsilon+1}^{\infty} (x, e_i) Te_i \right\| \leqslant \left(\sum_{i=n_\varepsilon+1}^{\infty} \| Te_i \|^2 \right)^{1/2} \cdot \| x \| \leqslant \varepsilon \cdot \| x \|.$$

The result is

$$\| Tx \| \leqslant \sup_{1 \leqslant i \leqslant n_\varepsilon} |\langle x, a_i \rangle| + \varepsilon \cdot \| x \|$$

with elements $a_i \in l_2$, $1 \leqslant i \leqslant n_\varepsilon$, depending on the positive number ε. But since ε can be chosen arbitrarily small, we may conclude that T is compact (see Proposition 2.3.1). Finally, any compact operator acting between Hilbert spaces is approximable (see Proposition 2.4.1 or Proposition 2.4.4). Therefore we make use of a Schmidt representation

$$Tx = \sum_{i=1}^{\infty} \sigma_i (x, u_i) v_i \tag{2.1.9'}$$

where (u_i) and (v_i) are finite or countable orthonormal systems in l_2 and $\sigma_1 \geqslant \sigma_2 \geqslant \cdots \geqslant 0$ is a non-increasing sequence, which is finite or tends to zero, respectively. The members of this sequence are nothing other than the approximation numbers of the operator T, that is

$$\sigma_n = a_n(T), \quad n = 1, 2, 3, \ldots \tag{2.1.12}$$

Furthermore we obtain

$$Tu_n = \sigma_n v_n$$

from the representation formula (2.1.9)' and hence, in analogy to (6.6.10), may conclude that

$$\sum_{n=1}^{\infty} \| Tu_n \|^2 = \sum_{n=1}^{\infty} \sigma_n^2 = \sum_{n=1}^{\infty} a_n^2(T) < \infty. \qquad (6.6.11)$$

This proves $T \in L_{2,2}^{(a)}(l_2, l_2)$. To obtain an estimate for $\lambda_{2,2}^{(a)}(T)$ we notice that

$$\sum_{n=1}^{\infty} |(u_n, a)|^2 \leqslant \|a\|^2 \quad \text{for } a \in l_2$$

and then extend the equality (6.6.11) to the inequality

$$\sum_{n=1}^{\infty} a_n^2(T) \leqslant \pi_2^2(T) \cdot \sup_{\|a\| \leqslant 1} \sum_{n=1}^{\infty} |(u_n, a)|^2$$

$$\leqslant \pi_2^2(T),$$

which makes it clear that

$$\lambda_{2,2}^{(a)}(T) \leqslant \pi_2(T).$$

On the other hand, starting from an operator $T \in L_{2,2}^{(a)}(l_2, l_2)$ we make use of a Schmidt representation (2.1.9)' with (2.1.12) from the very beginning. If we then consider an arbitrary finite system of elements $x_1, x_2, \ldots, x_n \in l_2$, we see that

$$\left(\sum_{j=1}^{n} \| Tx_j \|^2 \right)^{1/2} = \left(\sum_{i=1}^{\infty} \sigma_i^2 \sum_{j=1}^{n} |(x_j, u_i)|^2 \right)^{1/2}$$

$$\leqslant \left(\sum_{i=1}^{\infty} \sigma_i^2 \right)^{1/2} \cdot \sup_{\|a\| \leqslant 1} \left(\sum_{j=1}^{n} |(x_j, a)|^2 \right)^{1/2}$$

which proves $T \in P_2(l_2, l_2)$ and, simultaneously,

$$\pi_2(T) \leqslant \left(\sum_{n=1}^{\infty} a_n^2(T) \right)^{1/2} = \lambda_{2,2}^{(a)}(T).$$

This completes the proof of (6.6.8).

To verify (6.6.9) for $T \in P_2(l_2, l_2)$ we go back to the Schmidt representation (2.1.9)' of T, which gives

$$Te_n = \sum_{i=1}^{\infty} \sigma_i(e_n, u_i)v_i$$

and thus

$$\sum_{n=1}^{\infty} \| Te_n \|^2 = \sum_{i=1}^{\infty} \sigma_i^2 \left(\sum_{n=1}^{\infty} |(e_n, u_i)|^2 \right) = \sum_{i=1}^{\infty} \sigma_i^2 = \pi_2^2(T)$$

by (2.1.12) and (6.6.8). ∎

Proposition 6.6.3. *Over the Banach space l_2 the ideal P_2 of absolutely 2-summing operators coincides with the ideal P_1 of absolutely 1-summing operators, that is to say*

$$P_2(l_2, l_2) = P_1(l_2, l_2)$$

and, moreover,

$$\pi_2(T) \leqslant \pi_1(T) \leqslant c_K \cdot \pi_2(T) \quad for \quad T \in P_2(l_2, l_2) \qquad (6.6.12)$$

where c_K is the universal constant in Khinchine's inequality (6.6.3).

Proof. By Proposition 6.6.2 every operator $T \in P_2(l_2, l_2)$ allows a Schmidt representation

$$Tx = \sum_{i=1}^{\infty} \sigma_i(x, u_i)v_i \qquad (2.1.9)'$$

with $(\sigma_i) \in l_2$. Introducing the operators

$$Sx = (\sigma_i(x, u_i))_{1 \leqslant i < \infty}$$

from l_2 into l_1 and

$$R((\xi_i)) = \sum_{i=1}^{\infty} \xi_i v_i$$

from l_2 into l_2 we achieve a factorization

$$T = RJS \qquad (6.6.13)$$

of T involving the identity map $J: l_1 \to l_2$. Since $J \in P_1(l_1, l_2)$ by Proposition 6.6.1, the operator T turns out to be absolutely 1-summing as well. Furthermore, the product formula (6.6.13) implies

$$\pi_1(T) \leqslant \|R\| \pi_1(J) \|S\|$$

which, in view of

$$\|S\| \leqslant \left(\sum_{i=1}^{\infty} \sigma_i^2 \right)^{1/2} = \lambda_{2,2}^{(a)}(T) = \pi_2(T), \quad \|R\| = 1,$$

and

$$\pi_1(J) \leqslant c_K, \qquad (6.6.7)$$

provides the right-hand part of the estimate (6.6.12). For the inclusion $P_1(l_2, l_2) \subseteq P_2(l_2, l_2)$ and $\pi_2(T) \leqslant \pi_1(T)$, we only need to refer to (6.1.2). ∎

6.7. Tensor product techniques and the little Grothendieck theorem

Having proved the main global properties of absolutely 2-summing operators in Propositions 6.6.2 and 6.6.3, we now turn to a *local property of the absolutely 2-summing norm* π_2 concerning the so-called tensor product $S \otimes S$ of an operator $S: l_1^n \to l_2^m$. This requires some preliminaries.

We consider the n-dimensional linear space $\mathbb{K}^n$ with a basis $e_1, e_2, \ldots, e_n$ and the m-dimensional linear space $\mathbb{K}^m$ with a basis $f_1, f_2, \ldots, f_m$. The *algebraic tensor product $\mathbb{K}^n \otimes \mathbb{K}^m$ of $\mathbb{K}^n$ and $\mathbb{K}^m$ is defined to be a linear* space of dimension $n \cdot m$. The formal products $e_i \otimes f_j$, $1 \leqslant i \leqslant n$ and $1 \leqslant j \leqslant m$, are regarded as a basis of $\mathbb{K}^n \otimes \mathbb{K}^m$, so that every $z \in \mathbb{K}^n \otimes \mathbb{K}^m$ can be represented uniquely as

$$z = \sum_{i=1}^{n} \sum_{j=1}^{m} \zeta_{ij} e_i \otimes f_j \qquad (6.7.1)$$

with coefficients $\zeta_{ij} \in \mathbb{K}$. In addition the elements of the tensor product $\mathbb{K}^n \otimes \mathbb{K}^m$ are related to the elements of the factors $\mathbb{K}^n$ and $\mathbb{K}^m$ by the *laws of distributivity*

$$(x + y) \otimes u = x \otimes u + y \otimes u$$

and

$$x \otimes (u + v) = x \otimes u + x \otimes v$$

valid for arbitrary $x, y \in \mathbb{K}^n$ and $u, v \in \mathbb{K}^m$ and by the *law of associativity*

$$\alpha(x \otimes y) = (\alpha x) \otimes y,$$

and the law of commutativity

$$(\alpha x) \otimes y = x \otimes (\alpha y)$$

with respect to the scalars $\alpha \in \mathbb{K}$.

In correspondence with the tensor product $\mathbb{K}^n \otimes \mathbb{K}^m$ of the two linear spaces $\mathbb{K}^n$ and $\mathbb{K}^m$ we introduce the *tensor product* $S_1 \otimes S_2 : \mathbb{K}^{n_1} \otimes \mathbb{K}^{n_2} \to \mathbb{K}^{m_1} \otimes \mathbb{K}^{m_2}$ *of two linear operators* $S_1 : \mathbb{K}^{n_1} \to \mathbb{K}^{m_1}$ *and* $S_2 : \mathbb{K}^{n_2} \to \mathbb{K}^{m_2}$ by demanding that $S_1 \otimes S_2$ be linear and

$$(S_1 \otimes S_2)(x \otimes y) = S_1 x \otimes S_2 y. \qquad (6.7.2)$$

In the following we shall be concerned with the *tensor product* $S \otimes S$ *of operators* $S : l_2^m \to l_1^n$ on the one hand *and* $S : l_1^n \to l_2^m$ on the other. In this situation it is obvious to equip the algebraic tensor products $\mathbb{K}^m \otimes \mathbb{K}^m$ and $\mathbb{K}^n \otimes \mathbb{K}^n$ of the underlying linear spaces $\mathbb{K}^m$ and $\mathbb{K}^n$ with the $l_2^{m^2}$ norm and the $l_1^{n^2}$ norm, respectively, so that $S \otimes S$ appears as an operator from $l_2^{m^2}$ into $l_1^{n^2}$ or from $l_1^{n^2}$ into $l_2^{m^2}$, respectively. First we shall consider the case $S : l_2^m \to l_1^n$ and estimate the norm of the tensor product $S \otimes S : l_2^{m^2} \to l_1^{n^2}$ by the norm of the original operator $S : l_2^m \to l_1^n$.

Lemma 6.7.1. *If* $S : l_2^m \to l_1^n$ *is an operator from* l_2^m *into* l_1^n *and* $S \otimes S : l_2^{m^2} \to l_1^{n^2}$ *the corresponding tensor product, then*

$$\| S \otimes S \| \leqslant c_K \cdot \| S \|^2, \qquad (6.7.3)$$

where c_K *is the constant from Khinchine's inequality* (6.6.3).

Proof. Let $e_1, e_2, \ldots, e_m$ be the unit vector basis of l_2^m. Then any element $x \in l_2^m \otimes l_2^m$ can be represented uniquely as

$$x = \sum_{i,j=1}^{m} \xi_{ij} e_i \otimes e_j \qquad (6.7.1)'$$

with respect to the basis $e_i \otimes e_j$ of $l_2^m \otimes l_2^m$, so that

$$(S \otimes S)x = \sum_{i,j=1}^{m} \xi_{ij} Se_i \otimes Se_j$$

by (6.7.2). Referring to the unit vector basis $f_1, f_2, \ldots, f_n$ of l_1^n we have unique representations

$$Se_i = \sum_{k=1}^{n} \sigma_{ki} f_k \qquad (6.7.4)$$

for the images Se_i of the unit vectors $e_i \in l_2^m$, $1 \leqslant i \leqslant m$. Using these representations we can immediately write down the $l_1^{n^2}$ norm of $(S \otimes S)x$, namely

$$\|(S \otimes S)x\| = \sum_{k,l=1}^{m} \left| \sum_{i,j=1}^{n} \xi_{ij} \sigma_{ki} \sigma_{lj} \right|. \qquad (6.7.5)$$

If we assign to $x \in l_2^m \otimes l_2^m$ the operator $X: l_2^m \to l_2^m$ whose matrix representation with respect to the basis $e_1, e_2, \ldots, e_m$ of l_2^m is given by (ξ_{ij}), we can rewrite (6.7.5) as

$$\|(S \otimes S)x\| = \sum_{l=1}^{n} \|SXS'g_l\|$$

with $g_1, g_2, \ldots, g_n$ as the unit vector basis of l_∞^n and $S': l_\infty^n \to l_2^m$ as the dual matrix operator

$$(\sigma'_{jl}) = (\sigma_{lj})$$

of S. Furthermore, the definition of the absolutely 1-summing norm implies

$$\sum_{l=1}^{n} \|SXS'g_l\| \leqslant \pi_1(SXS') \cdot \sup_{\|a\| \leqslant 1} \sum_{l=1}^{n} |\langle g_l, a \rangle| = \pi_1(SXS').$$

Because

$$\pi_1(SXS') \leqslant \|S\| \pi_1(X) \|S'\| = \|S\|^2 \pi_1(X)$$

the problem of estimating $\|(S \otimes S)x\|$ from above amounts to the problem of estimating $\pi_1(X)$ from above. Proposition 6.6.3, tells us that

$$\pi_1(X) \leqslant c_K \cdot \pi_2(X).$$

Moreover, since the operator X acts in a Hilbert space, Proposition 6.6.2 applies. The m-dimensional version of formula (6.6.9) says that

$$\pi_2(X) = \left(\sum_{j=1}^{m} \|Xe_j\|^2 \right)^{1/2}$$

for the operator $X: l_2^m \to l_2^m$. But because of the relation between the matrix representation of the operator X and the representation (6.7.1)' of the element $x \in l_2^m \otimes l_2^m$ we have

$$\sum_{j=1}^{m} \|Xe_j\|^2 = \sum_{i,j=1}^{m} |\xi_{ij}|^2 = \|x\|^2$$

and hence

$$\pi_2(X) = \left(\sum_{j=1}^{m} \| Xe_j \|^2 \right)^{1/2} = \| x \|.$$

Going back to the starting point (6.7.5) of our estimates we see that

$$\| (S \otimes S)x \| \leqslant c_K \cdot \| S \|^2 \| x \|,$$

which completes the proof of the assertion (6.7.3). ∎

In the case of an operator $S : l_1^n \to l_2^m$ the norm $\| S \otimes S \|$ of the tensor product $S \otimes S : l_1^{n^2} \to l_2^{m^2}$ even coincides with the square $\| S \|^2$ of the norm of S.

Lemma 6.7.2. *If $S : l_1^n \to l_2^m$ is an operator from l_1^n into l_2^m and $S \otimes S : l_1^{n^2} \to l_2^{m^2}$ the corresponding tensor product, then*

$$\| S \otimes S \| = \| S \|^2. \tag{6.7.6}$$

Proof. In accordance with the notation used for the proof of Lemma 6.7.1 we refer to $f_1, f_2, \ldots, f_n$ as the unit vector basis of l_1^n. It can easily be verified that

$$\| S \| = \sup_{1 \leqslant k \leqslant n} \| Sf_k \|.$$

Correspondingly, the norm of the tensor product $S \otimes S : l_1^{n^2} \to l_2^{m^2}$ is given by

$$\| S \otimes S \| = \sup_{1 \leqslant k,l \leqslant n} \| S(f_k \otimes f_l) \|.$$

To calculate the norm of $S(f_k \otimes f_l)$ in $l_2^{m^2}$ we take account of

$$S(f_k \otimes f_l) = Sf_k \otimes Sf_l$$

and of the fact that the $l_2^{m^2}$ norm of an element $y \otimes z$ with $y, z \in l_2^m$ is subject to the law

$$\| y \otimes z \| = \| y \| \ \| z \|. \tag{6.7.7}$$

This gives

$$\| S \otimes S \| = \sup_{1 \leqslant k,l \leqslant n} (\| Sf_k \| \cdot \| Sf_l \|) = \| S \|^2$$

and thus completes the proof of (6.7.6). ∎

Next we estimate the absolutely 2-summing norm $\pi_2(S \otimes S)$ of the tensor product $S \otimes S : l_1^{n^2} \to l_2^{m^2}$ of an operator $S : l_1^n \to l_2^m$ from below by the absolutely 2-summing norm $\pi_2(S)$ of the operator S itself. This estimate is the *local property* of the absolutely 2-summing norm π_2 already mentioned.

Lemma 6.7.3. *If $S : l_1^n \to l_2^m$ is an operator from l_1^n into l_2^m and $S \otimes S : l_1^{n^2} \to l_2^{m^2}$*

is the corresponding tensor product, then

$$\pi_2^2(S) \leqslant c_K \cdot \pi_2(S \otimes S), \tag{6.7.8}$$

where c_K is the constant from Khinchine's inequality (6.6.3).

Proof. Given a system of elements $x_1, x_2, \ldots, x_k \in l_1^n$ we refer to the Banach space l_2^k and, in a similar way to that in (6.1.9), define an auxiliary operator $A : l_2^k \to l_1^n$ by prescribing the values of A on the unit vector basis $e_1, e_2, \ldots, e_k$ of l_2^k, namely

$$Ae_i = x_i \quad \text{for } 1 \leqslant i \leqslant k. \tag{6.1.9}'$$

In this way the sum $\sum_{i=1}^{k} \| Sx_i \|^2$, which has to be estimated for estimating the absolutely 2-summing norm $\pi_2(S)$ of S, takes the form $\sum_{i=1}^{k} \| SAe_i \|^2$. Since the assertion to be proved involves the tensor product $S \otimes S$, we take the square

$$\left(\sum_{i=1}^{k} \| Sx_i \|^2 \right)^2 = \left(\sum_{i=1}^{k} \| SAe_i \|^2 \right)^2 = \sum_{i,j=1}^{k} \| SAe_i \|^2 \| SAe_j \|^2.$$

Writing the product $\| SAe_i \| \cdot \| SAe_j \|$ as $\| (SAe_i) \otimes (SAe_j) \|$ (see (6.7.7)) we arrive at

$$\left(\sum_{i=1}^{k} \| Sx_i \|^2 \right)^2 = \sum_{i,j=1}^{k} \| (SAe_i) \otimes (SAe_j) \|^2.$$

Furthermore, by (6.7.2) the tensor product $(SAe_i) \otimes (SAe_j)$ of the images of $e_i, e_j \in l_2^k$ under SA arises from the image of the element $e_i \otimes e_j \in l_2^k \otimes l_2^k$ under $(S \otimes S)(A \otimes A)$, that is

$$(SAe_i) \otimes (SAe_j) = (S \otimes S)(Ae_i \otimes Ae_j)$$
$$= (S \otimes S)(A \otimes A)(e_i \otimes e_j).$$

Hence we obtain

$$\left(\sum_{i=1}^{k} \| Sx_i \|^2 \right)^2 = \sum_{i,j=1}^{k} \| (S \otimes S)(A \otimes A)(e_i \otimes e_j) \|^2$$

and thus may conclude that

$$\left(\sum_{i=1}^{k} \| Sx_i \|^2 \right)^2 \leqslant \pi_2^2((S \otimes S)(A \otimes A)) \cdot \sup_{\|a\| \leqslant 1} \sum_{i,j=1}^{k} |(e_i \otimes e_j, a)|^2.$$

Since the elements $e_i \otimes e_j$, $1 \leqslant i, j \leqslant k$, are an orthonormal basis of the Hilbert space $l_2^k \otimes l_2^k$ we have

$$\sup_{\|a\| \leqslant 1} \sum_{i,j=1}^{k} |(e_i \otimes e_j, a)|^2 = 1.$$

If we make use of the ideal property

$$\pi_2((S \otimes S)(A \otimes A)) \leqslant \pi_2(S \otimes S) \cdot \| A \otimes A \|$$

of the π_2 norm and then apply Lemma 6.7.1 to the operator $A:l_2^k \to l_1^n$ we get

$$\sum_{i=1}^{k} \|Sx_i\|^2 \leqslant c_K \cdot \pi_2(S \otimes S) \|A\|^2. \qquad (6.7.9)$$

The norm $\|A\|$ of the operator $A:l_2^k \to l_1^n$ in question is determined by

$$\|A\| = \sup_{b \in U_\infty^n} \left(\sum_{i=1}^{k} |\langle x_i, b \rangle|^2 \right)^{1/2}. \qquad (6.1.11)'$$

Therefore (6.7.9) amounts to the assertion

$$\pi_2^2(S) \leqslant c_K \cdot \pi_2(S \otimes S). \qquad \blacksquare$$

Little Grothendieck Theorem. *Every operator $T:l_1 \to l_2$ from l_1 into l_2 is absolutely 2-summing and*

$$\pi_2(T) \leqslant c_K \cdot \|T\| \quad \text{for } T \in L(l_1, l_2) \qquad (6.6.2)$$

with the constant c_K from Khinchine's inequality (6.6.3).

Proof. We first prove the inequality (6.6.2) for operators from l_1^n into l_2^m. So let $S:l_1^n \to l_2^m$ be fixed arbitrarily and let $x_1, x_2, \ldots, x_k$ be a system of elements in l_1^n. We again use the operator $A:l_2^k \to l_1^n$ defined by (6.1.9)' and give the sum $\sum_{i=1}^{k} \|Sx_i\|^2$ the form

$$\sum_{i=1}^{k} \|Sx_i\|^2 = \sum_{i=1}^{k} \|SAe_i\|^2.$$

Using the standard notation $P_k:l_2 \to l_2^k$ for the orthogonal projection of l_2 onto l_2^k and $I_m:l_2^m \to l_2$ for the natural embedding of l_2^m into l_2 and putting

$$T = I_m S A P_k, \qquad (6.7.10)$$

we can write

$$\sum_{i=1}^{k} \|Sx_i\|^2 = \sum_{i=1}^{k} \|Te_i\|^2. \qquad (6.7.11)$$

Because $e_1, e_2, \ldots, e_k$ also represent an orthonormal system in l_2 we have

$$\sum_{i=1}^{k} \|Te_i\|^2 \leqslant \pi_2^2(T). \qquad (6.7.12)$$

Moreover, the finite rank operator $T:l_2 \to l_2$ is subject to Proposition 6.6.2, in particular to

$$\pi_2(T) = \lambda_{2,2}^{(a)}(T), \qquad (6.6.8)$$

which means

$$\pi_2(T) = \left(\sum_{j=1}^{n} a_j^2(T) \right)^{1/2}$$

since $\text{rank}(T) \leqslant n$. By Theorem 3.4.1 we have

$$a_j(T) \leqslant g_j(T)$$

in the Hilbert space situation. Hence the universal inequality

$$g_j(T) \leqslant 2e_j(T)$$

yields

$$\pi_2(T) \leqslant 2\left(\sum_{j=1}^{n} e_j^2(T)\right)^{1/2}. \tag{6.7.13}$$

Finally, the relation (6.7.10) between the operator $T:l_2 \to l_2$ and the operator $S:l_1^n \to l_2^m$ implies

$$e_j(T) \leqslant e_j(S) \cdot \|A\| = e_j(S) \cdot \sup_{b \in U_\infty^n} \left(\sum_{i=1}^{k} |\langle x_i, b \rangle|^2\right)^{1/2} \tag{6.7.14}$$

(see (6.1.11)'). Combining (6.7.11), (6.7.12), (6.7.13), and (6.7.14) we get

$$\pi_2(S) \leqslant 2\left(\sum_{j=1}^{n} e_j^2(S)\right)^{1/2}$$

in analogy to (6.7.13). The estimate

$$e_j(S) \leqslant c \cdot \|S\| \cdot j^{-1/2} \log^{1/2}\left(\frac{n}{j} + 1\right) \tag{6.4.3}'$$

for the entropy numbers of operators $S:l_1^n \to H$ or $S:l_1^n \to l_2^m$ now serves us for estimating $\pi_2(S)$, namely

$$\pi_2(S) \leqslant 2c \cdot \|S\| \left(\sum_{j=1}^{n} j^{-1} \log\left(\frac{n}{j} + 1\right)\right)^{1/2} \leqslant 2c \cdot \|S\| \left(\log(n+1) \sum_{j=1}^{n} j^{-1}\right)^{1/2}.$$

Since

$$\sum_{j=1}^{n} j^{-1} \leqslant 1 + \log n \leqslant 2 \cdot \log(n+1)$$

we obtain

$$\pi_2(S) \leqslant \sqrt{8}c \cdot \|S\| \cdot \log(n+1), \tag{6.7.15}$$

a result rather far from the desired one. We have still to remove the logarithmic term. For this reason we compare the growth of $\log(n+1)$ with the growth of an arbitrary positive power n^ε of n and infer from (6.7.15) that there is a constant $c_1(\varepsilon)$ such that

$$\pi_2(S) \leqslant c_1(\varepsilon) \cdot \|S\| \cdot n^\varepsilon \tag{6.7.16}$$

for all $S:l_1^n \to l_2^m$; $n, m = 1, 2, 3, \ldots$ In particular, if we apply (6.7.16) to the tensor product $S \otimes S:l_1^{n^2} \to l_2^{m^2}$ of an operator $S:l_1^n \to l_2^m$ we see that

$$\pi_2(S \otimes S) \leqslant c_1(\varepsilon) \cdot \|S \otimes S\| \cdot n^{2\varepsilon} = c_1(\varepsilon) \cdot \|S\|^2 \cdot n^{2\varepsilon} \tag{6.7.17}$$

by Lemma 6.7.2. On the other hand, Lemma 6.7.3 says that

$$\pi_2^2(S) \leqslant c_K \cdot \pi_2(S \otimes S) \tag{6.7.8}$$

for arbitrary $S:l_1^n \to l_2^m$. We combine the two inequalities (6.7.8) and (6.7.17)

and then take the square roots, which gives
$$\pi_2(S) \leqslant (c_K \cdot c_1(\varepsilon))^{1/2} \cdot \|S\| \cdot n^\varepsilon.$$
Hence we have reproduced the inequality (6.7.16), but with the constant
$$c_2(\varepsilon) = (c_K \cdot c_1(\varepsilon))^{1/2}$$
instead of $c_1(\varepsilon)$. Repeating this procedure we arrive at
$$\pi_2(S) \leqslant c_k(\varepsilon) \cdot \|S\| \cdot n^\varepsilon \tag{6.7.18}$$
after $k-1$ steps, where
$$c_k(\varepsilon) = (c_K \cdot c_{k-1}(\varepsilon))^{1/2}.$$
By induction it immediately follows that
$$c_k(\varepsilon) = c_K^{1/2 + 1/4 + \cdots + 1/2^{k-1}} \cdot c_1(\varepsilon)^{1/2^{k-1}}.$$
This makes it clear that
$$\lim_{k \to \infty} c_k(\varepsilon) = c_K.$$
By using this limit relation the inequality (6.7.18) turns into
$$\pi_2(S) \leqslant c_K \cdot \|S\| \cdot n^\varepsilon$$
as k tends to infinity. But since c_K is independent of ε, we can take the limit $\varepsilon \to 0$. The result is the desired inequality
$$\pi_2(S) \leqslant c_K \cdot \|S\| \tag{6.6.2)'}$$
for operators $S: l_1^n \to l_2^m$.

If in contrast to $S: l_1^n \to l_2^m$ the operator S in question acts from l_1^n into l_2, the inequality (6.6.2)' obviously remains true. So let $T: l_1 \to l_2$ be an arbitrary operator from l_1 into l_2 and let $T_n: l_1 \to l_2$ be the rank n operator derived from T by
$$T_n = T I_n P_n$$
where $P_n: l_1 \to l_1^n$ is the natural projection from l_1 onto l_1^n and $I_n: l_1^n \to l_1$ is the corresponding natural embedding. Then we obtain
$$\sum_{i=1}^k \|T I_n(P_n x_i)\|^2 \leqslant c_K^2 \cdot \|T I_n\|^2 \cdot \sup_{b \in U_\infty^n} \sum_{i=1}^k |\langle P_n x_i, b \rangle|^2 \tag{6.7.19}$$
for any finite system of elements $x_1, x_2, \ldots, x_k \in l_1$ by taking account of (6.6.2)' with $S = T I_n$. In a similar way to the proof of Proposition 6.6.1 we have
$$\sup_{b \in U_\infty^n} \sum_{i=1}^k |\langle P_n x_i, b \rangle|^2 = \sup_{b \in U_\infty^n} \sum_{i=1}^k |\langle x_i, P_n' b \rangle|^2 \leqslant \sup_{a \in U_\infty} \sum_{i=1}^k |\langle x_i, a \rangle|^2.$$
Thus the inequality (6.7.19) turns into
$$\sum_{i=1}^k \|T I_n P_n x_i\|^2 \leqslant c_K^2 \cdot \|T\|^2 \cdot \sup_{a \in U_\infty} \sum_{i=1}^k |\langle x_i, a \rangle|^2 \tag{6.7.20}$$

since $\| T I_n \| \leqslant \| T \|$. Taking the limit $n \to \infty$ on the left-hand side of (6.7.20) and making use of

$$\lim_{n \to \infty} I_n P_n x_i = x_i$$

we arrive at

$$\sum_{i=1}^{k} \| T x_i \|^2 \leqslant c_K^2 \cdot \| T \|^2 \cdot \sup_{a \in U_\infty} \sum_{i=1}^{k} |\langle x_i, a \rangle|^2.$$

This proves $T \in P_2(l_1, l_2)$ and (6.6.2). ∎

The simple but striking tensor product trick used in the proof of the little Grothendieck theorem can be rephrased in the following *abstract setting*: let M be a set of real or complex $n \times n$ matrices, where n varies arbitrarily in the set of natural numbers, and $\alpha : M \to [0, \infty)$, $\beta : M \to [0, \infty)$ two non-negative functions on M such that for every $\varepsilon > 0$ there is a constant $c(\varepsilon) > 0$ which guarantees

$$\alpha(S) \leqslant c(\varepsilon) \cdot n^\varepsilon \cdot \beta(S) \qquad (6.7.21)$$

for all $n \times n$ matrices $S \in M$, $n = 1, 2, 3, \ldots$ If in addition M is known to be tensor stable with respect to the pair α, β in the sense that $S \in M$ implies $S \otimes S \in M$, as well as

$$\alpha^2(S) \leqslant a \cdot \alpha(S \otimes S)$$

and

$$\beta(S \otimes S) \leqslant b \cdot \beta^2(S)$$

with appropriate constants $a, b \geqslant 1$, then (6.7.21) can be strengthened to

$$\alpha(S) \leqslant ab \cdot \beta(S) \quad \text{for all } S \in M.$$

This tensor product concept combined with entropy estimates has been profitably employed by Carl and Defant (1988) to improve various inequalities in the theory of absolutely summing operators. A similar device was used by Russo (1977) and Pietsch (1986) for improving eigenvalue estimates of operators (see also König 1986; Pietsch 1987).

References

Allakhverdiev, D.Eh. (1957). On the rate of approximation of completely continuous operators by finite dimensional operators (Russian). *Azerbajzhan. Gos. Univ. Uchen. Zap.* **2**, 27–37.

Astala, K. and Tylli, O. (1987). On the bounded compact approximation property and measures of noncompactness. *J. Funct. Anal.* **70**, 388–401.

Bourgain, J. and Milman, V.P. (1987). New volume ratio properties for symmetric convex bodies in R^n. *Invent. Math.* **88**, 319–40.

Caradus, S.R., Pfaffenberger, W.E. and Yood, B. (1974). *Calkin algebras of operators in Banach spaces.* New York: Dekker.

Carl, B. (1981). Entropy numbers, s-numbers and eigenvalue problems. *J. Funct. Anal.* **41**, 290–306.

——— (1982). Inequalities between geometric quantities of operators in Banach spaces. *Integral Equations and Operator Theory* **5**, 759–773.

——— (1984). Entropy numbers, entropy moduli, s-numbers, and eigenvalues of operators in Banach spaces. Berlin: Seminar Analysis 1983/84, *Akad. Wiss DDR.*

——— (1985). Inequalities of Bernstein-Jackson-type and the degree of compactness of operators in Banach spaces. *Ann. Inst. Fourier. Grenoble* **35**(3), 79–118.

Carl, B. and Defant, A. (1988). Tensor products, entropy estimates and Grothendieck type inequalities of operators in L_p-spaces. (preprint).

Carl, B., Heinrich, S. and Kühn, T. (1988). s-numbers of integral operators with Hölder continuous kernels over metric compacta. *J. Funct. Anal.* **81**(1), 54–73.

Carl, B. and Pajor, A. (1988). Gelfand numbers of operators with values in Hilbert spaces. *Inventiones Math.* **94**, 459–504.

Carl, B. and Pietsch, A. (1978). Some contributions to the theory of s-numbers. *Comment. Math. Prace Mat.* **21**, 65–76.

Carl, B. and Triebel, H. (1980). Inequalities between eigenvalues, entropy numbers, and related quantities of compact operators in Banach spaces. *Math. Ann.* **251**, 129–33.

Dunford, N. and Schwartz, J.T. (1958). *Linear Operators*, vol. 1. New York, London: Interscience.

Edmunds, D.E. and Tylli, O. (1986). On the entropy numbers of an operator and its adjoint. *Math. Nachr.* **126**, 231–9.

Enflo, P. (1973). A counterexample to the approximation problem in Banach spaces. *Acta Math.* **130**, 309–17.

Figiel, T. and Johnson, W.B. (1980). Large subspaces of l_∞^n and estimates of the Gordon Lewis constant. *Israel J. Math.* **37**, 92–112.

Gantmacher, F.R. (1958). *Matrizenrechnung*, vol. 1. Berlin: Verlag der Wissenschaften.

Garling, D.J.H. and Gordon, Y. (1971). Relations between some constants associated with finite dimensional Banach spaces. *Israel J. Math.* **9**, 346–61.

Gohberg, I.C. and Krein, M.G. (1969). *Introduction to the theory of linear non-self-adjoint operators in Hilbert space.* Providence: American Mathematical Society.

Gordon, Y., König, H. and Schütt, C. (1987). Geometric and probabilistic estimates for entropy and approximation numbers of operators. *J. Approx. Theory* **49**, 219–39.

Grothendieck, A. (1956). Résumé de la théorie métrique des produits tensoriels topologiques. *Bol. Soc. Mat. Sao-Paulo* **8**, 1–79.

Halmos, P.R. (1955). *Finite dimensional vector spaces*. Princeton University Press. (1967). *A Hilbert space problem book*. Princeton, Toronto, London: Van Nostrand.

Hardy, G.H., Littlewood, J.E. and Polya, G. (1964). *Inequalities*. Cambridge University Press.

Heinrich, S. and Kühn, T. (1985). Embedding maps between Hölder spaces over metric compacta and eigenvalues of integral operators. *Indag. Math.* **47**, 47–62.

Heuser, H. (1975). *Funktionalanalysis*. Stuttgart: Teubner Verlag.

Hutton, C.V. (1974). On the approximation numbers of an operator and its adjoint. *Math. Ann.* **210**, 277–80.

Jarchow, H. (1981). *Locally convex spaces*. Stuttgart: Teubner Verlag.

John, F. (1948). Extremum problems with inequalities as subsidiary conditions. Courant Anniversary Volume, New York: Interscience, pp. 187–204.

Johnson, W.B., König, H., Maurey, B. and Retherford, J.R. (1979). Eigenvalues of p-summing and l_p-type operators in Banach spaces. *J. Funct. Anal.* **32**, 329–36.

Johnson, W.B., Rosenthal, H.P. and Zippin, M. (1971). On bases, finite dimensional decompositions and weaker structures in Banach spaces. *Israel J. Math.* **9**, 488–506.

Jörgens, K. (1970). *Lineare Integraloperatoren*. Stuttgart: Teubner Verlag.

Kadec, M.J. and Snobar, M.G. (1971). Certain functionals on the Minkowski compactum (Russian). *Mat. Zametki* **10**, 453–8.

Kantorowitsch, L.W. and Akilow, G.P. (1978). *Funktionalanalysis in normierten Räumen*. Berlin: Akademie Verlag.

Kolmogorov, A.N. (1936). Über die beste Annäherung von Funktionen einer gegebenen Funktionenklasse. *Ann. Math.* **37**, 107–10.

Kolmogorov, A.N. and Tichomirov, V.M. (1959). ε-entropy and ε-capacity of sets in function spaces (Russian). *Uspeki Mat. Nauk* **14**(2), 3–86.

König, H. (1977). *s-Zahlen und Eigenwertverteilungen von Operatoren in Banachräumen*. Habilitationsschrift. Universitat Bonn. (1979). A formula for the eigenvalues of a compact operator. *Studia Math.* **65**, 141–6. (1980). *Weyl-type inequalities for operators in Banach spaces*. Proceedings of the Conference on Functional Analysis, Paderborn 1979. North Holland, pp. 297–317. (1986). *Eigenvalue distributions of compact operators*. Basel, Boston, Stuttgart: Birkhäuser Verlag.

König, H. and Lewis, D. (1988). A strict inequality for projection constants. *J. Funct. Anal.* **74**, 328–32.

König, H. and Milman, V.D. (1987). *On the covering numbers of convex bodies*. Israel Functional Analysis Seminar GAFA. Springer Lecture Notes 1267, pp. 82–95.

König, H., Retherford, J.R. and Tomczak-Jaegermann, N. (1980). On the eigenvalues of $(p, 2)$-summing operators and constants associated to normed spaces. *J. Funct. Anal.* **37**, 88–126.

Köthe, G. (1960). *Topological vector spaces*, vol. 1. Berlin, Heidelberg, New York: Springer Verlag. (1979). *Topological vector spaces*, vol. 2. Berlin, Heidelberg, New York: Springer Verlag.

Krein, M.G., Krasnoselskij, M.A. and Milman, D.P. (1948). On the defect numbers of linear operators and on some geometric questions (Ukrainian). *Sbornik trud. inst. mat. AN USSR* **11**, 97–112.

Lewis, D.R. (1979). Ellipsoids defined by Banach ideal norms. *Matematika* **26**, 18–29.

Lindenstrauss, J. and Pełczyński, A. (1968). Absolutely summing operators in L_p spaces and their applications. *Studia Math.* **29**, 275–326.

Lindenstrauss, J. and Rosenthal, H.P. (1969). The L_p-spaces. *Israel J. Math.* **7**, 325–49.

Lindenstrauss, J. and Tzafriri, C. (1977/79). *Classical Banach spaces*, vols. 1, 2. Berlin, Heidelberg, New York: Springer Verlag.

Lorentz, G.G. (1966). *Approximation of functions.* New York, Toronto, London: Academic Press.

Makai, E. and Zemanek, J. (1982). Geometrical means of eigenvalues. *J. Oper. Theory* **7**, 173–8.

Mangoldt, H. and Knopp, K. (1973). *Einführung in die höhere Mathematik*, vol. 4. Leipzig: S. Hirzel Verlag.

Milman, V.D. (1985). Almost euclidean quotient spaces of subspaces of finite-dimensional normed spaces. *Proc. Am. Math. Soc.* **94**, 445–9.

(1986). Inégalité de Brunn-Minkowski inverse et applications a la théorie locale des éspaces normes. *C. R. hebd. Acad. Sci., Paris* **302**(1), 25–8.

Mitjagin, B.S. (1961). Approximative dimension and bases in nuclear spaces (Russian). *Uspeki Mat. Nauk* **16**(4) 63–132.

Mitjagin, B.S. and Pełczyński, A. (1966). *Nuclear operators and approximative dimension.* Proceedings of the International Congress of Mathematics, Moscow. pp. 366–72.

Natanson, I.P. (1955). *Konstruktive Funktionentheorie.* Berlin: Akademie Verlag.

Pełczyński, A. (1980). *Geometry of finite dimensional spaces and operator ideals.* Notes in Banach spaces. Texas University Press.

Persson, A. and Pietsch, A. (1969). *p*-nukleare und *p*-integrale Abbildungen in Banachräumen. *Studia Math.* **33**, 19–62.

Pietsch, A. (1965). *Nukleare lokalkonvexe Räume.* Berlin: Akademie Verlag.

(1967). Absolut *p*-summierende Abbildungen in normierten Räumen. *Studia Math.* **28**, 333–53.

(1974). *s*-numbers of operators in Banach spaces. *Studia Math.* **51**, 201–23.

(1978). *Operator ideals.* Berlin: Verlag der Wissenschaften.

(1980). Weyl numbers and eigenvalues of operators in Banach spaces. *Math. Ann.* **247**, 149–68.

(1986). Eigenvalues of absolutely *r*-summing operators. Aspects of mathematics and its applications. Amsterdam: Elsevier, pp. 607–17.

(1987). *Eigenvalues and s-numbers.* Leipzig: Akademische Verlagsgesellschaft Geest & Portig K.-G.

Pinkus, A. (1985). *n-widths in approximation theory.* Berlin, Heidelberg, New York: Springer Verlag.

Pisier, G. (1980/81). Remarques sur un resultat non publié de B. Maurey. Séminaire d' Analyse Fonctionnelle, Ecole Polytechnique-Palaiseau, expose 5.

(1983). Counterexamples to a conjecture of Grothendieck. *Acta Math.* **151**, 181–208.

(1986). *Factorization of linear operators and the geometry of Banach spaces.* CBMS Regional Conference Series in Mathematics 60.

(1988). Volume inequalities in the geometry of Banach spaces. Texas A&M University. Cambridge University Press, 1989.

(1989). A new approach to several results of V. Milman. *J. Reine angew. Math.* **393**, 115–31.

Pontrjagin, L.S. and Schnirelman, L.G. (1932). Sur une propriété métrique de la dimension. *Ann. Math.* **33**, 152–62.

Przeworska-Rolewlcz, D. and Rolewicz, S. (1968). *Equations in linear spaces.* Warsaw: PWN.

Riesz, F. and Sz.-Nagy, B. (1982). *Vorlesungen über Funktionalanalysis.* Berlin: Verlag der Wissenschaften.

Russo, B. (1977). On the Hausdorff–Young theorem for integral operators. *Pac. J. Math.* **68**, 241–53.

Saint Raymond, J. (1980/81). Sur le volume des corps convexes symétriques. Séminaire Initiation à l'Analyse. Exp. no. 11. Université P. et M. Curie, Paris.

Santaló, L. (1949). Un invariante afin para los cuerpos convexos del espacio de n dimensiones. *Portugal Math.* **8**, 155–61.

Stephani, I. (1970). Injektive Operatorenideale über der Gesamtheit aller Banachräume und ihre topologische Erzeugung. *Studia Math.* **38**, 105–24.

(1972). Surjektive Operatorenideale über der Gesamcheit aller Banachräume. *Wiss. Z. Univ. Jena* **21**, 187–206.

(1973). Surjektive Operatorenideale über der Gesamtheit aller Banachräume und ihre Erzeugung. *Beiträge zur Analysis* **5**, 75–89.

(1987). Injectively *A*-compact operators, generalized inner entropy numbers, and Gelfand numbers. *Math. Nachr.* **133**, 247–72.

Szarek, S. (1983). The finite dimensional basis problem with an appendix on nets of Grassmann manifolds. *Acta Math.* **151**, 153–80.

Taylor, A.E. and Lay, D.C. (1980). *Introduction to functional analysis.* New York, Chichester, Brisbane, Toronto: Wiley & Sons.

Timan, A.F. (1964). On the order of growth of the ε-entropy of spaces of real continuous functionals defined on a connected compactum (Russian). *Uspekhi Mat. Nauk* **19**. 173–7.

(1977). On the exact order of growth of the ε-entropy and ε-capacity of Arzela compacta of arbitrary big size. In *Metriceskie voprosy teorii funkcii i otobrashenii*, **8**, 131–42. Naukova Dumka. Kiev. (Russian).

Tomczak-Jaegermann, N. (1989). *Banach-Mazur distances and finite-dimensional operator ideals.* Harlow: Pitman, and New York: John Wiley.

Triebel, H. (1970). Interpolationseigenschaften von Entropie- und Durchmesseridealen kompakter Operatoren. *Studia Math.* **34**, 89–107.

Weyl, H. (1949). Inequalities between the two kinds of eigenvalues of a linear transformation. *Proc. Nat. Acad. Sci. USA* **35**, 408–11.

Whitley, R. (1966). Projecting *m* onto *c₀*. *Am. Math. Monthly* **73**, 285–6.

Yamamoto, T. (1967). On the extreme values of the roots of matrices. *J. Math. Soc. Japan* **19**(2), 173–8.

Zemanek, J. (1983). The essential spectral radius and the Riesz part of the spectrum. *Coll. Math. Soc. J. Bolyai* **35**, 1275–89.

List of symbols

Throughout this book, $\mathbb{R}$ and $\mathbb{C}$ denote the real and complex fields, respectively.

1. Metric, Banach and Hilbert spaces

Unless the contrary is explicitely stated, (X, d) denotes a metric space, E and F denote Banach spaces, while H and K are Hilbert spaces. Elements are usually denoted by x and y, functionals by a and b.

$\|x\|$, $\langle x, a \rangle$, $\|a\|$ 1, 3
E/N, E', $E'' = (E')'$ 3, 4
$U(x_0; \varepsilon)$, $\mathring{U}(x_0; \varepsilon)$, U_E, $\mathring{U}_E$, U_p^k 6, 8, 17

2. Geometrical parameters of spaces

$\lambda(M, E)$, $\lambda_k(E)$ 64, 239
$p(M; E, F)$, $q(E, F; N)$, $p_n(E, F)$, $q_n(E, F)$ 63, 68, 65, 70
$d(E, F)$, $\delta_n(E)$, $\delta^{(n)}(E)$ 106

3. Operators

Operators are mostly denoted by S and T

$\|T\|$,   2
T', T^*, C_H, $|T|$,   72, 115, 116
rank (T),   4
$N(T)$, $R(T)$, $\overline{R(T)}$, $N(T)^\perp$, $R(T)^\perp$ 2, 117
$N_\infty(\lambda I_E - T)$, $R_\infty(\lambda I_E - T)$,   131, 132
$d(T)$, $m(T; \lambda)$, $p(T; \lambda)$ 131
$\lambda_n(T)$, $\rho(T)$, $\sigma(T)$ 4, 129, 130
$r(\bar{T})$, $\rho(\bar{T})$, $\sigma(\bar{T})$, $r_{\text{ess}}(T)$, $\|\bar{T}\|$, $\alpha(T)$,   134
Q, Q_N^E, Q_E 3, 12, 53
K_E, I_E, I_M^E, J, J_F 4, 13, 24, 56, 61
I_0, I_∞, J_0 73, 74, 75

$D, D^{(k)}, D_k$ 1, 16, 17, 18
S_+, D_+ 136, 137
$T_{K,\mu}, T_K$ 214, 226

4. Entropy and approximation quantities of sets and operators

$\varepsilon_n(M), \varphi_n(M), g_n(M)$ 7, 10
$\varepsilon_n(T), \varphi_n(T), e_n(T), f_n(T), g_n(T)$ 11, 21, 22, 24
$a_n(T), c_n(T), d_n(T), t_n(T), s_n(T)$ 41, 56, 49, 82, 155
$\hat{a}_n(T), \hat{c}_n(T), \hat{d}_n(T), \hat{t}_n(T)$ 85, 90, 91, 94
$e(T), a(T), d(T), g(T), \beta(T)$ 76, 124, 126

5. Modulus of continuity of functions, kernels and operators

$\omega(f; \delta), \omega(K; \delta)$ 95, 170, 215
$\omega(T; \delta), \omega(T_{K,\mu}; \delta), \omega(T_K; \delta)$ 161, 174, 215, 226
$\Omega_p(K; \delta), \Omega_Z(K; \delta)$ 161, 215, 225

6. Norms, semi-norms and quasi-norms of sequences, functions, kernels and operators

$\lambda_{p,q}(x)$ 27
$\lambda_{p,q}^{(e)}(T), \lambda_{p,q}^{(a)}(T), \lambda_{p,q}^{(d)}(T), \lambda_{p,q}^{(c)}(T), \lambda_{p,q}^{(t)}(T)$ 35, 47, 53, 61, 83
$|f|_\alpha, \|f\|_\alpha, \|f\|, \|f\|_p, \|f\|_\infty$ 180, 213
$|T|_\alpha, \|T\|_\alpha$ 196, 198
$\|K\|, \|\|K\|\|_p, |K|_{p,\alpha}, \|\|K\|\|_{p,\alpha}, |K|_\alpha, \|K\|_\alpha$ 214, 215, 221, 224
$\|K\|, |K|_{Z,\alpha}, \|K\|_{Z,\alpha}$ 225, 226
$\pi_p(T), \nu_2(T)$ 231, 236

7. Operator classes

$L(E, F)$ 2
$A, K, G, A^S, A^I, G^S, G^I$ 35, 40, 54, 62
$L_{p,q}^{(e)}, L_{p,q}^{(a)}, L_{p,q}^{(d)}, L_{p,q}^{(c)}, (L_{p,q}^{(a)})^S, (L_{p,q}^{(a)})^I$ 27, 35, 37, 47, 53, 54, 61, 62
$L(E, E)/K(E, E)$ 133
$K(E, C(X)), L^\alpha(E, C(X))$ 169, 184
P_p, N_2 231, 236

8. Sequence and function spaces

$l_p, l_\infty, c_0, l_{p,q}, l_p^k$ 1, 9, 17, 27
$l_1(\Gamma), l_1(U_E), l_\infty(\Gamma), l_\infty(U_{F'})$ 51, 53, 60

$C(X)$, $L_p(X,\mu)$, $L_\infty(X,\mu)$ 1, 159, 213
$\mathscr{L}_{\infty,\lambda}$, $\mathscr{L}_{1,\lambda}$ 167
$C^\alpha(X)$, $C_Z(X)$, $C_Z^\alpha(X)$ 179, 225

9. Miscellaneous

vol_m, $E_n(f)$, $\mathrm{conv}(x_1,\ldots,x_n)$ 9, 95, 245
M^0, $\mathscr{E}$ 249, 250

Index

absolutely p-summing operator, 231
absolutely 2-summing norm, 232
absolutely 2-summing operator, 229, 231, 234
absolute value of an operator, 116
abstract kernel, 161, 226, 227
adjoint and self-adjoint operator, 115, 116
algebraic tensor product, 259
approximable operator, 46
approximation numbers
  local, 89, 168
  local symmetrized, 94
  of a function, 160
  of an operator, 41
  symmetrized, 82
Arzelà–Ascoli theorem, 175
Auerbach basis, 44
Auerbach's lemma, 43

ball, 249
Banach algebra, 134
Banach–Mazur distance, 106, 229
Banach space
  bidual of, 75
  dual, 3
  metric extension property of a, 59
  metric lifting property of a, 51
  of bounded number families, 60
  of Hölder continuous functions, 179
  of summable number families, 51
  reflexive, 75
Bernstein's inequality for functions, 95
Bernstein type inequality for operators, 96, 230
bidual operator, 76
Borel measure, 213
Borel set, 213
Borel σ-algebra, 212
bounded subset, 6

Calkin algebra, 133
canonical embedding, 4, 75
canonical factorization of an operator, 2
canonical surjection, 3
Cantor's ternary set, 194
classical spectral radius formula, 138

convex hull, 245
  volume of a, 245
coset, 3
covering, 6

degree of approximation, 178
degree of compactness, 178
diagonal operator, 16
Dirac functional, 226
distance, local injective and surjective, 106
distance of a point from a subset, 162

ε-chain, 189
ε-net, 6
eigenvalue, 130
  algebraic and geometric multiplicity
  of an, 131
  sequence, 4, 137
eigenvector, 131
ellipsoid, 250
entropy function, 8
entropy moduli, 10, 12
entropy numbers, 7, 11
  dyadic and inner dyadic, 21
  inner, 7, 11
  outer, 13
equicontinuity, 176
equivalent quasi-norm, 4
essential spectral radius, 134
essential spectrum, 134
extension of an operator, 63
extension constant, 63, 65
extension property of a Banach space, metric, 59

factorization theorem for absolutely-2 summing operators, 231
finite-dimensional operator, 2
finite-dimensional subspace, geometrical properties of a, 162
finite rank operator, 2
Fredholm operator, 133
functional, 2

Gelfand numbers, 56
  local, 90
  injectivity of, 61

generalized sequence, 87
generalized integral operator, 161
generating kernel, 214
generating operator, 198
geometrical parameter, 63
geometrical properties of
    finite-dimensional subspaces, 162
Grothendieck's theorem, 230, 264
Grothendieck's inequality, 254, 264

Hahn–Banach theorem, 4
Hardy's inequality, 31
Hölder classes of continuous functions, 160
Hölder continuous function of type α, 179
Hölder continuous operator of type α, 196
Hölder continuous Z-valued function
    of type α, 225

ideal quasi-norm, 37
    injective and surjective, 39
identity map I_∞, 180
integrable function, 213
integral operator, 161, 212, 221
    generalized, 161
isomorphic Banach spaces, 1
isomorphism, 1
iterated kernel, 131

Jackson's inequality for functions, 96, 160
Jackson type inequality for operators, 101

kernel, 2
    abstract, 161, 226
    generating, 214
    modulus of continuity of a, 215
    operator defined by an abstract, 225, 227
Khintchin's inequality, 254, 255
Kolmogorov numbers, 49
    local, 91
    surjectivity of, 53

Lebesgue measure, 9
lifting, 51, 68
lifting constant, 68, 70, 240
$\mathscr{L}_{\infty,\lambda}$-, $\mathscr{L}_{1,\lambda}$-space, 162
local approximation numbers, 89, 168
local Gelfand numbers, 90
local injective and surjective distance, 106
local Kolmogorov numbers, 91
local quantities, 85
local reflexive, 75
Lorentz sequence spaces, 37
    lexicographical order of, 29

μ-equivalent measurable functions, 213
μ-essential bounded, 213
μ-essential supremum, 213
μ-null set, 213

measurable function and set, 213
measures of non-approximation and
    non-compactness, 124
metric injection and surjection, 12, 13
metric lifting and extension property, 51, 59
metric space, connected and
    ε-connected, 188, 189
modulus of continuity, 95, 96, 160, 161
    of a kernel, 215
    of a Z-valued function, 225
    of an operator, 174, 178
    surjectivity of the, 175

2-nuclear norm and operator, 236
null space, 2

operator, 1, 11
    absolutely-summing, 229, 231, 234
    absolute value of an, 116
    adjoint and self-adjoint, 115, 116
    ascent of an, 131
    canonical factorization of an, 2
    compact, 159
    dual, 72
    finite rank (finite-dimensional), 2
    Fredholm, 133
    generating, 198
    Hilbert-Schmidt, 231
    Hölder continuous, 196
    ideal, 37: injective and surjective, 39;
        injective and surjective hull, 54, 62;
        quasi-normed and complete
        quasi-normed, 37
    induced, 2, 3
    integral (generalized integral), 161,
        212, 221
    modulus of continuity of an, 174, 178
    norm, 2
    nuclear, 236
    of type A, 37
    polar decomposition of an, 117
    positive (root of a positive), 116
    resolvent and spectrum of an, 130, 134
    Riesz, 134
    spectral radius of an, 130, 134

packing, 1
    functions subordinate to a, 181
partial isometry, 117
partition of unity, 162, 163, 200
    subordinate to an open covering, 163
Pisier's theorem, 250
polar decomposition of an operator, 117
polar set, 249
positive (and root of a positive)
    operator, 116
precompact, 8
principal vector, 132

principle of local reflexivity, 75, 76
projection, 42
    spectral, 132
projection constant, 229, 239
    relative, 64, 239

quasi-norm
    equivalent, 4
    injective and surjective, 39, 40
quasi-normed operator ideal, 37
    complete, 39
quotient map, 3, 49
quotient norm, 49
quotient space, 2, 49

ρ-distant set, 6
range, 2
resolvent set, 130, 134
Riesz operator, 134

Santaló inequality, 230, 253
    inverse of, 230, 253
Schauder's theorem, 84
Schmidt's representation theorem, 46
semi-norm, 214
singular numbers, 155
spectral projection, 132

spectral radius, 130, 134
    classical, 138
    essential, 134
    formula, 148
spectrum, 130
    essential, 134
support of a function, 162
symmetric subset, 249
symmetrized approximation numbers, 82
    local, 94

tensor product
    algebraic, 259
    of operators, 260

universality of the Banach space
    $C[a, b]$, 159
universality of $C[a, b]$-valued compact
    operators, 159

volume of a convex body, 230
volume of a convex hull, 230
volume ratio, 230

Weyl's inequalities, 157
Weyl type inequality, 146